Troposcatter Radio Links

Troposcatter Radio Links

by
Giovanni Roda

Artech House
Boston and London

British Library Cataloguing in Publication Data
Roda, Giovanni
 Troposcatter radio links.
 1. Radio waves. Tropospheric propagation
 I. Title
 621.3841′1

 ISBN 0–89006–293–5

Library of Congress Cataloging-in-Publication Data
Roda, Giovanni. 1929–
 Troposcatter radio links.
 Includes bibliographies and index.
 1. Tropospheric scatter communication systems.
 2. Radio relay systems. I. Title.
 TK6570.T76R63   1988   621.3841′56   88–6166

 ISBN 0–89006–293–5

Copyright © 1988

ARTECH HOUSE, INC.
685 Canton Street
Norwood, MA 02062

All rights reserved. Printed and bound in the United States of America. No part of this book may be reproduced or utilized in any form or by any means, electronic or mechanical, including photocopying, recording, or by any information storage and retrieval system, without permission in writing from the publisher.

International Standard Book Number: **0–89006–293–5**
Library of Congress Catalog Card Number: **88–6166**

10 9 8 7 6 5 4 3 2 1

Contents

Preface ... xi

Chapter 1 An Introduction to Tropospheric Scatter Radio links ... 1

1.1 Beyond-the-horizon and tropospheric scatter paths ... 2
1.2 Short historical outline of beyond-the-horizon radio communications ... 5
1.3 The place of troposcatter links among other transhorizon communications techniques ... 7
1.4 Main characteristics of troposcatter radio links ... 8
1.5 Design calculations and their reliability ... 10
1.6 Feasibility of a troposcatter path ... 11
1.7 Specific field of application of troposcatter systems ... 14
1.8 Types of troposcatter systems ... 14
1.9 Advantages of troposcatter systems ... 17
1.10 International recommendations and authorities ... 18
1.11 Comments and suggestions for further reading ... 20
References ... 20

Chapter 2 Review of Statistical Distributions used in Troposcatter ... 21

2.1 Random variables and their statistical distributions ... 21
2.2 Normal and log-normal distributions ... 23
  2.2.1 Normal distribution ... 24
  2.2.2 Log-normal distribution ... 26
2.3 Rayleigh distribution ... 27
2.4 Nakagami distributions ... 31
  2.4.1 Nakagami m distribution ... 31
  2.4.2 Nakagami n distribution ... 32
2.5 Composed distributions—convolutions ... 35
  2.5.1 General remarks ... 35
  2.5.2 Convolution of normal distributions ... 36
  2.5.3 Convolution of log-normal distributions ... 36
  2.5.4 Convolution of Rayleigh distributions ... 36
  2.5.5 Convolution of log-normal and Rayleigh distributions ... 37
2.6 Correlation between random variables—autocorrelation ... 37
2.7 Comments and further reading ... 38
References ... 39

Chapter 3 The Geometry of the Troposcatter Path and Related Problems ... 41

3.1 Profile of the troposcatter path ... 42

3.2 Definition of geometrical data and parameters
    3.2.1 Basic data
    3.2.2 Angular parameters
    3.2.3 Length parameters
    3.2.4 Geometrical parameters of antennas
3.3 Derivation of the formulae
    3.3.1 Calculation of the effective radius of the Earth
    3.3.2 Calculation of the distance and elevation of the horizon over sea (or flat land)
    3.3.3 Calculation of the elevation angle of the horizon
    3.3.4 Calculation of angles α and β
    3.3.5 Calculation of the angular length and the scatter angle
    3.3.6 Calculation of the distance along the horizon rays to their crosspoint
    3.3.7 Calculation of $EH = h$
    3.3.8 Calculation of $EG = h_0$
    3.3.9 Calculation of path differences
3.4 Comments

Chapter 4   The Tropospheric Scatter Mode of Propagation

4.1 Tropospheric refractivity and its characteristics
    4.1.1 The refractive index and its gradient
    4.1.2 The effective radius of the Earth
    4.1.3 Structure of the atmospheric refractivity
    4.1.4 Radiometeorological parameters
4.2 Troposcatter multipath transmission
4.3 Representations of the main propagation phenomena
4.4 Degradation of antenna gain
4.5 The troposcatter multipath signal
4.6 Spread and bandwidth parameters of the medium
4.7 Variability in the level of the tropscatter signal
4.8 Anomalous propagation
4.9 The CCIR climates
4.10 Comments and suggestions for further reading
References

Chapter 5   Diversity Techniques

5.1 Decorrelated paths and diversity effects
5.2 Diversity systems
    5.2.1 Space diversity
    5.2.2 Frequency diversity
    5.2.3 Polarization diversity
    5.2.4 Angle diversity
    5.2.5 Time diversity

	Linear combination of diversity branches	90
	Predetection or postdetection combination	91
	Combining methods	93
5.5.1	Signal selection	94
5.5.2	Equal-gain combination	96
5.5.3	Maximal-ratio combination	96
	Combined signal and diversity gain	98
	Implicit diversity techniques to combat multipath distortion	100
	Combiners and adaptive modems	103
5.8.1	A simple baseband combiner	103
5.8.2	An i.f. combiner	106
5.8.3	Multiple-subband moden for digital transmission	107
5.8.4	Independent sideband diversity modem for digital transmission	107
5.8.5	Distortion adaptive modem for megabit transmission	108
	Comments and suggestions for further reading	111
References		111

Chapter 6 Path Loss and Path Distortion   113

	Path loss between isotropic antennas	114
	Method of calculation	116
	Path geometry calculations	117
	CCIR methods of calculating path loss	117
6.4.1	Method I	121
6.4.2	Method II	125
6.4.2	Radiometeorological method	132
6.4.4	Chinese method	132
	Calculation of instantaneous path loss with fast fading and diversity	133
	Dependence of path loss on the main parameters and its actual variability	140
	Path loss for line-of-sight links	141
	Diffraction path loss	142
6.8.1	Knife-edge diffraction	143
6.8.2	Smooth-earth diffraction	145
	Path loss in particular propagation conditions	146
	Evaluation of path distortion and the multipath delay spectrum	148
	Comments and suggestions for further reading	153
References		153

Chapter 7 Tropospheric Scatter Equipment   155

	Composition of a troposcatter terminal	155
	Transmitting system	158
7.2.1	Transmitters (exciters)	158
7.2.2	Power amplifiers	160

7.3	Receiving system	16
	7.3.1 Receivers	16
	7.3.2 Input thermal noise	16
7.4	Design parameters for analog radio equipment	16
	7.4.1 Analog receiver threshold for telephony	16
	7.4.2 Telegraph failure point (telegraphic threshold)	16
	7.4.3 Thermal signal-to-noise ratio in a telephone channel	16
	7.4.4 Overall performance characteristics of analog radio equipment	17
7.5	Design parameters of digital radio equipment	17
	7.5.1 Bit error rate curve of a digital system	17
	7.5.2 Threshold of a digital receiver	17
	7.5.3 Overall performance characteristics of digital radio equipment	18
7.6	Typical troposcatter radio equipment	18
	7.6.1 Analog radio equipment	18
	7.6.2 Digital radio equipment	18
	7.6.3 Radio frequency filters and duplexers—isolators	18
	7.6.4 Coaxial and waveguide feeders	18
7.7	Antennas	18
	7.7.1 General characteristics of fixed antennas	18
	7.7.2 Mobile antennas	19
	7.7.3 Gain and gain degradation: path gain	19
7.8	Multiplex equipment	19
	7.8.1 Analog multiplex	19
	7.8.2 Digital multiplex	19
7.9	Auxiliary equipment	19
7.10	Ability of combined equipment to overcome path loss	19
7.11	Systems with adaptive capability	19
7.12	Comments and suggestions for further reading	20
References		20

Chapter 8 Path Design and Prediction of Performance for Analog Troposcatter Links 20

8.1	Performance objectives	20
8.2	Noise	20
	8.2.1 Thermal noise	20
	8.2.2 Intermodulation noise	21
8.3	Path calculation	22
	8.3.1 Conventional troposcatter links	22
	8.3.2 High quality troposcatter links	22
8.4	Telegraphic transmission performance	22
8.5	Comments and suggestions for further reading	23
References		23

CONTENTS

Chapter 9 Path Design and Prediction of Performance for Digital Troposcatter Radio Links ... 233

9.1 Digital transmission on a troposcatter channel ... 233
    9.1.1 Signal-to-noise ratio in telephone channels ... 234
    9.1.2 Transmission degradation: digital errors ... 234
    9.1.3 Multipath delay spread and pulse distortion ... 235
    9.1.4 Expedients against multipath delay dispersion ... 236
    9.1.5 Behaviour of diversity reception ... 238
9.2 Quality of digital transmission ... 242
    9.2.1 Outages and quality parameters ... 244
    9.2.2 Hypothetical reference circuit ... 245
    9.2.3 Performance objectives ... 245
9.3 Design of a digital troposcatter link ... 248
    9.3.1 Path calculation ... 248
    9.3.2 Path equation ... 249
    9.3.3 High quality troposcatter links ... 250
    9.3.4 Conventional troposcatter links ... 250
9.4 Comments and suggestions for further reading ... 252
References ... 253

Chapter 10 Topographic and Survey Problems ... 255

10.1 Preliminary work using maps ... 256
    10.1.1 Calculation of the antenna azimuth ... 256
    10.1.2 Drawing the connection line on the maps ... 259
    10.1.3 Plotting the horizon skyline profile ... 260
10.2 Surveying problems on site ... 263
    10.2.1 Basic principles ... 264
    10.2.2 Fundamental parameters of geographical astronomy ... 267
    10.2.3 Coordinate systems ... 269
    10.2.4 Determination of the direction of the distant station ... 270
    10.2.5 Determination of the horizon skyline ... 271
    10.2.6 Determination of latitude and longitude ... 272
10.3 Topographic problems in antenna installation and line-up ... 274
    10.3.1 Orienting the antenna in the horizontal plane ... 275
    10.3.2 Orienting the antenna in the vertical plane ... 276
    10.3.3 Assessment of the profiles of large billboard antennas ... 277
10.4 Comments and suggestions for further reading ... 280
References ... 280

Chapter 11 From Planning to Implementation and Maintenance ... 281

11.1 Planning the system ... 282
11.2 Site surveys ... 284

11.3   Siting the stations
11.4   Choice of frequency band
11.5   Frequency plan
11.6   Procedure for design calculations
11.7   Choice of equipment
       11.7.1   Radio equipment
       11.7.2   Multiplex equipment
       11.7.3   Antennas
       11.7.4   Power equipment
       11.7.5   Air conditioning
11.8   Installation of the system
       11.8.1   Fixed systems
       11.8.2   Installation of transportable systems
       11.8.3   Deployment of mobile systems
11.9   Tests and measurements
       11.9.1   Path propagation tests
       11.9.2   Tests on the installed system
11.10 Radiation hazards
11.11 System operation
11.12 Maintenance problems
11.13 Cost considerations
11.14 Comments and suggestions for further reading
References

Appendix 1   **CCIR Recommendations and Reports Concerning Transhorizon Systems**

Appendix 2   **Computer Calculations**

Bibliography

Index

Preface

This book is the result of the author's many years of experience in system design in the field of troposcatter radio links. It is an expanded version of an earlier internal publication of the company Applicazioni Radio Elettroniche (ARE) that was aimed at potential customers and clients who were already involved in this field in some way. The present book is directed at a wider range of readers who do not already have some experience of the subject. Since troposcatter radio links are a particular case of radio relay links, it is assumed that the reader has some background in this field and therefore some subjects are introduced without explanation or derivation.

At present many good books dealing with conventional line-of-sight radio links are available to design engineers. In contrast, apart from two books published some years ago which only partially serve our purposes (F. du Castel, *Propagation Troposphérique et Faisceaux Hertziens Transhorizon*, Chiron, Paris, 1961; P. F. Panter, *Communication Systems Design—Line-of-sight and Troposcatter Systems*, McGraw-Hill, New York, 1972), little information is available in the troposcatter field unless reference is made to the original papers in specialized journals and technical reviews. I hope that this book will fill the gap. It has been written specifically to describe the principles and procedures necessary to plan, design, survey and implement a troposcatter radio link circuit with a clear understanding of what has to be done and why. By following the procedures suggested, the reader should be able to set up a system providing reliable communication circuits using this mode of propagation.

Elementary explanations of the theories and their use and applications are given, and are supported by data, figures, diagrams, tables, work forms and examples of calculation provided or taken from actual experience in the field. The reader should study the examples carefully and work through the calculations in order to become completely familiar with practical design procedures.

As in the case of conventional radio links, the International Radio Consultative Committee (CCIR) has produced several Recommendations and Reports specifically concerning the characteristics of transhorizon radio links. They are listed in Appendix 1, and many of them are cited in the main text. The engineer planning to work in this field should familiarize himself with these Recommendations as well as with others concerning radio links in general.

This book is divided into 11 chapters. The first three are introductory. Chapter 1 provides an overview of the subject which is a useful preliminary to more detailed considerations. The statistical distributions required to understand the mathematics used in this book and in the original papers are described as simply as possible in Chapter 2. The geometry of the troposcatter path is introduced in Chapter 3 and many formulae which are not easily accessible in the literature are given. The troposcatter mode of propagation, together with its characteristics and consequences, is described in Chapter 4. One of the main consequences of this is the development of diversity techniques, which are analyzed in detail in Chapter 5 together with some basic theory and a description of diversity combiners. The study and prediction of path loss and path distortion, which are of fundamental interest to the designer, are described in Chapter 6. By using the equipment and arrangements described in Chapter 7 the designer should be able to dimension an analog path, as explained in Chapter 8, or a digital path, as explained in Chapter 9, with the desired characteristics. In practice the work does not follow such a linear path, as the first tentative investigation will lead to a feasibility study, which will be followed by site surveys and measurements, collection of data etc. Problems encountered in this phase and during the installation of antennas are described in Chapter 10. The contents of the preceding chapters are summarized in Chapter 11, in which outstanding problems are also discussed. The book is completed by two appendices which include a collection of data in various forms which are not easily available elsewhere.

The author hopes that his work will be appreciated by the general reader and that it will be of value to engineers involved in the design, implementation, operation and maintenance of troposcatter links.

GIOVANNI RODA
Busto Arsizio, Italy

Dedication
To my wife

Acknowledgments

The author thanks the companies and organizations for whom he worked or whom he contacted, who gave him the opportunity to encounter, discuss and solve the problems which have helped to develop his experience in this field. Thanks are also due to the ITU, who gave permission to reproduce several figures and diagrams from CCIR Volumes 5 and 6, 1986, and the definition of climates given in Chapter 4, Section 4.9. The author is also grateful to ARE, Castellanza, Italy, GTE-COMELIT, Milan, Italy, and Raytheon, Lexington, USA, who provided data and photographs of their equipment which appear in the book.

Chapter 1

An Introduction to Tropospheric Scatter Radio Links

The concepts of beyond-the-horizon (or transhorizon) and tropospheric scatter (or "troposcatter") radio paths are introduced in this chapter. A brief history of this type of communications system is given and its main characteristics are described. The specific field of application, the advantages, the utilization and other features of this mode of propagation are discussed. A general introduction to the various topics discussed in detail in subsequent chapters is also provided.

Troposcatter radio relay links are a particular type of radio link which exploit the transmission possibilities of the inhomogeneities of the troposphere for establishing communications with points well beyond the horizon. The reader should therefore have some familiarity with the techniques of the conventional line-of-sight (LOS) radio links, as the new techniques are extensions of these.

1.1 Beyond-the-horizon and tropospheric scatter paths

The postwar period has produced many new developments in communications technology and system design, of which the most important are the extensive use of microwave radio relay point-to-point links, the worldwide use of satellites and the utilization of troposcatter propagation.

The discovery of radio wave propagation by tropospheric scatter was almost accidental, as unexplained signals from remote stations were received at sites well beyond the predicted LOS paths and at stations hundreds of kilometers from the transmitter of origin. The mechanism by which this type of propagation takes place is still not completely understood, but much experimental work has been carried out from which sufficient data have been obtained to enable circuit behavior to be predicted and to provide a sound basis for fabricating practical and reliable communications systems. The technique is now used extensively over difficult terrain where the terminals are separated by a distance well beyond the radio horizon.

In order to introduce the concept of troposcatter path let us consider a radio path like that shown in Fig. 1.1, with the two terminal antennas reciprocally in line of sight, and let us then increase the distance between the antennas until the link is obstructed by the horizon. Given the necessary clearance, the r.f. signal received is relatively stable in LOS conditions and its level decreases (at least theoretically) by 6 dB when the path length is doubled if the antennas remain reciprocally visible. This is indicated as the "LOS zone" in Fig. 1.1. As the first obstructions appear in the path the received r.f. signal exhibits some (diffraction) fringes and then rapidly decreases, but it does not disappear completely. Propagation between the two antennas still takes place through the mechanism of diffraction, with a reasonably stable signal. This is indicated as the "diffraction zone" in Fig. 1.1. When the antennas are moved a long way beyond the horizon the diffracted signal vanishes but low r.f. signals are still received. Although they are continuously variable, they can be used for the transmission of information provided that certain conditions are fulfilled. This is indicated as the "troposcatter zone" in Fig. 1.1, and it can extend for hundreds of kilometers beyond the horizon before the signal can no longer be used. The atmosphere, which introduces unexpected diffusion or scattering of the radio waves, is responsible for the propagation beyond the diffraction zone.

The atmosphere is divided into three zones.

(a) The *troposphere* is the zone where the air is in permanent motion and meteorological phenomena take place. The influence of the weather prevails, the temperature decreases with height, and cloud formation and convection predominate. There is no ionization of the air. This zone extends from ground level to a height of about 8–10 km.

(b) In the *stratosphere* the air is still and stratified. The temperature is

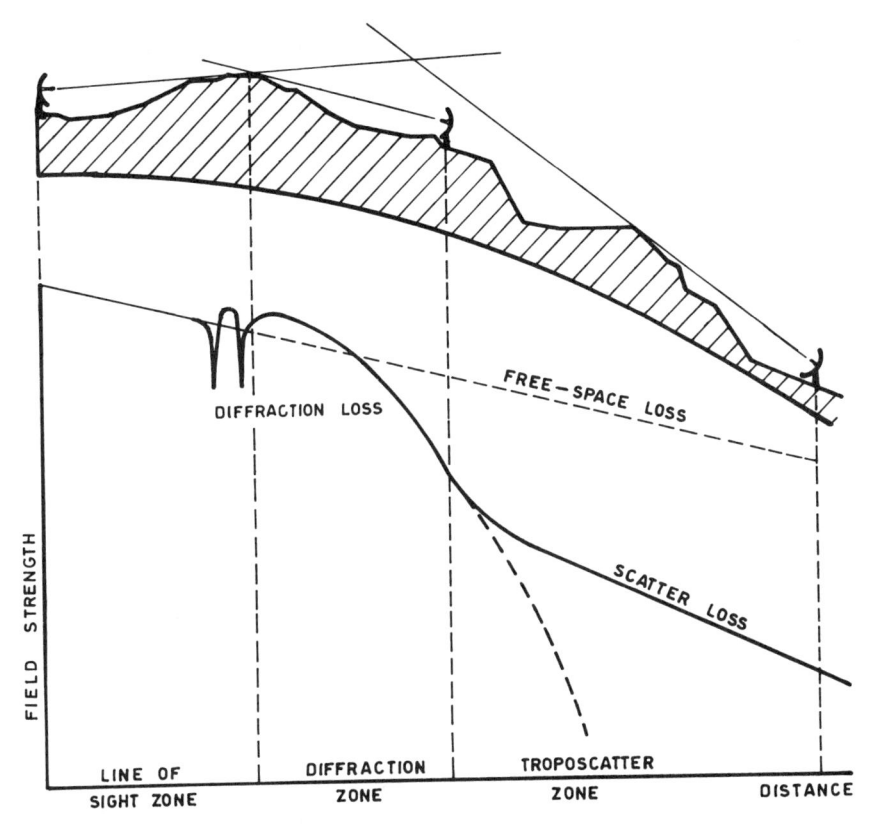

Fig. 1.1 From line of sight to troposcatter.

almost constant, humidity is almost absent and there are almost no perturbations. This zone is located above the troposphere and extends to a height of 40–50 km.

(c) The *ionosphere* is located about 50–60 km from the surface of the Earth where the ionization of the air is appreciable. The reflection properties of this region at certain high frequencies are used for very long distance communications. Scattering of the radio waves (ionoscatter) occurs at certain v.h.f. bands, and this is utilized for some narrow-band transmissions.

The irregularities of the troposphere produced by the irregular motion of the air enable r.f. signals to be transmitted beyond the horizon by means of troposcatter techniques. Figure 1.2 illustrates how the inhomogeneities in the troposphere, when "illuminated" by the radio beam from a transmitting antenna pointing at the horizon, diffuse or scatter the electromagnetic energy in all directions (including backwards), but mainly forwards within a cone whose axis is the original direction of the beam. Therefore some scattered energy reaches the receiving antenna (which is also pointing at the horizon) and the transmission of information is still possible with a high degree of circuit continuity. The nature of the inhomogeneities is not well known, but they are generally treated as a mixture of blobs, small layers, "feuillets" etc. which continuously vary in number, shape, position, velocity and physical characteristics.

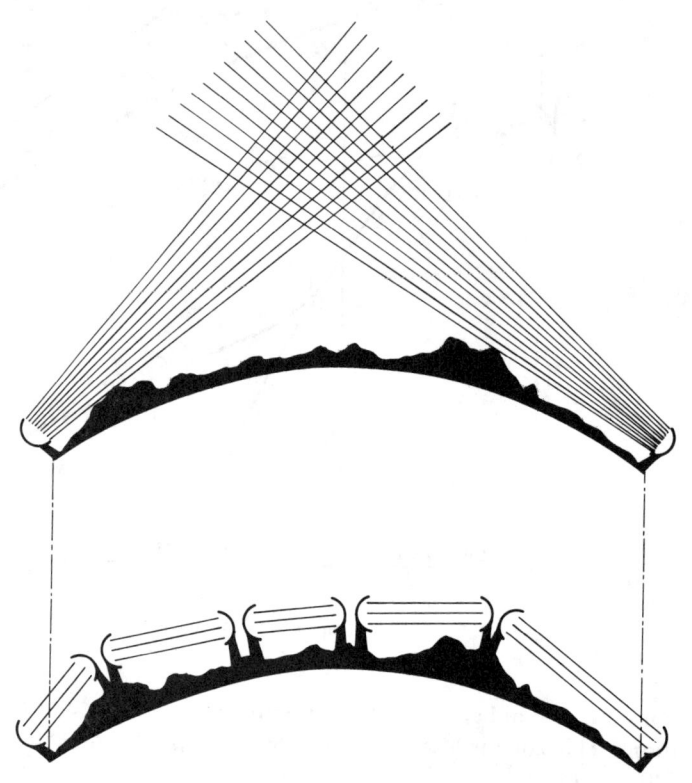

Fig. 1.2 Troposcatter and LOS paths.

This technique can be used to connect the two terminal sites in a single hop. If conventional LOS radio links were used to obtain the same connection (Fig. 1.2) several intermediate repeaters would be required to maintain reciprocal visibility between corresponding antennas, but the terrain conditions could be unfavorable for the installation of repeaters.

In subsequent chapters we will study in detail the techniques utilized for exploiting this type of propagation which permits transmission well beyond the horizon. However, we must remember that, as we saw in Fig. 1.1, propagation just beyond the horizon may take place by diffraction. Therefore diffraction paths are also considered to be transhorizon and are briefly discussed in Chapter 6. For the moment we just note that there are two types of diffraction path: (a) the knife-edge diffraction path (Fig. 1.3); (b) the smooth-earth diffraction path (Fig. 1.4). Both types have been studied and solved theoretically, but they are ideal cases which generally represent a simplification of the real situation.

Fig. 1.3 Knife-edge diffraction path.

Fig. 1.4 Smooth-earth diffraction path.

1.2 Short historical outline of beyond-the-horizon radio communications

Since radio waves are part of the electromagnetic spectrum they were initially believed to propagate in the same way as optical waves, and therefore certainly not beyond the horizon or any other kind of obstacle. LOS propagation was considered as the only possibility for radio waves. Thus it was rather astonishing to discover that waves with frequencies of the order of megahertz or tens of megahertz could apparently ignore the curvature of the Earth and be received hundreds or thousands of kilometers from the transmitter. This strange phenomenon was finally explained by the existence of ionospheric reflecting layers, and the first transatlantic commercial link was achieved in the 1920s.

However, it was soon recognized that the ionosphere was no longer reflecting at frequencies beyond say 30 MHz, and it was verified that the limit for transmission at these and higher frequencies was actually the horizon, as expected from the optical laws. Research on long-range communications was therefore abandoned by most workers, who ignored the experiments and ideas of some courageous investigators, including Marconi, who had not left the field but were performing tests at higher frequencies. However, the power available at that time was not sufficient to allow the weak transhorizon effects to be detected. About 12 years passed before the ideas of these pioneers gained acceptance.

Studies of diffraction over a spherical surface were conducted around 1935, and in 1937 the first complete and applicable theory of this subject was published by Van Der Pool and Bremmer. The predictions of the theory were somewhat pessimistic with respect to the experimental data, and therefore refraction by the atmosphere was included in the calculations. This is the origin of the concept of the "effective radius of the Earth", as it was recognized that the decrease in the refractive index of the atmosphere with height produced the equivalent of an increase in the Earth's radius.

The problem of propagation beyond the horizon appeared to be completely solved and no further research was carried out until the Second World War. The advent of radar, which operated at higher frequencies and higher power, revealed the existence of reflections well beyond the horizon where the theory predicted that transmission was impossible. Similar observations were made at ultra-short-wave stations. These results were explained by developing a theory of diffraction in an atmosphere where the refractive index varied nonlinearly with height. In 1946 some researchers discovered the existence of reflecting layers and ducts in the atmosphere. These were more common over the sea, where radar was most frequently used.

The rapid development of frequency modulation (FM) radio stations, television stations and radio relay links after the Second World War resulted in reports of the presence of much inexplicable interference, over land as well as over the sea. The conventional explanation of this phenomenon would have required a permanent condition of super-refraction in the atmosphere, which certainly could not be the case over land. In 1949 the US Government was obliged to restrict the issue of licenses for new television stations because of cochannel interference caused by propagation beyond expected boundaries.

The first theory of propagation beyond the horizon as a result of atmospheric turbulence was proposed in about 1950. It demonstrated the possibility of permanent communications at long distances. Other theories of scattering as a result of partial reflection, diffusion, dispersion, terrain irregularities etc. were proposed subsequently. At the same time the technology had developed sufficiently to allow the transmission of r.f. powers of the order of kilowatts at frequencies of up to 1000 MHz using power klystrons and low noise amplifiers for receivers, and parabolic antennas of large diameter were available. Many experimental links were implemented between 1950 and 1955, and they showed that the received signal, although continuously variable, was permanently present and gave promise of reliable communications well beyond the horizon in hops of more than 500 km.

Theoretical and experimental work was continued, and by about 1960 the theory was well established. The first military and commercial troposcatter links were installed from about 1953 onwards. These links were in the 400–900 MHz band and had a capacity of up to 60 telephone channels. The path length was up to 300 km. Thus the new troposcatter technique was born.

1.3 The place of troposcatter links among other transhorizon communication techniques

Present-day techniques for achieving single-hop long-range transhorizon radio communications are as follows.

(a) *H.f. (short-wave) communications*: this technique utilizes ionospheric reflections for transmitting signals at distances of up to thousands of kilometers. The bandwidth of the medium permits transmission of only one or two telephone channels, and there are other major physical limitations on the use of h.f. subbands which depend on the time of day. Furthermore, the h.f. spectrum is very crowded. The technique has been used extensively, particularly, before the advent of satellites.

(b) *Ionospheric scatter*: this technique utilizes the scattering of signals in the ionosphere (a phenomenon similar to troposcatter) at very high frequencies of up to say 100 MHz with hops reaching some thousands of kilometers. The bandwidth of the medium is very limited, permitting transmission of only a few telephone channels, with other limitations due to fading etc. The technique is rarely used.

(c) *Meteor bursts*: this technique utilizes reflections from the ionized tails of micrometeors that are always present in the high atmosphere and permits hops of many hundreds or some thousands of kilometers. Continuity of transmission is not ensured by the physics of the phenomenon, and the signals have to be transmitted in bursts. The technique is under investigation but is not currently utilized.

(d) *Troposcatter*: this technique, which is the subject of this book, permits the transmission of more than 100 telephone channels over hops of some hundreds of kilometers.

(e) *Diffraction*: this technique permits transmission of a large number of channels over short distances beyond the horizon. We will study it in this book.

(f) *Satellites*: this is the most suitable method for very long hops (e.g. intercontinental communications), but it has proved uneconomical for the substitution of local transhorizon hops or networks and of insufficient capacity in some cases.

It therefore appears that the troposcatter technique is most suitable for single transhorizon hops or "local" networks with hop lengths of the order of hundreds of kilometers where other techniques are unsuitable or too expensive.

1.4 Main characteristics of troposcatter radio links

As an introduction to the topics that will be discussed in detail in subsequent chapters we give here a general review of the main characteristics of troposcatter links (Table 1.1). A troposcatter path can vary in length from about 100–150 km to almost 1000 km. However, the path length of most of the systems currently in use is between 150 and 400 km. The maximum path attenuation including deep fades is very high, of the order of 190–240 dB. Consequently high r.f. transmitted powers, low noise receivers and high gain parabolic antennas are normally required.

Table 1.1 Main characteristics of troposcatter radio links

Very long paths (with terminal sites reciprocally beyond visibility)
Very high path attenuation
High r.f. radiated power
High gain antennas (large paraboloids)
Very sensitive low noise receivers
Frequency range 300–5000 MHz
Frequency modulation
Traffic capacity of up to 120 telephone channels (analog) and up to 8 Mbit s^{-1} (digital)
Use of diversity techniques
Strong independence of terrain profile
Limited bandwidth

Normal r.f. power outputs are 100 W, 1 kW and 10 kW. In some cases lower or higher powers have also been used. However, it seems that the trend is in favor of 1 kW. Owing to the variation in path loss throughout the year, the r.f. power can sometimes be decreased with a consequent reduction in power consumption and interference. The antennas are normally of the parabolic type with diameters ranging from a few meters up to 30 or even 40 m. Antenna gain increases with increasing diameter, but gain degradation also increases owing to the incoherence of the r.f. field and this limits the maximum diameter that can be used. Receivers with low noise characteristics are normally employed. From a cost-effectiveness standpoint, it is sometimes better to extend the receiver's threshold by means of low noise amplifiers and bandwidth compression devices rather than to increase the r.f. power or the diameter of the antennas.

The frequency bands of troposcatter systems generally lie between 300 and 5000 MHz. These limits are not determined by particular theoretical or physical requirements but by practical considerations. At the lower end of this band the antennas would be too large to provide the necessary gain, while at the upper end the gain degradation would be too high. Some frequency bands available for troposcatter systems are listed in Table 1.2. The data are taken from the ITU *Radio Regulations* and represent bands allocated for fixed services on a primary basis. For mobile services the same bands are used with some restrictions. It will be seen later that each system has an optimum frequency band for economic operation.

Table 1.2 Frequency bands allocated to fixed services on a primary basis

Region 1 (Europe, Asia excluding zones south of the USSR)	Region 2 (North and South America)	Region 3 (India, China, Iran and zones south of them, Australia)
335.4–399.9 MHz	335.4–399.9 MHz	335.4–399.9 MHz
406.1–430.0 MHz	406.1–430.0 MHz	406.1–430.0 MHz
440–470 MHz	440–470 MHz	440–470 MHz
790–960 MHz	806–960 MHz	470–960 MHz
1427–1530 MHz	1427–1525 MHz	1427–1530 MHz
1668.4–1690 MHz	1668.4–1690 MHz	1668.4–1690 MHz
1700–2690 MHz	1700–2690 MHz	1700–2690 MHz
3400–4200 MHz	3400–4200 MHz	3400–4200 MHz
4400–5000 MHz	4400–5000 MHz	4400–5000 MHz
5850–8500 MHz	5850–8500 MHz	5850–8500 MHz

The equipment used must match the characteristics of the propagation, which produces a continuously variable r.f. field at the receiver input with frequent deep fades. This is done by using "diversity techniques". The receiving section of the equipment is composed of a number of radio receivers which receive independent samples of the r.f. signal and combine them in an appropriate manner by means of a "diversity combiner", thus obtaining an output signal with much more stable characteristics. The order of diversity, i.e. the number of receivers used, is normally either two or four. Good economic systems can be obtained with dual diversity, but the variations in the r.f. field are not always smoothed sufficiently to permit satisfactory transmission of the telegraphy channels carried by some of the telephone channels. Quadruple diversity appears to be an optimum technical solution which permits good transmission of both telephone and telegraph channels.

As in LOS radio relay links, both analog and digital techniques are used in troposcatter links. Originally only analog links were considered, and most of the links in operation at present are of this type. Digital techniques were studied theoretically and experimentally for many years before the first links were implemented. The characteristics of both types of link will be studied in detail in this book. For the moment we can anticipate that a subjective comparison of the performances of the two types of link by the normal user will favor the digital link. Objectively, the digital link performs better under poor conditions. In fact during a telephone call the user of an analog link may hear a variable background noise, whereas with a digital link he would hear a perfect signal except for some minor brief interruptions under poor propagation conditions. The main advantages of digital over analog troposcatter systems are summarized in Table 1.3. However, these advantages may be outweighed by other features (e.g. wider bandwidths), and all aspects of the problem must be considered before a system is chosen.

Table 1.3 Advantages of digital troposcatter

Much better control of the transmission quality owing to the fixed value of the signal-to-noise ratio which is almost independent of the propagation characteristics and the length of the system
Better protection against interference, jamming and other disturbances
Possibility of the introduction of error-correcting techniques
Possibility of encryption for ensuring secrecy or privacy
Easier interfacing with other digital systems

Analog troposcatter systems have a traffic capacity, in terms of 4 kHz FDM telephone channels, of 6–120 channels. Higher capacities are generally not recommended because the propagation mechanism generates nonlinear noise in the channels, thus limiting the transmissible bandwidth. Systems carrying 240 channels have also been implemented at the cost of accepting a lower quality. However, for TV transmission the performance would normally be rather poor. Frequency modulation is always used. Experiments with amplitude modulation–single-sideband modulation (AM–SSB) have been carried out, but they did not prove to be practicable.

Digital troposcatter systems can transmit a bit stream of up to 8–12 Mbit s^{-1}. These systems are more sensitive to propagation distortions and their bandwidth becomes too large above 2–3 Mbit s^{-1}, so that transmission even at an acceptable path loss is impossible unless special complex modem techniques are used. Modulation can be of the FSK or PSK type.

1.5 Design calculations and their reliability

We will see in subsequent chapters that all parameters necessary for the design of a troposcatter system can be stated with sufficient accuracy except those related to propagation, which are statistical in nature and difficult to predict.

Troposcatter propagation can take place because of the presence of inhomogeneities in the atmosphere. These are of various types and can vary in unpredictable ways with time and from region to region of the world. A new link could be designed with a maximum of confidence by previously performing propagation tests or sounding the troposphere for sufficiently long periods of time. However, this is generally impossible or too expensive. Prediction methods are practical only if they are based on simple parameters characterizing the behavior of the troposphere, but the complexity of the physical situation is such that a solution which is both simple and reliable cannot be found. The only parameters available in practice, which are measured at least daily at many meteorological stations around the world, are the refractivity of air and its gradient. These parameters have been used in semi-empirical prediction formulae. In the absence of better solutions, they can be considered as approximate descriptions of the behavior of the troposphere and they are currently used with caution in the predictions.

The various prediction methods may give different results for the same path, and the behavior of a newly designed link is always subject to some uncertainties. This situation is recognized by the CCIR. Therefore the design calculations are only approximate, and their validity depends on the method used and the data available for the zone of interest. Various climatic regions have been studied and an adequate amount of propagation data is available for them. The designer must make estimations for other regions.

1.6 Feasibility of a troposcatter path

The design of a troposcatter path requires much less information than is necessary for an LOS radio link, and this is of great importance particularly in those regions of the world for which accurate topographic maps are not available. In such a case the planning of an LOS microwave system could require expensive aerial surveys which are not necessary for a troposcatter system. It is necessary to know only the path length and the elevation or takeoff angle of the horizon from each site to design a troposcatter path. The essential data required for different levels of study are listed in Table 1.4.

Table 1.4 Data required to design a troposcatter path

Feasibility study	Complete study (accurate maps not available)	Complete study (accurate maps available)
Region and available climatic data Path length Measured elevation angle of radio horizon or (a) altitude of sites (b) altitude of radio horizon (c) distance of radio horizon If the radio horizon is the sea, (b) and (c) are not required	Available climatic data Coordinates of the sites Altitude of the sites above sea level Measured elevation angle of radio horizon	Available climatic data Coordinates of the sites

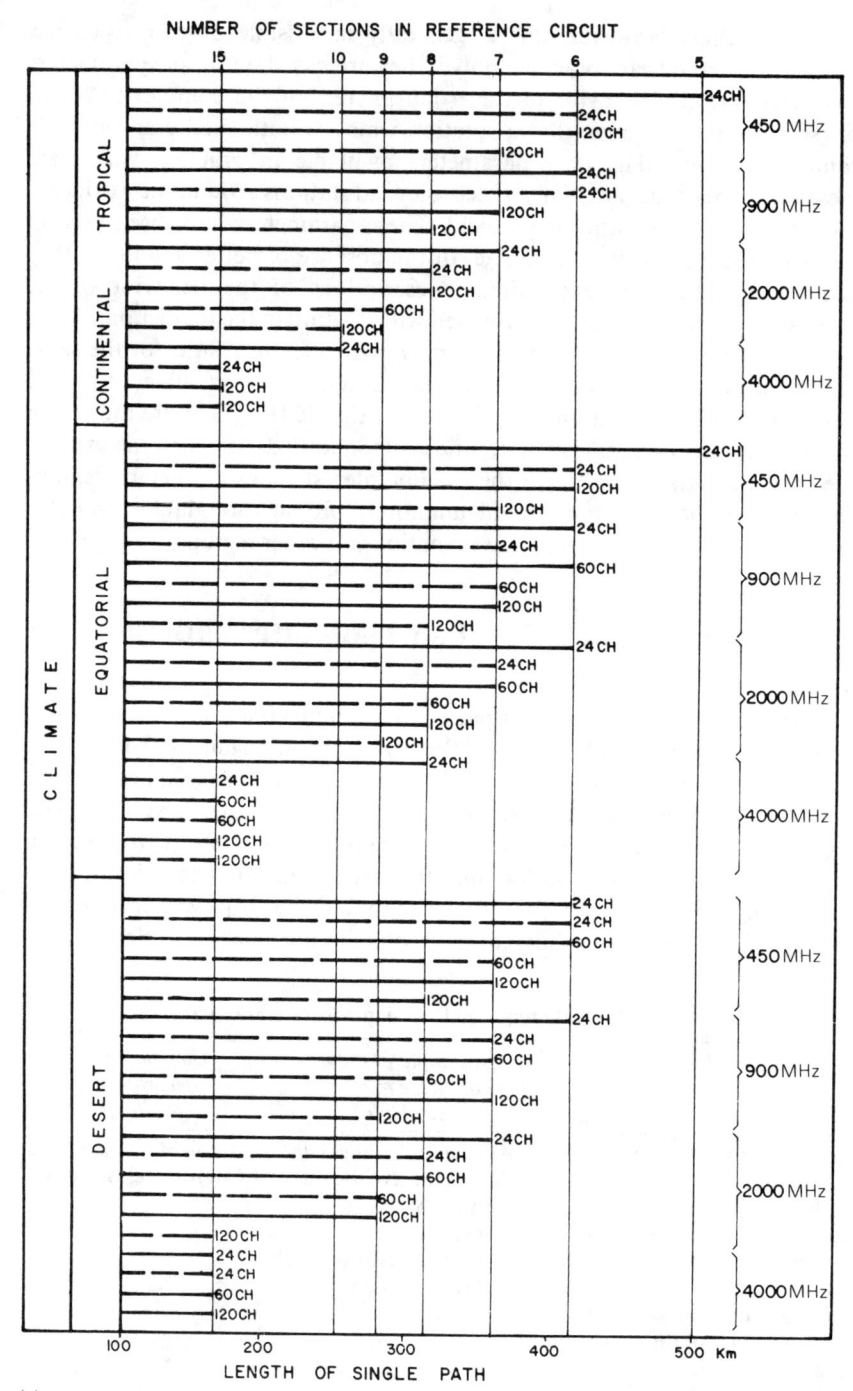

(a)

[Figure: bar chart showing feasibility of troposcatter paths by climate (Temperate Sea, Temperate Land, Desert, Equatorial and Sub Tropical) versus length of single path in km]

(b)

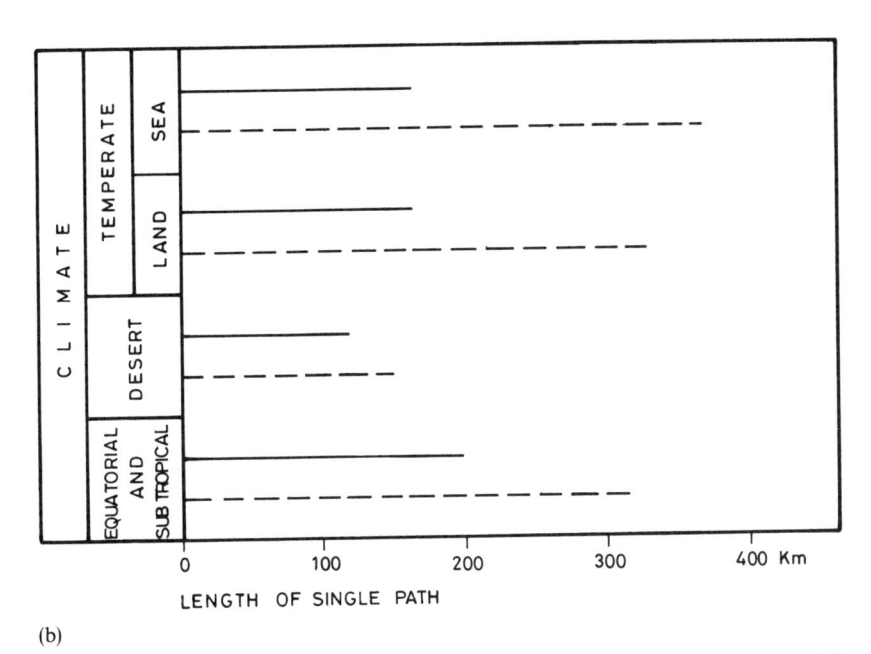

Fig. 1.5 Feasibility of troposcatter paths: (a) analog paths with CCIR characteristics for second-class systems (- - -, dual diversity; ———, quadruple diversity); (b) digital paths (———, 2 Mbit s^{-1}; - - -, 512 kbit s^{-1}; 4 GHz quadruple diversity).

The design of an analog link can be based on the CCIR Recommendations when the link is integrated with a public telecommunications network. In this case the performance may be first class or second class. If the link is only part of a small local system the performance requirements are normally lower and the cost is also lower.

The feasibility of analog troposcatter paths in accordance with the CCIR Recommendations for some typical climatic regions and for second-class systems is given in Fig. 1.5(a). The results are only indicative and are based on zero elevation angles at the terminal sites, an r.f. power no greater than 10 kW and antenna diameters not exceeding 30 m (or 10 m for 4 GHz). Second-class systems are feasible for distances of up to 400 or 500 km for a small number of channels and a low r.f. band and up to a few hundred kilometers for higher capacities, depending on the frequency band and the type of climate. Other conditions being equal, quadruple diversity systems can generally operate over longer distances than dual diversity systems. The feasibility range is broader for troposcatter systems of lower performance. As the above is only indicative, actual paths should be assessed individually using the methods explained in this book.

Similar considerations apply to Fig. 1.5(b) which shows the feasibility of some digital paths.

1.7 Specific field of application of troposcatter systems

From an application point of view the most interesting characteristic of troposcatter radio links is the great distance over which reliable communication can be obtained without the need for intermediate repeaters (Fig. 1.2). This feature is particularly useful for cases where the terrain is difficult such as the following:

(a) connection between sites in the desert or the jungle;

(b) connection of a remote island to the mainland or another island;

(c) military networks, in which the possibility of sabotage at unattended repeater stations should be avoided at the design stage;

(d) connection of an oil drilling platform far out at sea to offices on shore.

The possibility of avoiding repeater stations, with their equipment, antennas, buildings, roads and problems of accessibility and maintenance, may lead to a more economical solution with a troposcatter link instead of a conventional LOS link. Extensive use of troposcatter radio systems is expected in the future, particularly in developing countries. It is expected that troposcatter links will provide a substantial part of their national networks, particularly in the first expansion phase when only a moderate number of channels is required.

1.8 Types of troposcatter system

The following types of troposcatter radio link have been developed.

(a) *Fixed systems*, which are the most common, in which the electronic equipment is installed in buildings and the antennas are fixed in the terrain. Examples of troposcatter stations of this type are shown in Fig. 1.6.

(b) *Transportable systems*, such as complete stations mounted in shelters which are prepared at the factory and then transported to the sites by truck, ship and/or helicopter (Fig. 1.7). The antennas are often of the fixed type.

(c) *Mobile systems*, in which the radio station is mounted in a small shelter or container that is either fixed on a truck or can be easily loaded/unloaded from the truck. The antennas and power generators can also be mounted and dismounted easily and are transportable by truck. Figure 1.8 shows an example of a truck-mounted quadruple diversity mobile terminal.

Transportable systems are particularly useful in those cases in which conventional installation of the equipment at the sites would be difficult

Fig. 1.6 Fixed troposcatter stations: (a) Lampedusa, Italy (quadruple diversity); (b) Al Bayda, Yemen (dual frequency diversity).

Fig. 1.7 Transportable troposcatter station.

Fig. 1.8 Quadruple diversity mobile troposcatter station.

because of local conditions. Examples of such cases are installations in the desert or on mountain tops and when there is a limit on the time available for installation (because of short-term visas, difficult meteorological conditions etc.). Installation in a shelter can be done more easily at the factory where any problem can be solved immediately without the difficulties and the waste of time often experienced at inhospitable sites.

Mobile systems are often required by military authorities for strategic communications and may also be useful for emergency connections. Since mobile antennas cannot exceed a certain size (e.g. a diameter of 7 m) because of the requirements of flexibility in mounting/dismounting, stability against wind, transportability etc., mobile systems are used mainly on shorter paths and/or with low traffic capacities.

1.9 Advantages of troposcatter systems

The main advantages of troposcatter systems can be summarized as follows.

(a) They permit a long path: a single hop can have a length of hundreds of kilometers, i.e. up to five or six times the usual length of an LOS hop (Fig. 1.2).

(b) They can be used on difficult terrain. The only requirement for a troposcatter path is that sites are chosen such that the antennas are if possible pointed horizontally or even slightly downward. The type and topography of the terrain have little influence on the system design.

(c) Coverage of very large areas can be obtained with a small number of hops.

(d) Few repeaters are required because of the long hop length.

(e) There is a saving of frequencies because of the reduced number of hops and stations. This saving is substantial and is very important in congested areas.

(f) Reduced maintenance is required because of the reduction in the number of stations.

(g) Security against sabotage or catastrophic failure is easier to achieve as there are fewer stations to protect.

(h) Costs are reduced because of the lower number of repeaters with associated buildings, access roads, power plants, spares, test equipment, maintenance personnel etc.

(i) There is a high immunity to interception because of the use of very narrow beam antennas.

The main disadvantages are as follows.

(a) The cost is high. However, this factor may be superseded by the advantages given above and the troposcatter solution may be the best choice. Under certain circumstances the overall cost (investment, operation and maintenance during the life of the plant) may be no higher than the costs of alternative solutions.

(b) There is a risk of interference over a wide area if the same frequencies are used at other stations. Therefore the CCIR advises against troposcatter when other methods can be used without excessive difficulty.

Experience shows that when troposcatter is one of a number of solutions under consideration, a technical–economical evaluation together with consideration of the factors discussed above may easily favor troposcatter. However, it is difficult to give a general rule, as each case is different and requires individual analysis.

1.10 International recommendations and authorities

The field of telecommunications is coordinated by national and international agencies, which also issue recommendations, regulations etc. for ordering, standardizing and making compatible the various types of equipment, systems, networks and so on. These recommendations also serve as an important reference for designers and purchasers, who can base their specifications on widely accepted standards. In the specific field of troposcatter reference is often made to such agencies, particularly in international requests for bids and in the evaluation of acceptable performances of equipment and systems. Therefore we review the most important of these organizations here.

The United Nations has a specialized agency for the planning, coordination, regulation and standardization of telecommunications worldwide. This agency is the International Telecommunications Union (ITU). The work of the ITU is performed by permanent committees, two of which are described below.

The International Radio Consultative Committee (CCIR) studies and issues recommendations on technical and operating questions relating to radiocommunications.

The International Telegraph and Telephone Consultative Committee (CCITT) studies and issues recommendations on technical, operating and tariff questions relating to telegraphy and telephony.

We are particularly interested in the work of the CCIR. This committee authorizes study groups composed of international experts who examine practical problems in the field of radio communications and propose solutions in the form of recommendations and reports. These documents are submitted to a Plenary Assembly held every 4 years. If the documents are accepted, they are published in the CCIR Books which are issued in an updated edition after the Assembly.

The CCIR have issued several recommendations and reports concerning troposcatter links (see Appendix 1). When a troposcatter link is intended for public service its design, characteristics and performance should be in agreement with the CCIR Recommendations. This is often specifically requested in the Technical Specifications annexed to the Request for Bid. However, the CCIR has so far considered mainly analog troposcatter links, and there are still few references to digital troposcatter links.

High quality, which is also expensive, is not required for troposcatter links that are not intended to be connected to public networks but are part of local limited networks, and thus international recommendations can be ignored. The only civil authority that has issued international recommendations in this field appears to be the ITU.

In the military field very good standards for troposcatter (and LOS) links have been issued by the American Defense Communications Agency (DCA) under the auspices of the Department of Defense. They cover both analog and digital links to be used for global communications in the Defense Communications System (DCS) and are very stringent. In the absence of CCIR Recommendations for the dimensioning of digital troposcatter radio links we will refer to the DCA Standards in Chapter 9.

In 1967 the National Bureau of Standards (NBS) issued a revised edition, which is still valid, of *Technical Note 101, Transmission Loss Predictions for Tropospheric Communication Circuits* (two volumes). It outlines a method that has been very widely used by designers and that has also been accepted by the CCIR in *Recommendation 617*.

In addition the ITU has issued the well-known *Radio Regulations* which, amongst many other things, list the subdivisions of the radio spectrum in the frequency bands reserved for the various types of services (fixed, mobile, broadcasting, radio navigation etc.). There is no specific mention of troposcatter in these *Regulations*, except for two notes. One appears in *Recommendation 100*, in which it is suggested that the frequency bands of the fixed service to be used in preference for troposcatter systems should be established at a future world administrative conference. The second note concerns the 2500–2690 MHz band in which development of new troposcatter systems should be avoided, except for Region 1 where it is allowed only after agreement between administrations and if antennas do not point towards the orbit of geostationary satellites.

1.11 Comments and suggestions for further reading

The only existing books dedicated to troposcatter appear to be refs 1.1 and 1.2. They were published some time ago but can still be read with profit. Other books may include a chapter on this subject but do not deal with it in detail. The CCIR Books [1.3, 1.4] should be studied carefully by the designer, who should be familiar with the main Recommendations and Reports. It is also advisable to look through the GAS 3 Manuals [1.6], which have been developed to help administrations in the design of transmission systems. Reference 1.7 is a classic publication on troposcatter, and most of its content is still valid. Many systems which have been implemented, their performance, the results of measurements, methods of calculating various parameters, theories, new equipment etc. are described in the literature. The most important papers are listed in the bibliography after the CCIR Recommendations and Reports.

References

1.1 du Castel, F. *Propagation Troposphérique et Faisceaux Hertziens Transhorizon*, Chiron, Paris, 1961.
1.2 Panter, P. F. *Communication Systems Design—Line-of-sight and Troposcatter Systems*, McGraw-Hill, New York, 1972.
1.3 *Recommendations and Reports of the CCIR, 16th Plenary Assembly, Dubrovnik, 1986*, Vol. 5, *Propagation in Non-ionized Media*, ITU, Geneva, 1986.
1.4 *Recommendations and Reports of the CCIR, 16th Plenary Assembly, Dubrovnik, 1986*, Vol. 9, *Fixed Service Using Radio Relay Systems*, ITU, Geneva, 1986.
1.5 ITU General Secretariat, *Radio Regulations*, Vols 1 and 2, ITU, Geneva, 1982.
1.6 Transmission systems—economic and technical aspects of the choice of transmission systems, *GAS 3 Manual*, Vols 1 and 2, ITU, Geneva, 1976.
1.7 Transmission loss predictions for tropospheric communication circuits, *NBS Tech. Note 101*, Vols 1 and 2, January 1967.

Chapter 2

Review of Statistical Distributions used in Troposcatter

As discussed in Chapter 1 an originally steady signal transmitted through a troposcatter path presents widely varying characteristics at the receiver input. For example, its amplitude varies randomly, but it has been observed that to a good approximation this variation obeys well-known statistical laws. Furthermore, we also noted the use of diversity techniques to combine various random signals in an appropriate manner to decrease the overall variability and to obtain a more stable signal.

The main characteristics of the statistical distributions encountered in troposcatter are reviewed in this chapter. Detailed descriptions are not given, but all properties mentioned in this book and used in practice are noted. The reader is assumed to have a basic understanding of statistical distributions.

2.1 Random variables and their statistical distributions

The random variables of greatest interest in troposcatter are generally signals, noise and their associated parameters such as attenuation, signal-to-noise ratio etc. However, statistical variability may also affect some other characteristic troposcatter parameters such as the correction factor for the

Earth's radius, the transmissible bandwidth, the multipath spread, the duration of fading etc. These variables can be discrete (e.g. a set of 1-min-averaged signal strengths over a period of 1 month) or continuous (e.g. the r.f. signal level received during a period of 5 min). For simplicity we will consider only continuous variables here; the expressions for the discrete cases can easily be derived. Furthermore, we will sometimes take probability to imply frequency of occurrence. We will express probabilities in either numerical values (0–1) or percentages (0%–100%).

The statistical behavior of a continuous random variable x can be represented by the following functions.

The *probability density function* $p(x)$ yields the probability dp that the random variable x has a value between x and $x+dx$:

$$dp = p(x)\,dx \tag{2.1}$$

The integral of (2.1) between $-\infty$ and $+\infty$ is obviously equal to unity.

The *(cumulative) distribution function* $P(x)$ yields the probability P that the variable x does not exceed the value X:

$$P(x<X) = \int_{-\infty}^{X} p(x)\,dx \tag{2.2}$$

The corresponding probability $P(x>X)$ that x exceeds X can be obtained by integrating between X and $+\infty$ or from the relation $1 - P(x<X)$.

Cumulative distributions are currently used in troposcatter. They are often presented in diagrammatic form on probability (Gaussian) paper and indicate for any value of the variable its probability of being exceeded or not exceeded. The adjective "cumulative" is redundant, although it is frequently used, because a distribution is automatically cumulative.

The following parameters are used to characterize the statistical distribution of a continuous random variable x.

(a) The *mean* or *average* value is given by

$$M = \int_{-\infty}^{\infty} x\,p(x)\,dx \tag{2.3}$$

(b) The *median* value is exceeded (or not exceeded) by 50% of the values. It is given by X in (2.2) when $P = 0.5$.

(c) The *variance* σ^2 is given by

$$\sigma^2 = \int_{-\infty}^{\infty} (x-M)^2 p(x)\,dx$$
$$= \int_{-\infty}^{\infty} x^2 p(x)\,dx - M^2 \tag{2.4}$$

If the variable is a voltage with zero mean (e.g. a noise voltage), its variance is proportional to its power. If M is nonzero, the signal power is proportional to its variance plus M^2.

(d) The *standard deviation* σ is the square root of the variance. If the variable is a voltage with zero mean, its standard deviation is proportional to its root mean square (r.m.s.) value.

(e) The *spread* 2σ yields a general indication of the width of the probability density function about the mean M.

The variables can be measured in a variety of units which can be reduced to the following fundamental types:

numerical units of amplitude (V, mV, µV, mA, µA etc.)
numerical units of power (W, mW, pW, pW0, pW0p etc.)
logarithmic units of level which are derived from the logarithm or natural logarithm of the above units (dB, dBm, dBm0, dBm0p etc. or nepers (Np) etc. (1 Np = 8.686 dB))

The same random variable can be presented in either numerical or logarithmic units. For example, a signal strength can be indicated as S mW or S dBm, and this affects the shape of the distribution curves displayed on probability paper. We will see that a normal (Gaussian) distribution of levels corresponds to a log-normal distribution of amplitudes for the same variable. A Rayleigh distribution of amplitudes corresponds to an exponential distribution of powers. However, in practice the name of the basic distribution is often maintained.

Before proceeding we recall the following definitions (a is an amplitude ratio; ln, natural logarithm; log, logarithm to base 10):

$L(\text{Np}) = \ln a$ (level in nepers corresponding to the ratio a)
$a = e^L$ (reverse of the above formula (L in Np))
$L(\text{dB}) = 20 \log a$ (the same level expressed in dB)
$a = 10^{L/20}$ (reverse of the above formula (L in dB))
$r = e^{2L}$ (ratio of powers corresponding to $L(\text{Np})$)
$r = 10^{0.1L}$ (ratio of powers corresponding to $L(\text{dB})$)

2.2 Normal and log-normal distributions

We now consider a random variable such as signal strength that can be expressed in different units, i.e. x(mW) (signal power in mW) or X(dBm) (signal level in dBm) where

$$X(\text{dBm}) = 10 \log x(\text{mW}) = 4.34 \ln x(\text{mW}) \tag{2.5}$$

We use this variable to illustrate the two distributions.

2.2.1 Normal distribution

If X(dBm) has a normal (Gaussian) distribution we can write

$$dp = \frac{1}{\sigma(2\pi)^{1/2}} \exp\left\{-\frac{(X-M)^2}{2\sigma^2}\right\} dX \qquad (2.6a)$$

$$= \frac{1}{\sigma(2\pi)^{1/2}} \exp\left(-\frac{Y^2}{2\sigma^2}\right) dY \qquad (2.6b)$$

where dp is the elementary probability, M is the mean value (in dB) of the variable X (for a normal distribution the mean and the median coincide), σ is the standard deviation (in dB) of the distribution and $Y = X - M$ is the deviation from the mean (in dB).

The probability density function of the normal distribution (eqn (2.6b)), relative to its mean value and taking σ as unity, is shown in Fig. 2.1. Scale C represents the percentage probability that the value Y is not exceeded and scale E represents the probability that the absolute value $|Y|$ is exceeded (see eqn (2.8) below).

The (cumulative) distribution, which yields the probability P that the value Y_0(dB) is not exceeded, is given by

$$P(Y < Y_0) = \frac{1}{\sigma(2\pi)^{1/2}} \int_{-\infty}^{Y_0} \exp\left(-\frac{Y^2}{2\sigma^2}\right) dY \qquad (2.7)$$

Conversely, $1 - P$ is the probability that Y exceeds the value Y_0.

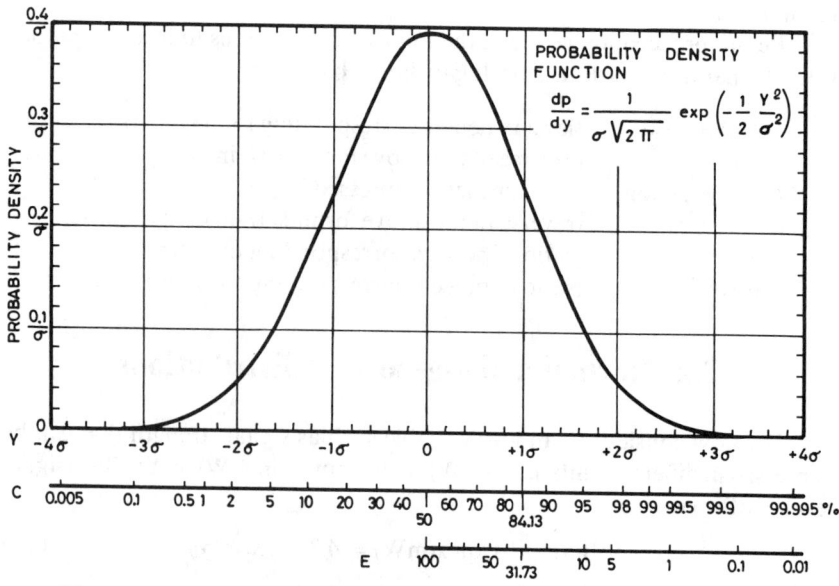

Fig. 2.1 Normal probability density function. (After *ITT Manual*, 1968.)

Statistical distributions are often drawn on special Gaussian paper with the probability scale suitably altered so that a normal distribution is represented by a straight line. For example, Fig. 2.2 shows the distribution (2.7), relative to its median value, plotted on Gaussian paper. In order to plot the line on this special paper it is necessary and sufficient to know only two points, e.g. the median M and the standard deviation σ. Table 2.1 gives the coordinates of some significant points of the normal distribution with reference to its median value. For example, if the median value is M dBm, the probability that the value $M(\text{dBm}) - 1.7\sigma(\text{dB})$ is not exceeded is 5% (and the probability that it will be exceeded is 95%).

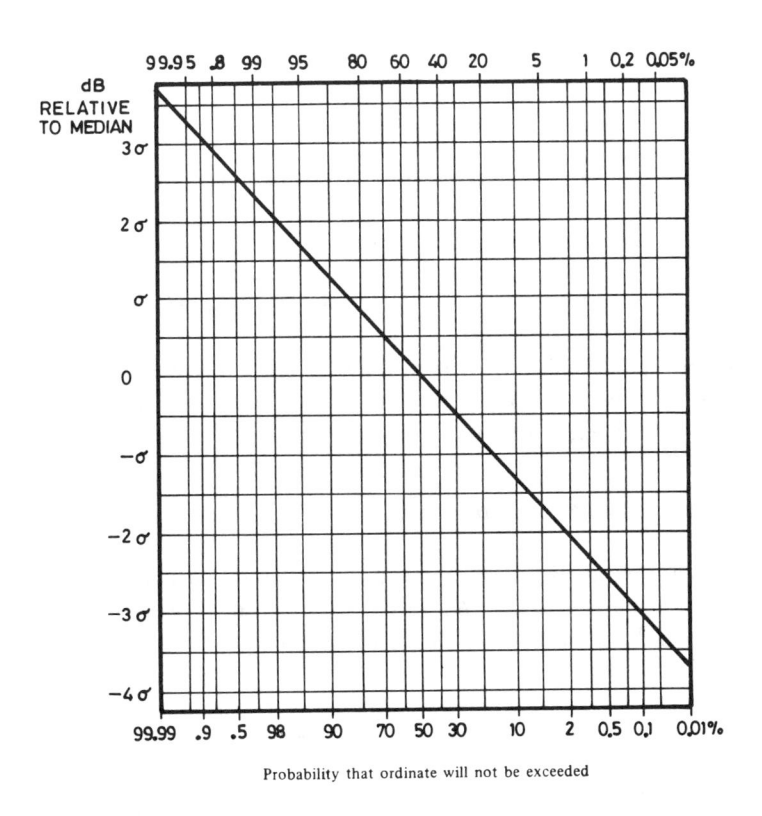

Fig. 2.2 Cumulative normal distribution.

Some random variables, e.g. the input r.f. signal to an FM radio receiver and the corresponding output noise, have the same cumulative distribution but with opposite variations. If X(dBm) is the level of the signal and N(dBm) is the level of the noise, an increase in X by 1 dB implies a decrease in N by 1 dB. Thus the same curve with an inverted scale can represent both distributions.

Table 2.1 Some significant points in the normal distribution

Per cent	Y	Per cent	Y
0.01	-3.7σ	99.99	$+3.7\sigma$
0.1	-3.1σ	99.9	$+3.1\sigma$
1	-2.3σ	99	$+2.3\sigma$
5	-1.7σ	95	$+1.7\sigma$
10	-1.3σ	90	$+1.3\sigma$
20	-0.8σ	80	$+0.8\sigma$
50	-0	50	$+0$
15.87	-1.0σ	84.13	$+1.0\sigma$
6.68	-1.5σ	93.32	$+1.5\sigma$
2.28	-2.0σ	97.72	$+2.0\sigma$
0.62	-2.5σ	99.38	$+2.5\sigma$
0.13	-3.0σ	99.87	$+3.0\sigma$
0.02	-3.5σ	99.98	$+3.5\sigma$

If the variable X is referred to its mean M and is measured in units of $2^{1/2}\sigma$, i.e. a new variable $t = (X-M)/2^{1/2}\sigma$ is chosen, the probability that $-t_0 \leqslant t \leqslant +t_0$ or that $|t| \leqslant t_0$ can be written

$$P(|t| \leqslant t_0) = \frac{2}{\pi^{1/2}} \int_0^{t_0} \exp(-t^2)\, \mathrm{d}t = \mathrm{erf}(t) \qquad (2.8)$$

which defines the well-known error function erf(t). The complementary error function is erfc(t) = 1 − erf(t) and it also can be obtained from (2.8) by integrating between t_0 and $+\infty$.

2.2.2 Log-normal distribution

The normal distribution of the X(dBm) values corresponds to the log-normal distribution of x(mW) values for the same variable. The density function can be obtained from (2.6a) by substituting (2.5) and its differential $\mathrm{d}X = 4.34\,\mathrm{d}x/x$:

$$\mathrm{d}p = \frac{4.34}{\sigma(2\pi)^{1/2}} \exp\left\{-\frac{(10\log x - M)^2}{2\sigma^2}\right\} \frac{\mathrm{d}x}{x} \qquad (2.9)$$

The cumulative distribution, which yields the probability that the value x_0(mW) is not exceeded, is given by

$$P(x \leqslant x_0) = \frac{4.34}{\sigma(2\pi)^{1/2}} \int_0^{x_0} \exp\left\{-\frac{(10\log x - M)^2}{2\sigma^2}\right\} \frac{\mathrm{d}x}{x} \qquad (2.10)$$

The diagram of the distribution can be the same as that of Fig. 2.2 but with a logarithmic ordinate scale in milliwatts.

The relations between parameters of the normal and log-normal distributions for the same variable are shown in Table 2.2.

Table 2.2 Relations between parameters of the normal and log-normal distributions for the same variable $X(\text{dBm}) = 10 \log x(\text{mW})$

Parameter	Distribution of X		Distribution of x	
Median value	$M = 10 \log m$	(dBm)	m	(mW)
Mean value $\begin{cases} \text{of } X \\ \text{of } x \end{cases}$	M $M + 0.115 \sigma^2$	(dBm) (dBm)	— $m \exp(\sigma^2/37.7)$	(mW)
Standard deviation	σ	(dB)	$m \left\{ \exp\left(\dfrac{\sigma^2}{9.42}\right) - \exp\left(\dfrac{\sigma^2}{18.83}\right) \right\}^{1/2}$	(mW)

2.3 Rayleigh distribution

The Rayleigh distribution is that of the modulus of a vector whose Cartesian components vary randomly and independently according to the Gaussian law with zero mean value and the same variance σ^2. It is also the distribution of the modulus of the sum of a large number of equal vectors with random equiprobable phases.

The probability density function of this two-dimensional Gaussian distribution can be obtained by using the transformation $dx\, dy = r\, dr\, d\theta$ and writing the expression for the probability that the end of the vector r is contained in the area $dx\, dy$ in both Cartesian and polar coordinates:

$$d^2 p = \frac{1}{2\pi\sigma^2} \exp\left(-\frac{x^2 + y^2}{2\sigma^2}\right) dx\, dy \tag{2.11}$$

$$= \frac{d\theta}{2\pi} \frac{r}{\sigma^2} \exp\left(-\frac{r^2}{2\sigma^2}\right) dr$$

The probability density function is shown in Fig. 2.3, and its mathematical expression derived from (2.11) by integration with respect to θ is

$$p(r) = \frac{r}{\sigma^2} \exp\left(-\frac{r^2}{2\sigma^2}\right) \tag{2.12}$$

The cumulative distribution of the probability $P(r)$ that the modulus exceeds the value r is obtained by integrating from r to infinity:

$$P(r) = \exp\left(-\frac{r^2}{2\sigma^2}\right) \tag{2.13}$$

If r is a voltage, r^2 is proportional to the instantaneous signal power W, and $2\sigma^2$, which is the sum of the mean powers of the two orthogonal components, is proportional to the mean signal power W_m. The probability

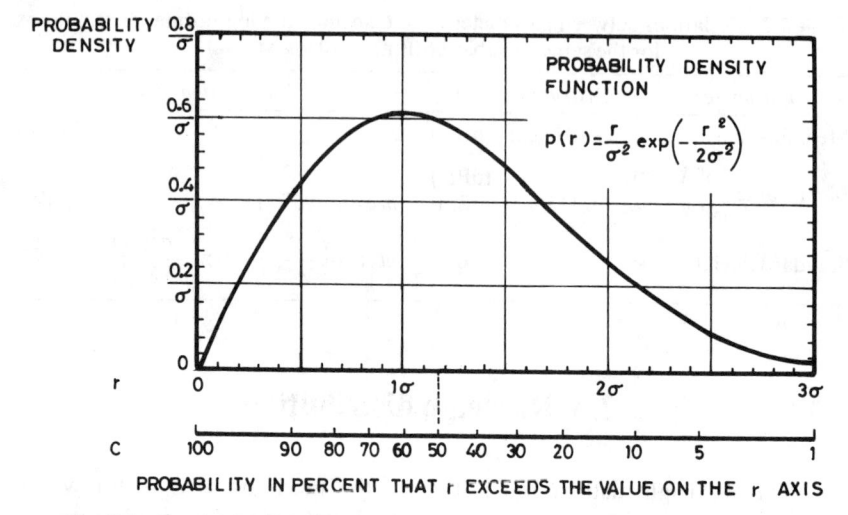

Fig. 2.3 Rayleigh probability density function. (After *ITT Manual*, 1968.)

that the power W is exceeded is then given by

$$P(W) = \exp\left(-\frac{W}{W_m}\right) \tag{2.14}$$

Thus the Rayleigh distribution of voltages becomes an exponential distribution of powers. Equation (2.14) can also be written in terms of the median value W_M, for which $P(W_M) = 0.5$:

$$P(W) = \exp\left\{-(\ln 2)\frac{W}{W_M}\right\} = \exp\left(-0.7\frac{W}{W_M}\right) \tag{2.15}$$

These functions are represented by straight lines on semilogarithmic paper. However, it is customary to draw them on Gaussian paper with values in decibels referred to the median value and to call them cumulative Rayleigh distributions. For example, Fig. 2.4 shows the probability that the signal is below the ordinate value. The probability that the signal will fade by more than -28 dB relative to the median level (i.e. a 28 dB fade) is 0.1%, which means that the signal is more than 28 dB below the median for 0.1% of the time.

This draws attention to the important practical problem of calculating the probability or the fraction of time that the level of the signal falls below a given value. Typically this is a problem of outages: the communications are interrupted each time the r.f. signal falls below the threshold of the receiver.

Consider signal levels X(dB) and U(dB) referred to the median or to the mean power as follows:

$$X(\text{dB}) = 10\log\left(\frac{W}{W_M}\right) \text{ dB above the median} \tag{2.16a}$$

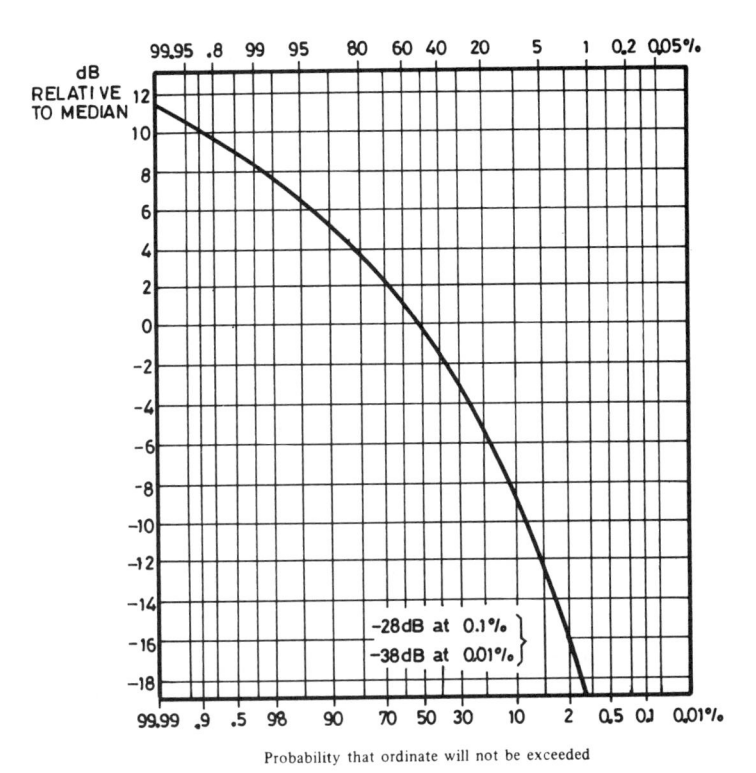

STATISTICAL DISTRIBUTIONS USED IN TROPOSCATTER 29

Fig. 2.4 Cumulative Rayleigh distribution.

$$U(\text{dB}) = 10 \log\left(\frac{W}{W_m}\right) \text{ dB above the mean} \quad (2.16\text{b})$$

Obviously the negative values $-X$ and $-U$ indicate that W is X dB or U dB below the median or the mean power. Note that the difference between the mean and median levels is

$$X - U = 10 \log\left(\frac{W_m}{W_M}\right) = 10 \log\left(\frac{1}{\ln 2}\right) = 1.6 \text{ dB} \quad (2.17)$$

The mean power corresponds to the r.m.s. amplitude and not to the mean amplitude. The dependence of the power ratio on X and U can now be written as follows:

$$s = \frac{W}{W_m} = 10^{0.1U} \quad (2.18\text{a})$$

$$s = (\ln 2)\frac{W}{W_M} = 0.7 \times 10^{0.1X} \quad (2.18\text{b})$$

Expressions (2.18) can be substituted in (2.14) and (2.15) and the probability that s is not exceeded can be rewritten as follows:

$$P(s) = 1 - \exp(-s) \tag{2.19}$$

$$P(X) = 1 - \exp(-0.7 \times 10^{0.1X}) \tag{2.20a}$$

$$P(U) = 1 - \exp(-10^{0.1U}) \tag{2.20b}$$

Figure 2.4 shows a plot of eqn (2.20a). The following data can be obtained from this figure:

(a) the percentage of time for which the signal level is below (or does not exceed) the ordinate value;

(b) the probability of fades deeper than the ordinate value;

(c) the outage probability, or the percentage of time for which the instantaneous signal level C(dBm) is more than X dB below (X negative) its median value M(dBm), so that

$$C(\text{dBm}) \leqslant M(\text{dBm}) - X(\text{dB}) = T_d(\text{dBm})$$

where T_d(dBm) is the threshold level;

(d) the minimum amount in decibels by which the median signal level must be above the threshold for ensuring a given outage probability.

Equation (2.20b) can be interpreted in a similar manner.

It should be noted that, if s is small, eqn (2.19) can be approximated by

$$P(s) = 1 - (1 - s) = s \tag{2.21}$$

In view of the definition of s (eqn (2.18a)) this means that, for small values of the power W, the probability of exceeding W/W_m is proportional to W and that the power varies by 10 dB for each decade of variation in probability.

For digital systems the outage probability can also be expressed as a function of the r.f. carrier-to-noise power ratio γ (numerical value) and of the error probability (or bit error rate (BER)) at threshold P_0. If we assume that the error probability P (or BER) is given by (see Chapter 7, Section 7.5.1)

$$P = \tfrac{1}{2} \exp(-\gamma) \tag{2.22}$$

and conversely $\gamma = -\ln(2P)$, so that at threshold

$$P_0 = \tfrac{1}{2} \exp(-\gamma_0)$$

and conversely $\gamma_0 = -\ln(2P_0)$, the ratio s can be written as

$$s = \frac{W_0}{W_m} = (\ln 2) \frac{W_0}{W_M}$$

$$s = \frac{\gamma_0}{\gamma} = -\frac{1}{\gamma} \ln(2P_0) = -\ln(2P_0)^{1/\gamma} \tag{2.23}$$

where W_0 is the threshold value, and therefore
$$\exp(-s) = (2P_0)^{1/\gamma} \tag{2.24}$$
Therefore the outage probability (2.19) becomes
$$P(\gamma) = 1 - (2P_0)^{1/\gamma} \tag{2.25}$$
From (2.16b),
$$U(\text{dB}) = 10 \log s = 10 \log \left(\frac{\gamma_0}{\gamma}\right) = 10 \log \left\{\frac{\ln(2P_0)}{\ln(2P)}\right\} \tag{2.26}$$

If $R(\text{dB}) = 10 \log \gamma$ is the r.f. carrier-to-noise ratio in decibels, eqn (2.25) can be rewritten as
$$P(R) = 1 - (2P_0)^{10^{-0.1R}} \tag{2.27}$$

2.4 Nakagami distributions

The Nakagami distributions can be considered as generalizations of the Rayleigh distribution. The following two distributions are of interest: the Nakagami m distribution and the Nakagami n distribution.

2.4.1 Nakagami m distribution

The Nakagami m distribution is useful because it fits various types of fade and is suitable for theoretical calculations. It includes as particular cases the normal and Rayleigh distributions.

If r is the modulus of a vector representing a signal amplitude for example, its density function is given by
$$\frac{dp}{dr} = 2\left(\frac{m}{\sigma^2}\right)^m \frac{r^{2m-1}}{\Gamma(m)} \exp\left(-\frac{mr^2}{\sigma^2}\right) \tag{2.28}$$
where $\sigma^2 = \overline{r^2}$ represents the average signal power (the bar indicates an average value),
$$m = \left(\frac{\sigma^2}{\overline{r^2 - \sigma^2}}\right)^2 \geq \frac{1}{2}$$
is a parameter, which never takes values of less than $1/2$, representing the inverse of the normalized variance of r^2 and $\Gamma(m)$ is the gamma function of m. If m is an integer, $\Gamma(m) = (m-1)!$, and if $m = 1/2$ we have $\Gamma(1/2) = \pi^{1/2}$. It can easily be verified that eqn (2.28) has the following characteristics.

(i) If $m = 1/2$ it becomes a Gaussian function like eqn (2.6). A factor of 2 remains because r is always positive (the negative portion of the Gaussian function is superimposed on the positive portion).

(ii) If $m = 1$ it becomes a Rayleigh function like (2.12).

The probability density function for various values of m is shown in Fig. 2.5 and the cumulative m distribution is shown in Fig. 2.6. This distribution has been used to calculate the effects of diversity for a generalized type of fading [2.1].

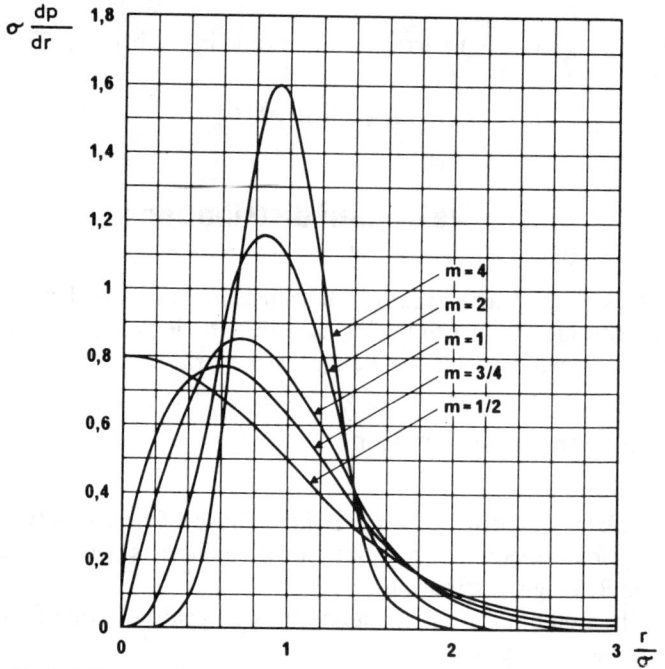

Fig. 2.5 Probability density function of the m distribution .(After *CCIR Report 762-1.*)

2.4.2 Nakagami *n* distribution

The Nakagami n distribution is also called the Nakagami–Rice distribution. It is the distribution of the modulus of the vector resulting from the sum of a fixed vector and a vector with a modulus which varies according to the Rayleigh distribution.

If a is the modulus of the fixed vector and σ is the most probable modulus of the random vector, the density function of the resultant vector r is given by

$$\frac{dp}{dr} = \frac{r}{\sigma^2} \exp\left(-\frac{r^2 + a^2}{2\sigma^2}\right) I_0\left(\frac{ar}{\sigma^2}\right) \qquad (2.29)$$

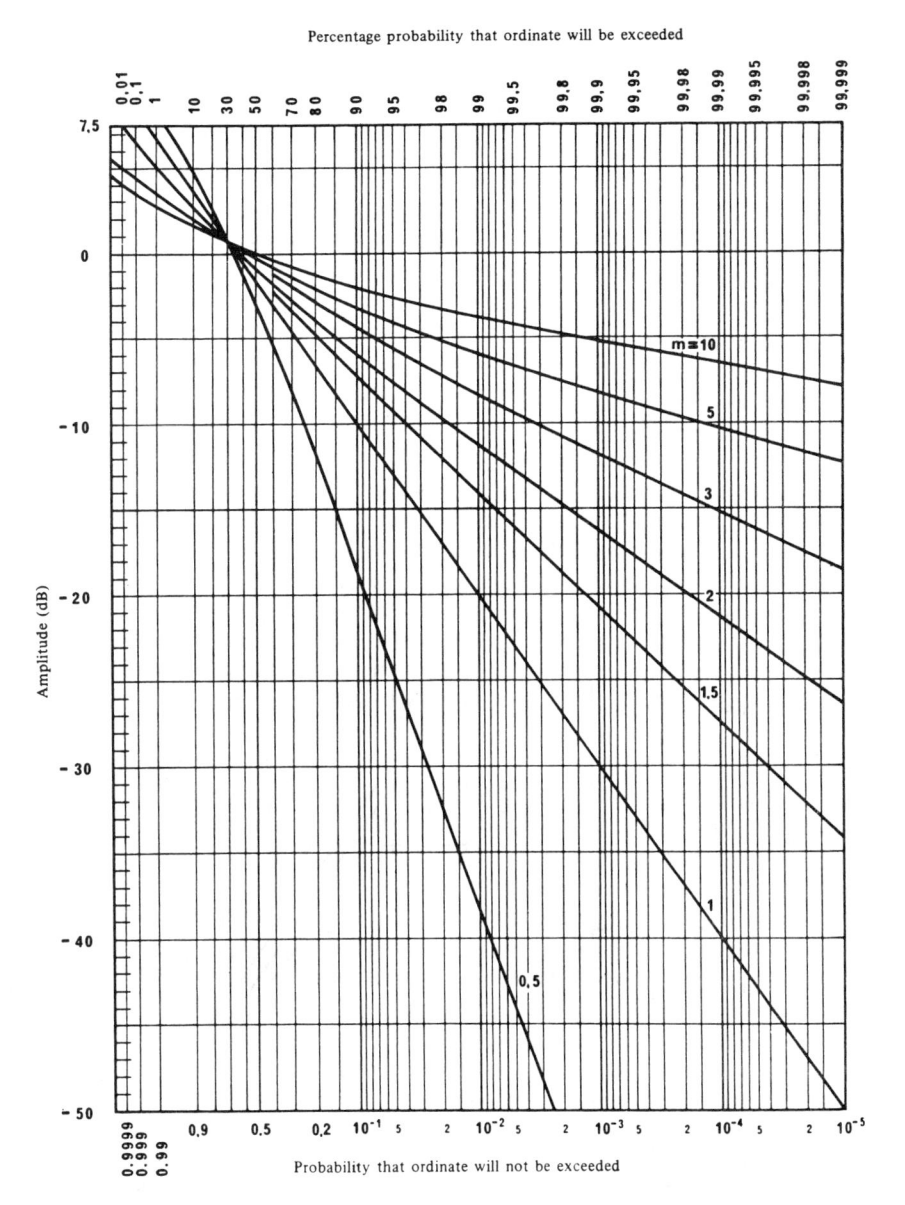

Fig. 2.6 Cumulative Nakagami m distribution ($\sigma = 1$). (From *CCIR Report 1007*.)

where a^2 and $2\sigma^2$ are proportional to the powers of the respective vectors and I_0 is a modified zero-order Bessel function of the first kind.

This distribution is suitable for the study of multipath propagation in which the transmitted power (which is obviously constant) is split into a fixed and a random component. The cumulative distributions obtained from (2.29) and drawn in a form suitable for this application are shown in Fig. 2.7 as functions of a parameter representing the fraction of power carried by the random vector. The scales are chosen such that a Rayleigh distribution is represented by a straight line.

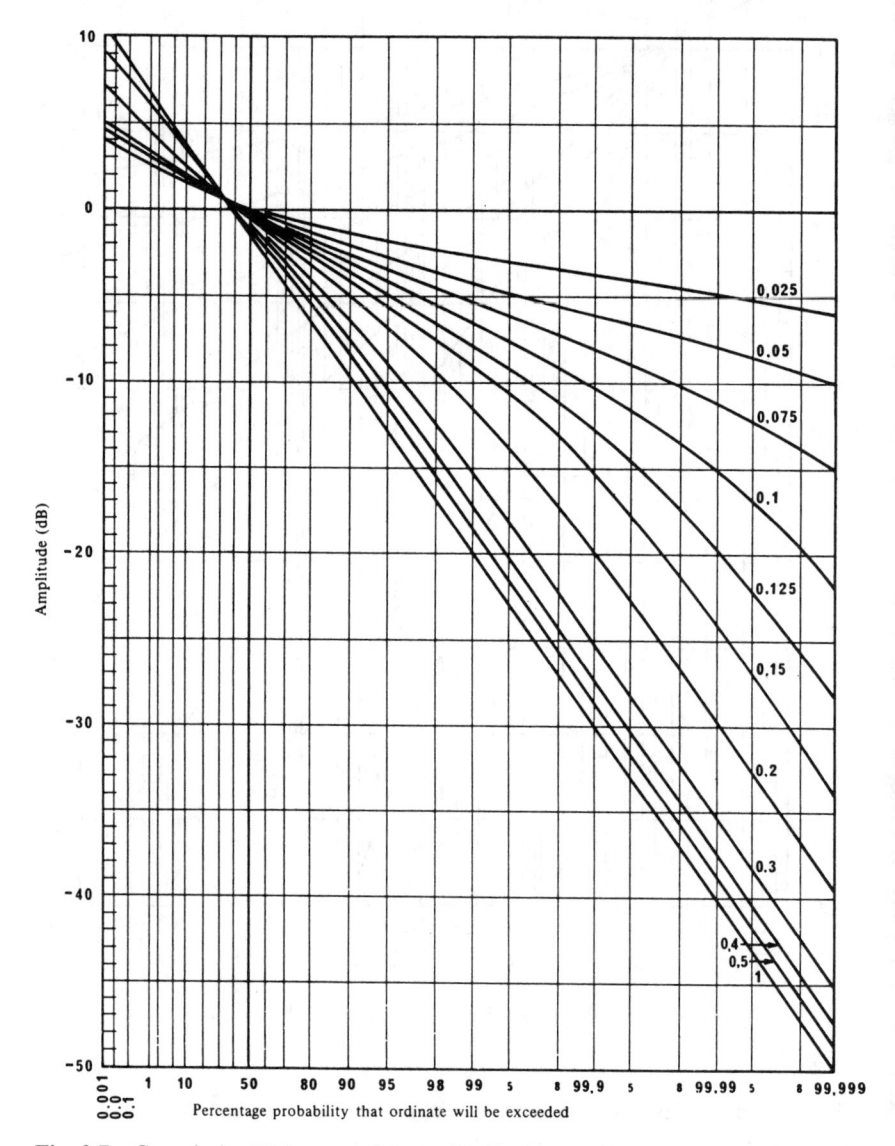

Fig. 2.7 Cumulative Nakagami–Rice n distribution for constant total power, with the fraction of power carried by the random vector as parameter. (From *CCIR Report 1007*.)

Consider for example a path in which the propagation occurs partly by troposcatter and partly by diffraction. If most of the propagation is by troposcatter the parameter approaches unity and a Rayleigh distribution is obtained. If the stable diffracted component increases, the parameter decreases and the distribution becomes uniform.

2.5 Composed distributions—convolutions

2.5.1 General remarks

In a number of cases a random signal is obtained as the sum of two or more uncorrelated random signals. When a random variable z is the sum of two (or more) statistically independent random variables x and y, i.e. $z = x+y$, it has the following properties:

(a) the mean value of z is the sum of the mean values of x and y;
(b) the variance of z is the sum of the variances of x and y;
(c) the mean of the product xy is equal to the product of the means of x and y.

It is necessary to know whether the resulting signal is the sum of numerical or logarithmic values. For example, if the variable is a troposcatter path loss Z dB resulting from the combination of a short-term loss X dB and a long-term loss Y dB, it is obvious that the overall loss will be the sum of the losses in decibels, i.e. $Z = X + Y$ dB. However, in the case where the variable is a noise level l_z dBm resulting from the combination of the noise in two cascaded channels with noise levels l_x dBm and l_y dBm respectively, it is not possible to sum the dBm values directly. They must first be transformed into power units (e.g. mW or pW) which can be added. The resulting power is then transformed back to dBm. This process can be expressed mathematically as follows:

$$l_z = 10 \log(10^{0.1 l_x} + 10^{0.1 l_y}) \tag{2.30}$$

Care must be taken to avoid errors that may occur when the variables of interest are not expressed in appropriate units.

The probability density function for a random variable $z = x+y$ with densities $p(x)$ and $q(y)$ is given by the convolution integral

$$p(z) = p(x) * q(y) = \int_{-\infty}^{\infty} p(z-y) q(y) \, dy \tag{2.31}$$

If $P(x)$ is the integral of $p(x)$, the cumulative distribution function for z is given by

$$P(z) = P(x)*q(y) = \int_{-\infty}^{\infty} P(z-y)\, q(y)\, \mathrm{d}y \qquad (2.32)$$

Sometimes it is not necessary to know the convolution function to solve a particular problem. For example, if the probabilities $P(x)$ and $P(y)$ of outage in two cascaded circuits are normally very small, the outage probability for the overall circuit can be written to a good approximation as

$$P(z) = P(x) + P(y) \qquad (2.33)$$

The probability $P(x)P(y)$ of outages occurring simultaneously in the two hops, which is extremely small, is ignored.

2.5.2 Convolution of normal distributions

The distribution of the sum of two or more statistically independent normal variables is also a normal distribution with a mean value equal to the sum of the individual mean values and a variance equal to the sum of the individual variances. This defines the resulting distribution completely.

If the original variables are perfectly correlated (fixed ratios between corresponding values), the resulting distribution is normal with mean equal to the sum of the means and standard deviation equal to the sum of the standard deviations.

2.5.3 Convolution of log-normal distributions

The distribution of the sum of two or more log-normal variables is not a log-normal distribution. According to well-known probability theorems, as the number of variables increases to infinity, the resulting distribution approaches a Gaussian form regardless of the form of the individual distributions. The law of the sum of means and variances always holds, but the exact form of the resulting distribution is not known. Approximate curves given in the literature [2.2, 2.3] will be used in this book for the summation of noise (see Chapter 8).

2.5.4 Convolution of Rayleigh distributions

Rayleigh distributions, which are normal in two dimensions, have the same properties as stated above for Gaussian distributions. The vectorial sum is a Rayleigh distribution of amplitudes or an exponential distribution of powers (see Section 2.3) completely defined by its mean power which is the sum of the individual mean powers.

2.5.5 Convolution of log-normal and Rayleigh distributions

It is important to know the cumulative distribution of the level of a signal in which a short-term Rayleigh variability is superimposed on a long-term log-normal variability of the amplitude.

Consider for example the case of a signal S dBm randomly oscillating by Z dB around its hourly median value M dBm, so that

$$S(\text{dBm}) = M(\text{dBm}) + Z(\text{dB})$$

Z dB is the sum of a Rayleigh variation of the signal by X dB around, say, its 1 min median and a Gaussian variation by Y dB of this median around the hourly median M dBm, so that

$$Z(\text{dB}) = X(\text{dB}) + Y(\text{dB})$$

The function $P(x)$ in (2.32) becomes $P(X) = P(Z-Y)$ and, using (2.14) and (2.18b), can be written as

$$P(Z-Y) = \exp(-0.7 \times 10^{0.1(Z-Y)})$$

Then, using (2.6b), function $q(y)$ of (2.32) becomes

$$q(Y) = \frac{1}{\sigma(2\pi)^{1/2}} \exp\left(-\frac{Y^2}{2\sigma^2}\right)$$

Thus the convolution of $P(Z-Y)$ and $q(Y)$ is given by

$$P(Z) = \frac{1}{(2\pi)^{1/2}} \int_{-\infty}^{\infty} \exp(-0.7 \times 10^{0.1(Z-Y)}) \exp\left(-\frac{Y^2}{2\sigma^2}\right) d\left(\frac{Y}{\sigma}\right) \quad (2.34)$$

It represents the probability that the signal level does not fall more than Z dB below (Z negative) the median M dBm of the normal distribution. If $M - Z$ dBm is the threshold level, the outage probability is $1 - P(Z)$.

2.6 Correlation between random variables—autocorrelation

In some cases the random variables are only partially independent, which means that their statistical variabilities correlate to some extent. The degree of correlation between two random variables x and y is measured by the "correlation coefficient" which is defined as

$$\rho = \frac{\sigma_{xy}^2}{\sigma_x \sigma_y} \quad (2.35)$$

where

$$\sigma_{xy}^2 = M(xy) - M(x)M(y)$$

$M(\ )$ is the mean value of the variable in parentheses and σ_x and σ_y are the standard deviations of the distributions of x and y. If x and y are statistically independent $M(xy) = M(x)M(y)$ and $\rho = 0$. Complete correlation is indicated by $\rho = 1$. The variance of the variable $z = x + y$ is given by

$$\sigma_z^2 = \sigma_x^2 + \sigma_y^2 + 2\rho\sigma_x\sigma_y \tag{2.36}$$

Sometimes different samples of the same random variable are not statistically independent. The variable is then said to be "autocorrelated". This may happen when, for example, the variable $f(t)$ is compared with the variable $f(t+\tau)$ for any t. In this case we define the "autocorrelation function"

$$F(\tau) = \int_{-\infty}^{\infty} f(t)f(t+\tau)\,dt \tag{2.37}$$

which is zero when there is no correlation and maximum in the case of correlation of a finite number of samples of sufficient length.

A similar case to that of autocorrelation is the correlation between a long digital stream of completely random bits and a copy of this stream. Suppose that the reference stream is $f(t) = a\,s(t-iT)$ where $a = +1, -1$ is the amplitude of the ith bit represented by $s(t-iT)$ and T is the duration of the bit. The copy of $f(t)$ can be written as $f(t+\tau)$, indicating that it is shifted or delayed by a time τ with respect to the reference stream. When $f(t)$ and $f(t+\tau)$ are substituted in (2.37), it is easily seen that the following hold:

(a) for $\tau = 0$ the integrand becomes a set of 1s and the integral reaches its maximum value;

(b) for $\tau > T$ the integral vanishes as the bits are random and there is no correlation;

(c) for $\tau \leqslant T$ the integral decreases linearly from the maximum value to zero and there is partial correlation.

The correlation function is therefore a triangle with a base 2 bits wide centered around the zero delay condition. A rectangular correlation function of width 1 bit can be obtained by sampling the bit polarity at the center of the reference bits.

These functions will be discussed further in Chapter 11, Section 11.9.1, with reference to measurement of the propagation delay spectrum.

2.7 Comments and further reading

Statistical distributions used in radio wave propagation are described in ref. 2.4, which also includes a bibliography. Most of them are also given in ref. 2.5.

References

2.1 Panter, P. F. *Communication Systems Design—Line-of-sight and Troposcatter Systems*, McGraw-Hill, New York, 1972.
2.2 Jacobsen, B. B. *Thermal Noise in Multi-section Radio Links, IEE Monograph 262 R*, IEE, London, 1957.
2.3 Sheffield, B. Nomograms for the statistical summation of noise in multihop communications systems, *IEEE Trans. Commun. Syst.*, 285–288, 1963.
2.4 *CCIR Rep. 1007.* In *Recommendations and Reports of the CCIR, 16th Plenary Assembly, Dubrovnik, 1986*, Vol. 5, *Propagation in Non-ionized Media*, ITU, Geneva, 1986.
2.5 Boithias, L. *Propagation des Ondes Radioélectriques dans l'Environnement Terrestre*, Dunod, Paris, 1984 (English translation (revised and updated): *Radiowave Propagation*, North Oxford Academic, London, 1987).

Chapter 3

The Geometry of the Troposcatter Path and Related Problems

The geometry of the troposcatter path plays an important role in design calculations. It includes several parameters which are required for the prediction of path loss and path distortion for example. The definitions of geometrical parameters and the formulae relating to such parameters are collected in this chapter. First we consider the fundamental data from which the design starts and then the parameters derived from the formulae.

The formulae are initially presented without derivation. In a subsequent section we show how they are deduced and also present some additional relationships which may be useful in particular cases. The antenna beamwidth and its geometrical effects are also taken into consideration. Note that angles are often measured in milliradians (mrad) rather than in degrees ($1° = 17.45$ mrad).

It is not necessary to go through all the formulae on the first reading and Section 3.3 can be ignored. Readers should familiarize themselves with the various parameters and should study Fig. 3.5 carefully.

3.1 Profile of the troposcatter path

The troposcatter path can be represented by a path profile diagram of the type used for ordinary radio links (Fig. 3.1). It is well known that such diagrams are constructed using an expanded vertical scale and that the Earth's curvature is altered so that an electromagnetic ray, which would normally be bent towards the Earth, can be represented by a straight line. Thus we define an "effective radius" $a = kR$ where $R = 6400$ km is the true radius of the Earth and k is a factor which depends on the atmospheric refractivity and has an average value of 4/3.

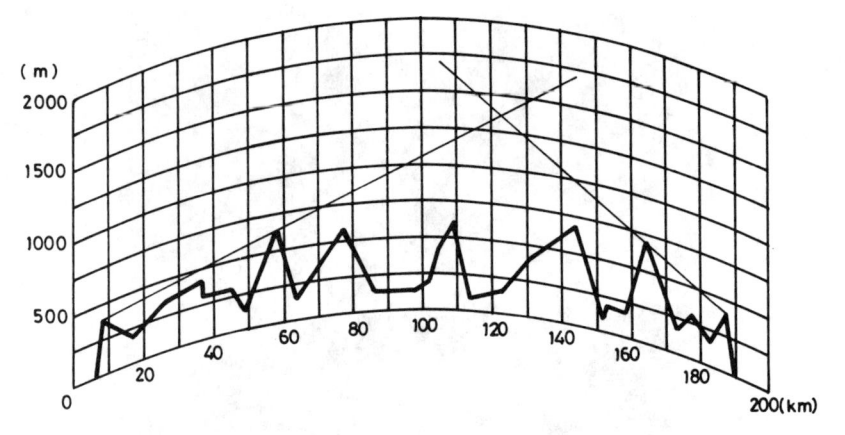

Fig. 3.1 Troposcatter path on a profile chart.

Profile charts are normally drawn for $k = 4/3$ and for a given ratio of horizontal to vertical length units. However, it is not necessary to draw a complete and exact profile for a troposcatter path; it is sufficient to determine the distance and height of the horizons, or alternatively the elevation angle of the horizons (see below). These data can be obtained from maps or from measurements on site. Calculations are sometimes made using the actual value of k instead of its average. Therefore in general only indicative path profiles are used for the visualization of troposcatter and for collecting the relevant parameters.

A typical troposcatter path profile which also shows the antenna beams is presented in Fig. 3.2. The following features should be noted.

(a) The antennas are pointing towards the horizon (or more correctly the "radio horizon" because of the alteration in the Earth's curvature which could give a different horizon), which is seen from the site with an elevation or takeoff angle that can be negative, zero or positive (Fig. 3.3). The ray to the radio

GEOMETRY OF THE TROPOSCATTER PATH

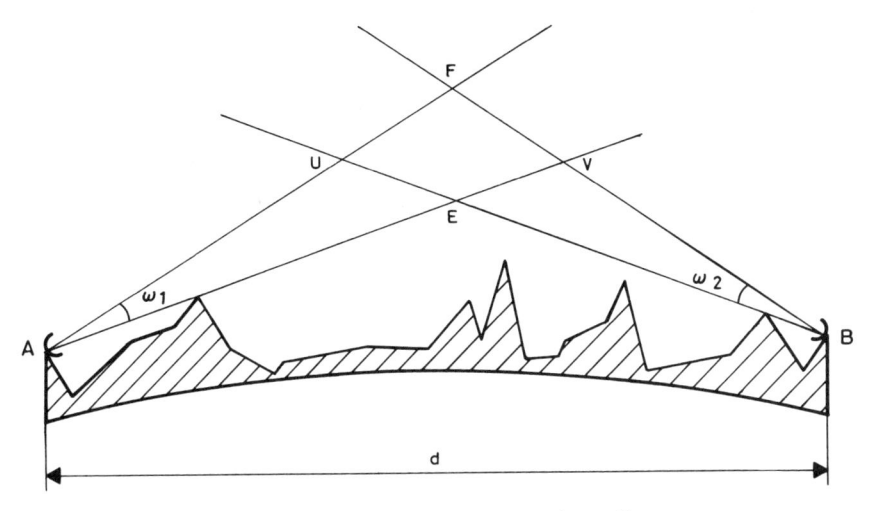

Fig. 3.2 Typical troposcatter path profile.

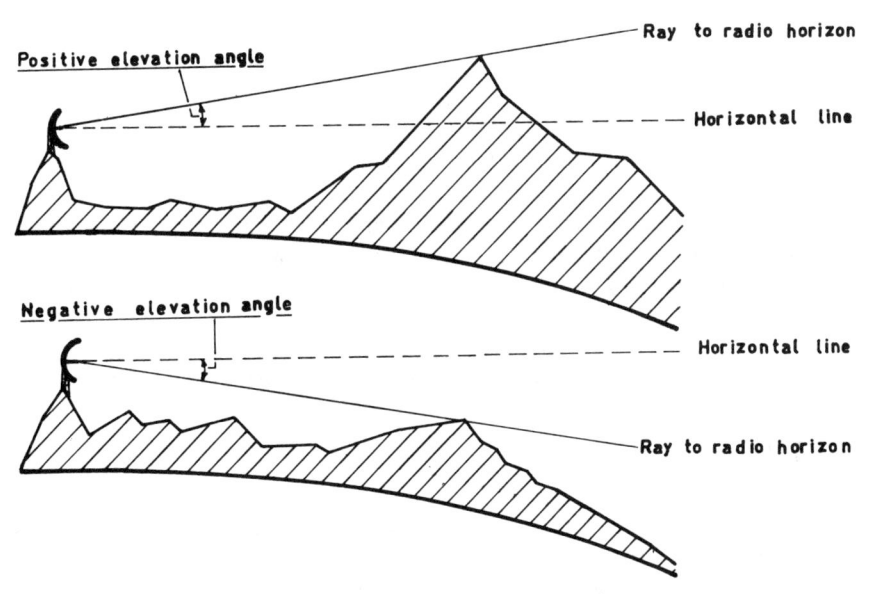

Fig. 3.3 Elevation or takeoff angle.

horizon has often been confused in the literature with the antenna axis. The real situation in practical cases is that of Fig. 3.4, which shows that the 3 dB beam is partially obstructed by the horizon. Therefore a "path beamwidth" should be taken into account in the calculations.

(b) The 3 dB beams of the antennas or, better, the path beams of Fig. 3.4 intersect in a common volume UFVE (Fig. 3.2).

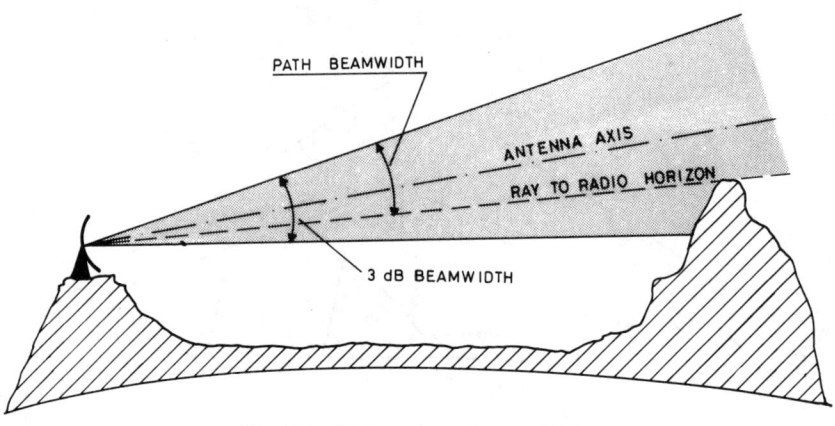

Fig. 3.4 Path antenna beamwidth.

(c) There is a shadow zone between the two horizons.

(d) Rays from antenna to antenna such as AFB and AEB are of different lengths, and this implies different transmission delays and signal distortions.

The study of the profile parameters starts from the pure geometrical characteristics. Some parameters are fundamental and must be known from the outset, whereas others are derived. Other factors such as the antenna beamwidth, multiple paths etc. are considered at a later stage.

We now define the various parameters mainly with reference to Fig. 3.5 or to other figures as indicated. We also give the numerical values of some parameters as found in practice.

3.2 Definition of geometrical data and parameters

3.2.1 Basic data

The following parameters must be known in advance for the calculation of a troposcatter path: path length d, heights h_1 and h_2 of the two sites or, more exactly, the heights of the centers of the two antennas above sea level; heights h_1' and h_2' of the two horizons above sea level (these parameters are not considered when the horizon is the sea); distances d_1 and d_2 from each site to its respective horizon (when the horizon is the sea these parameters are obtained by calculation); the air refractivity correction factor k for the Earth's radius R (an average value of R is 6400 km and of k is 4/3). All other geometrical path parameters can be calculated from these fundamental parameters as shown below. In some cases the horizon elevation angle (see Section 3.2.2) can be used instead of the height and the distance to the horizon.

GEOMETRY OF THE TROPOSCATTER PATH 45

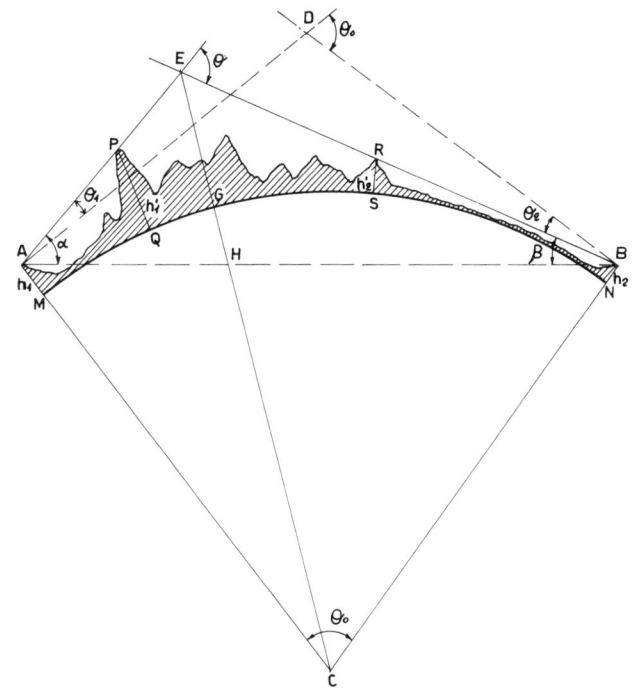

Fig. 3.5 Elements of path geometry.

Segments

AM	height h_1 of antenna 1 above sea level (a.s.l.)
BN	height h_2 of antenna 2 a.s.l.
EG	height h_0 a.s.l. of the crosspoint of the rays
EH	height h a.s.l. of the crosspoint above chord AB
PQ	height h_1' a.s.l. of the obstacle representing the horizon from A
RS	height h_2' a.s.l. of the obstacle representing the horizon from B
AB	path length d, assumed to be equal to the arc MN
MQ	distance d_1 of the horizon from site A
NS	distance d_2 of the horizon from site B
CM = CG = CN	effective radius a of the Earth
AD	horizontal line from site A
BD	horizontal line from site B
MG	distance of crosspoint from A
NG	distance of crosspoint from B

Angles (all small angles for which $\sin x = \tan x = x$, $\cos x = 1 - x^2/2$)

$\widehat{MCN} = \hat{D} = \theta_0$	angular length
$\hat{E} = \theta$	angular distance or scatter angle
$\widehat{EAD} = \theta_1$	elevation of the horizon from A (positive because it is above AD)
$\widehat{EBD} = \theta_2$	elevation of the horizon from B (negative because it is below BD)
$\widehat{EAB} = \alpha$	elevation above chord AB from A
$\widehat{EBA} = \beta$	elevation above chord AB from B

Relations

$s = \alpha/\beta$	path symmetry factor (if $\alpha/\beta \leq 1$; otherwise $s = \beta/\alpha$)

Current values of the path length range from 100 to 500 km, and exceptionally up to 1000 km. The heights of the sites and the horizons above sea level may reach several thousand meters. The distance from a site to its radio horizon normally ranges from a minimum of a few kilometers to a maximum of several tens of kilometers.

3.2.2 Angular parameters

Most angles used are normally so small that they can be substituted for their sines and tangents in the formulae without introducing appreciable errors. The angular parameters of interest are as follows (Fig. 3.5).

The *angular length* θ_0 is defined as

$$\theta_0 = \frac{d}{kR} \tag{3.1}$$

For a 500 km path for example the angular length θ_0 becomes 59 mrad corresponding to 3.4° for $k = 4/3$.

The *elevation angles* θ_1 and θ_2 of the two horizons from the horizontal plane at each site are given by

$$\theta_1 = \frac{h_1' - h_1}{d_1} - \frac{d_1}{2kR} \tag{3.2a}$$

$$\theta_2 = \frac{h_2' - h_2}{d_2} - \frac{d_2}{2kR} \tag{3.2b}$$

when the horizon is on land and by

$$\theta_1 = -\frac{d_1}{kR} \tag{3.3a}$$

$$\theta_2 = -\frac{d_2}{kR} \tag{3.3b}$$

when it is on the sea. These angles can also be measured in the field, but when the horizon is the sea this is not advisable because the error due to refraction may become unacceptably high. The elevation angle of the radio horizon may vary from -30 mrad ($-1.7°$) to $+35$ mrad (2°). The elevation of the horizon on the sea from the top of a mountain of height 3000 m is -27 mrad ($-1.52°$). The optical elevation of the horizon is slightly different (see Section 3.3.1).

The angles α and β between the ray to the radio horizon and the chord joining the two sites are given by

$$\alpha = \theta_1 + \frac{d}{2kR} + \frac{h_1 - h_2}{d} = \theta_1 + \frac{1}{2}\theta_0 + \frac{h_1 - h_2}{d} \tag{3.4a}$$

$$\beta = \theta_2 + \frac{d}{2kR} - \frac{h_1 - h_2}{d} = \theta_2 + \frac{1}{2}\theta_0 - \frac{h_1 - h_2}{d} \tag{3.4b}$$

The *scatter angle* or *angular distance* θ is defined as the angle between the two horizon rays and is given by

$$\theta = \theta_0 + \theta_1 + \theta_2 \tag{3.5}$$

or

$$\theta = \alpha + \beta \tag{3.6}$$

Scatter angles vary from a few milliradians to 80 mrad (4.6°), but current values do not generally exceed 35 mrad (2°).

The *symmetry factor* s is

$$s = \begin{cases} \alpha/\beta \text{ for } \alpha/\beta \leqslant 1 \\ \beta/\alpha \text{ for } \alpha/\beta > 1 \end{cases} \tag{3.7}$$

3.2.3 Length parameters

The following length parameters are of interest.
The distance d_h to the sea horizon is given by

$$d_h = (2kRh_1)^{1/2} \tag{3.8}$$

or

$$d_h = -kR\theta_h \tag{3.9}$$

The second formula should not be used because of the difficulty of measuring a reliable (negative) elevation angle θ_h on the sea.

The lengths r_0 and s_0 of the horizon rays from the site to their crosspoint are

$$r_0 = d\frac{\beta}{\theta} \tag{3.10a}$$

$$s_0 = d\frac{\alpha}{\theta} \tag{3.10b}$$

The height h_0 above sea level of the crosspoint of the horizon rays is given by

$$h_0 = h_1 + \frac{1}{2kR}\frac{d^2}{}\left(\frac{\beta}{\theta}\right)^2 + d\frac{\beta}{\theta}\theta_1 \tag{3.11a}$$

or

$$h_0 = h_2 + \frac{1}{2kR}d^2\left(\frac{\alpha}{\theta}\right)^2 + d\frac{\alpha}{\theta}\theta_2 \tag{3.11b}$$

This is also the height of the base of the common volume and ranges from several hundred to several thousand meters.

The difference Δd between the length of a path following the horizon rays and the distance between the two sites is

$$\Delta d = r_0 + s_0 - d \approx \tfrac{1}{2}\alpha\beta d \tag{3.12}$$

The height h of the crosspoint of the horizon rays above the chord joining the two sites is

$$h = \frac{ds\theta}{(1+s)^2} \tag{3.13}$$

The difference $\Delta d_{1,2}$ between the lengths of paths AFB and AEB in Fig. 3.2 for $\omega_1 = \omega_2 = \omega$, i.e. the difference between the upper and lower paths along the antenna beams, is

$$\Delta d_{1,2} = \frac{d}{2}(\omega^2 + \omega\theta) \tag{3.14}$$

This difference ranges from, say, 10 m to 200 m.

3.2.4 Geometrical parameters of antennas

The antenna beamwidth ω is defined as either the 3 dB beamwidth or the path beamwidth depending on the particular case. The 3 dB beamwidth for a parabolic antenna is (in degrees)

$$\omega = 70\frac{\lambda}{D} = \frac{21\,000}{fD} \tag{3.15}$$

where λ is the wavelength (in meters), D is the diameter of the parabolic reflector (in meters) and f is the frequency (in megahertz). This beamwidth ranges from several degrees (e.g. 8° for small antennas in the lower frequency range) to a fraction of a degree (e.g. 0.8° for large antennas in the upper frequency range).

3.3 Derivation of the formulae

The formulae given above are now derived in logical sequence. Some additional useful formulae are also included. Most of the angles are so small that we can consider $\sin x = \tan x = x$, $\cos x = 1 - x^2/2$. Unless stated otherwise, all derivations are made with reference to the geometry of Fig. 3.5.

3.3.1 Calculation of the effective radius of the Earth

We have

$$a = kR \tag{3.16}$$

where a is the effective radius of the Earth, R is the true radius of the Earth (6400 km) and k is the correction factor, which has an average value of $4/3 = 1.33$ for microwaves and 1.18 for light.

3.3.2 Calculation of the distance and elevation of the horizon over sea (or flat land)

With reference to Fig. 3.6 we can write

$$\overline{SC}^2 = \overline{TC}^2 + \overline{TS}^2$$

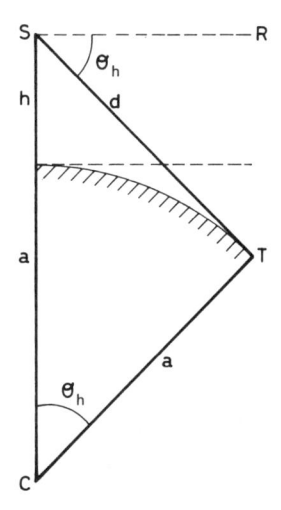

Fig. 3.6 Distance and elevation of the horizon over the sea (or a flat Earth).

Thus

$$(a+h_1)^2 = a^2 + d_h^2$$
$$a^2 + h_1^2 + 2ah_1 = a^2 + d_h^2$$

If h_1^2 is neglected this equation becomes

$$d_h = (2ah_1)^{1/2} \tag{3.17}$$

Furthermore

$$\theta_h = R\hat{S}T = S\hat{C}T = d_h/a$$

which, combined with eqn (3.17), gives the elevation angle

$$\theta_h = \left(\frac{2h}{a}\right)^{1/2} \tag{3.18}$$

This angle is assumed to be negative.

3.3.3 Calculation of the elevation angle of the horizon

Consider region APMQ in Fig. 3.5. If A and P were at the same height each one would see the other at an elevation angle $-MQ/2a$ or $-d_1/2a$, because the sum of these two equal angles is equal to the angle between the horizontal lines from A and P which is d_1/a. The difference in height adds another term so that eqn (3.2) is obtained.

3.3.4 Calculation of angles α and β

For a perfectly flat Earth $\alpha = \beta = \theta_0/2$. Since the two sites are at different heights h_1 and h_2, the chord is tilted by an angle $(h_1 - h_2)/d$ which must be added to α and subtracted from β. Furthermore angles θ_1 and θ_2 are added to α and β respectively, and thus eqn (3.4) is obtained.

3.3.5 Calculation of the angular length and the scatter angle

We have by definition

$$\theta_0 = d/a \qquad (3.19)$$

and thus the scatter angle is given by eqns (3.5) and (3.6):

$$\theta = \theta_0 + \theta_1 + \theta_2 = \frac{2d - (d_1 + d_2)}{2a} + \frac{h_1' - h_1}{d_1} + \frac{h_2' - h_2}{d_2}$$

$$\theta = \alpha + \beta$$

In a knife-edge diffraction path there is only one obstacle, PQ coincides with RS, $h_1' = h_2'$ and $d_1 + d_2 = d$. In this case

$$\theta = \frac{d}{2a} + \frac{h_1' - h_1}{d_1} + \frac{h_2' - h_2}{d - d_1} \qquad (3.20)$$

3.3.6 Calculation of the distance along the horizon rays to their crosspoint

According to the sine theorem

$$\frac{AB}{\sin(180° - \theta)} = \frac{AB}{\sin \theta} = \frac{AE}{\sin \beta} = \frac{BE}{\sin \alpha}$$

from which we obtain eqns (3.10a) and (3.10b):

$$r_0 = AE = d\frac{\sin \beta}{\sin \theta} \approx d\frac{\beta}{\theta}$$

$$s_0 = BE = d\frac{\sin \alpha}{\sin \theta} \approx d\frac{\alpha}{\theta}$$

These are also the distances to the crosspoint as, in practice, $AE \approx MG$ and $BE \approx NG$.

3.3.7 Calculation of $EH = h$

Using $s = \alpha/\beta$ and eqn (3.10a) we can obtain eqn (3.13):

$$h = AE \sin \alpha = d\alpha\frac{\beta}{\theta} = d\alpha\frac{\beta\theta}{\theta^2} = \frac{\theta d(\alpha/\beta)}{(1+\alpha/\beta)^2} = \frac{ds\theta}{(1+s)^2}$$

3.3.8 Calculation of $EG = h_0$

According to a well-known trigonometrical formula

$$\overline{CE}^2 = \overline{AC}^2 + \overline{AE}^2 - 2\,\overline{AC}\,\overline{AE}\cos E\hat{A}C$$

Since $\cos E\hat{A}C = \cos(90° + \theta_1) = -\sin \theta_1 \approx -\theta_1$ we can write, using eqn (3.10a),

$$(h_0 + a)^2 = (h_1 + a)^2 + d^2\frac{\beta^2}{\theta^2} + 2\theta_1(h_1 + a)\,d\frac{\beta}{\theta}$$

$$h_0^2 + a^2 + 2ah_0 = h_1^2 + a^2 + 2ah_1 + d^2\frac{\beta^2}{\theta^2} + 2\theta_1 h_1 d\frac{\beta}{\theta} + 2\theta_1 ad\frac{\beta}{\theta}$$

After simplifying and neglecting the smallest terms this becomes

$$2ah_0 = 2ah_1 + d^2\frac{\beta^2}{\theta^2} + 2\theta_1 ad\frac{\beta}{\theta}$$

Dividing all terms by $2a$ we obtain eqn (3.11a):

$$h_0 = h_1 + \frac{d^2}{2a}\left(\frac{\beta}{\theta}\right)^2 + \theta_1 d\frac{\beta}{\theta}$$

We can obtain eqn (3.11b) in a similar manner:

$$h_0 = h_2 + \frac{d^2}{2a}\left(\frac{\alpha}{\theta}\right)^2 + \theta_2 d\frac{\alpha}{\theta}$$

3.3.9 Calculation of path differences

Consider the path difference
$$\Delta d = AE + EB - AB$$
Using eqns (3.10a) and (3.10b) we can write
$$\Delta d = d\frac{\sin \alpha + \sin \beta}{\sin \theta} - d$$
$$= d\frac{\sin \alpha + \sin \beta}{\sin(\alpha + \beta)} - d$$
$$= d\frac{\sin \alpha + \sin \beta}{\sin \alpha \cos \beta + \cos \alpha \sin \beta} - d$$

and, since $\sin x = x$ and $\cos x = 1 - x^2/2$ for small x, we obtain eqn (3.12):
$$\Delta d = \frac{d(\alpha + \beta)}{\alpha(1 - \beta^2/2) + \beta(1 - \alpha^2/2)} - d$$
$$= \frac{d(\alpha + \beta)}{\alpha + \beta - \alpha\beta^2/2 - \beta\alpha^2/2} - d$$
$$= \frac{d}{1 - \alpha\beta/2} - d$$
$$= d\left(1 + \frac{\alpha\beta}{2}\right) - d$$
$$= \frac{1}{2}\alpha\beta d$$

The path difference between two rays with parameters α_1, β_1 and α_2, β_2 respectively can be written as
$$\Delta d_{12} = \frac{d}{2}(\alpha_2\beta_2 - \alpha_1\beta_1) \qquad (3.21)$$

In the important practical case in which the two antennas shown in Fig. 3.2 have equal path beamwidths ω,
$$\alpha_2 = \alpha_1 + \omega = \alpha + \omega$$
$$\beta_2 = \beta_1 + \omega = \beta + \omega$$
and eqn (3.21) becomes eqn (3.14):
$$\Delta d_{1,2} = \frac{d}{2}\{(\alpha + \omega)(\beta + \omega) - \alpha\beta\}$$

$$= \frac{d}{2}\{\omega^2 + \omega(\alpha+\beta)\}$$

$$= \frac{d}{2}(\omega^2 + \omega\theta)$$

An alternative expression for Δd which is valid for the case of knife-edge diffraction can be obtained by using $h = \alpha\beta d/\theta$ (from the derivation of eqn (3.13)), $\alpha = h/d_A$ and $\beta = h/d_B$, where $d_A = AH$ and $d_B = BH$ (Fig. 3.5). Then, substituting in (3.12), we can write

$$\Delta d = \frac{\alpha\beta d}{2}$$

$$= \frac{h^2}{2}\frac{d}{d_A d_B}$$

$$= \frac{(\alpha\beta d)^2}{2}\frac{1}{\theta^2}\frac{d}{d_A d_B}$$

$$= 2(\Delta d)^2 \frac{1}{\theta^2}\frac{d}{d_A d_B}$$

and we obtain the alternative expression

$$\Delta d = \frac{\theta^2}{2}\frac{d_A d_B}{d} \qquad (3.22)$$

3.4 Comments

Path geometry formulae and calculations are distributed sparsely throughout the literature. In this chapter they have been collected together in a more systematic way.

Chapter 4

The Tropospheric Scatter Mode of Propagation

The transmission of signals beyond the horizon by tropospheric scatter (troposcatter) propagation takes place because of the presence of the atmosphere with its peculiar large-scale and small-scale characteristics. In a vacuum propagation would take place only by diffraction, and we saw in Fig. 1.1 that the diffraction loss far beyond the horizon becomes extremely high, making transmission impossible. However, the presence of the atmosphere with its inhomogeneities permits an unexpected type of propagation with unusual characteristics that can be exploited for reliable communications.

In this chapter we will first examine the characteristics of the atmosphere or, more exactly, of the troposphere, which is the region from the ground to say 8–10 km where the air is continuously in motion and meteorological phenomena take place. In most cases the radio waves are scattered within the first kilometer above ground, where the phenomena are accentuated.

We will then examine the effect of the scattering process on the radio signal and categorize the various types of climate. The latter information is useful for predicting signal behavior in various regions of the world.

4.1 Tropospheric refractivity and its characteristics

Troposcatter propagation occurs because of the refraction characteristics of the troposphere and their variation with time. Both the spatial and temporal distributions of these characteristics show large-scale and small-scale structures. The large-scale structure can be considered as a statistical average over the individual small-scale structures.

First we examine the large-scale characteristics.

4.1.1 The refractive index and its gradient

The refractive index n of a medium is the ratio of the electromagnetic wave propagation velocity in a vacuum to that in the medium. An average value for air at radio frequencies is $n = 1.000\,315$. The radio wave velocity in air is therefore slightly lower than that in a vacuum, where $n = 1$. In order to facilitate dealing with problems involving air refraction, it is convenient to write

$$n = 1 + N \times 10^{-6} \tag{4.1}$$

which defines the refractivity N of air. The refractivity is not constant everywhere but varies in both time and space depending on the temperature, the humidity and the pressure of the air. Temporal variations may be fast (seconds) or slow (months). Spatial variations may involve large-scale (kilometers) and small-scale irregularities.

The most important large-scale average variation is the decrease in the refractive index with height, i.e. there is a negative refraction gradient with height. A vertical plane wave propagates faster at the top of the atmosphere than at the bottom, so that its trajectory is bent towards the ground. The variation in the average refractivity with height can be represented by an exponential law, at least for the lowest layers of the atmosphere where most of the propagation takes place. *CCIR Report 563* defines the "average exponential atmosphere" as that for which

$$N(h) = 315 \exp(-0.136h) \tag{4.2}$$

where $N(h)$ is the value of N at height h (kilometers) and 315 is the average value of radio refractivity at the surface of the Earth. This exponential model holds for temperate regions in average conditions, but its validity may be more questionable in other regions. In practice, a more complex situation may exist in the first 2 km. In any case we will see in Chapter 6 that the surface refractivity N_s is an important parameter in the evaluation of propagation effects, at least in some current methods.

World charts of radio refractivity have been prepared (Figs 4.1(a) and 4.1(b)). These charts show values N_0 reduced to sea level using the exponential model. We can see that N_0 varies between about 300 and 400 units. For a site h km above sea level we find N_0 on the charts and obtain

$$N_s = N_0 \exp(-0.136h) \tag{4.3}$$

The two charts represent the monthly mean values in February and August, when the refractivity is generally minimum or maximum. It should be noted, however, that the values shown may undergo major variations from year to year.

The refractivity gradient is also used in prediction calculations (see Chapter 6). By differentiating (4.2) we can see that the average gradient still follows an exponential law and its average value at the Earth's surface is $-43\ N$ units km^{-1}. In practice the average gradient for the lower layers of the troposphere in various regions of the world may vary from -20 to $-80\ N$ units km^{-1}. A knowledge of the refractivity gradient is also important in studies of ducting.

World charts of radio refractivity gradients ΔN, obtained as the difference between the refractivity at the Earth's surface and at a height of 1 km (therefore taken as positive instead of negative), are given in *CCIR Report 563*. Two of these charts are reproduced in Figs 4.1(c) and 4.1(d).

The refractivity of air can be measured using radiosondes, radar or refractometers carried by aircraft.

4.1.2 The effective radius of the Earth

In the propagation of radio waves it is known that ray bending is a function of the gradient dn/dh and that the rays can be considered as rectilinear above an Earth of "effective radius" $a = kR$ obtained from the following sum of curvatures:

$$\frac{1}{kR} = \frac{1}{R} + \frac{dn}{dh} \tag{4.4}$$

where k is the correction factor for the Earth's radius and R is the actual radius of the Earth (6400 km). It is then possible to derive a curve like that in Fig. 4.2 which yields both k and kR for each value of N_s.

An average value of k generally used for calculations and in particular for drawing path profile charts is $k = 4/3$. Most troposcatter path loss calculations are made assuming $k = 4/3$ or taking the k value derived from Fig. 4.2 as a function of N_s. This is firstly because of the difficulty of evaluating the average value of a factor variable in a generally unknown way along the path, and secondly because of the tacit inclusion in the path equations of the effects of values different from $k = 4/3$.

The (spatial) average value of k can be determined from propagation measurements or from the average of many simultaneous meteorological

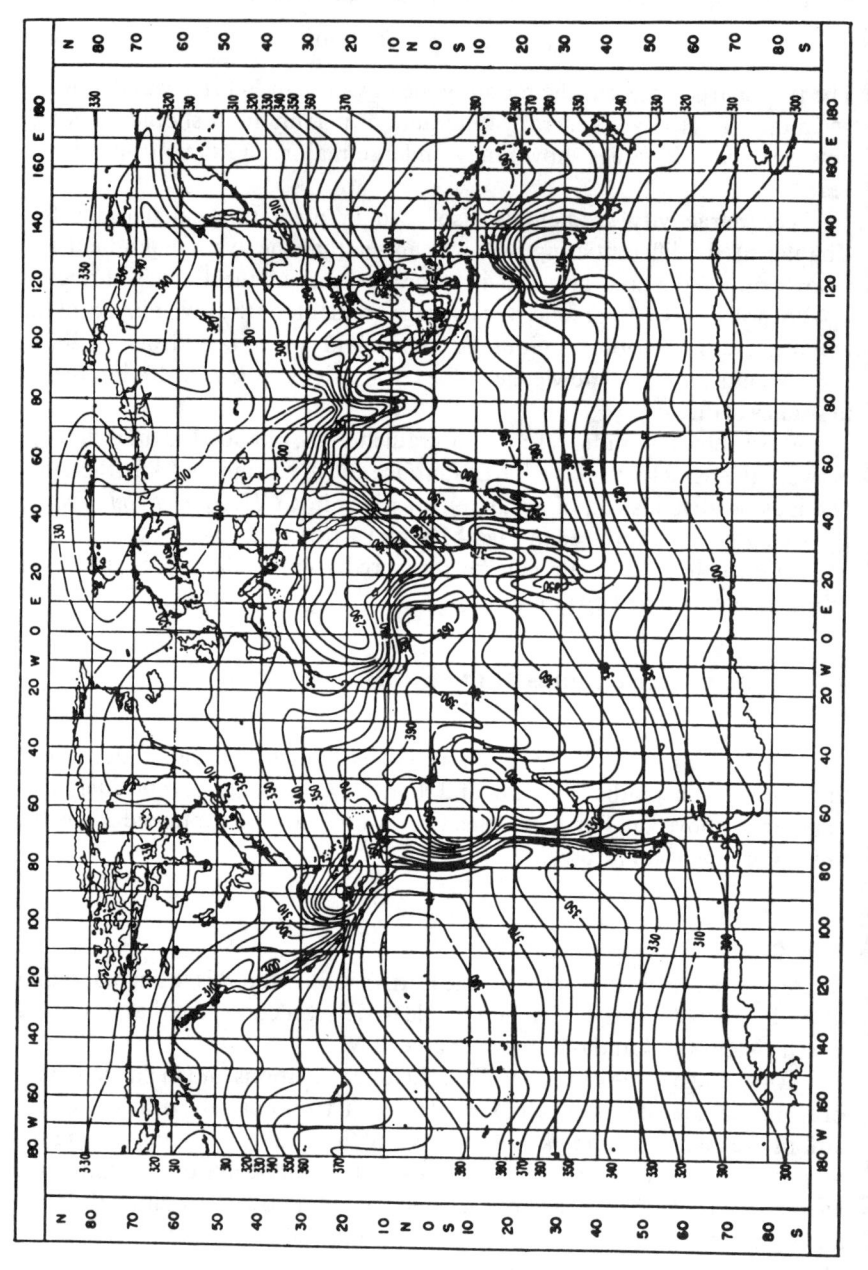

(a)

TROPOSPHERIC SCATTER MODE OF PROPAGATION

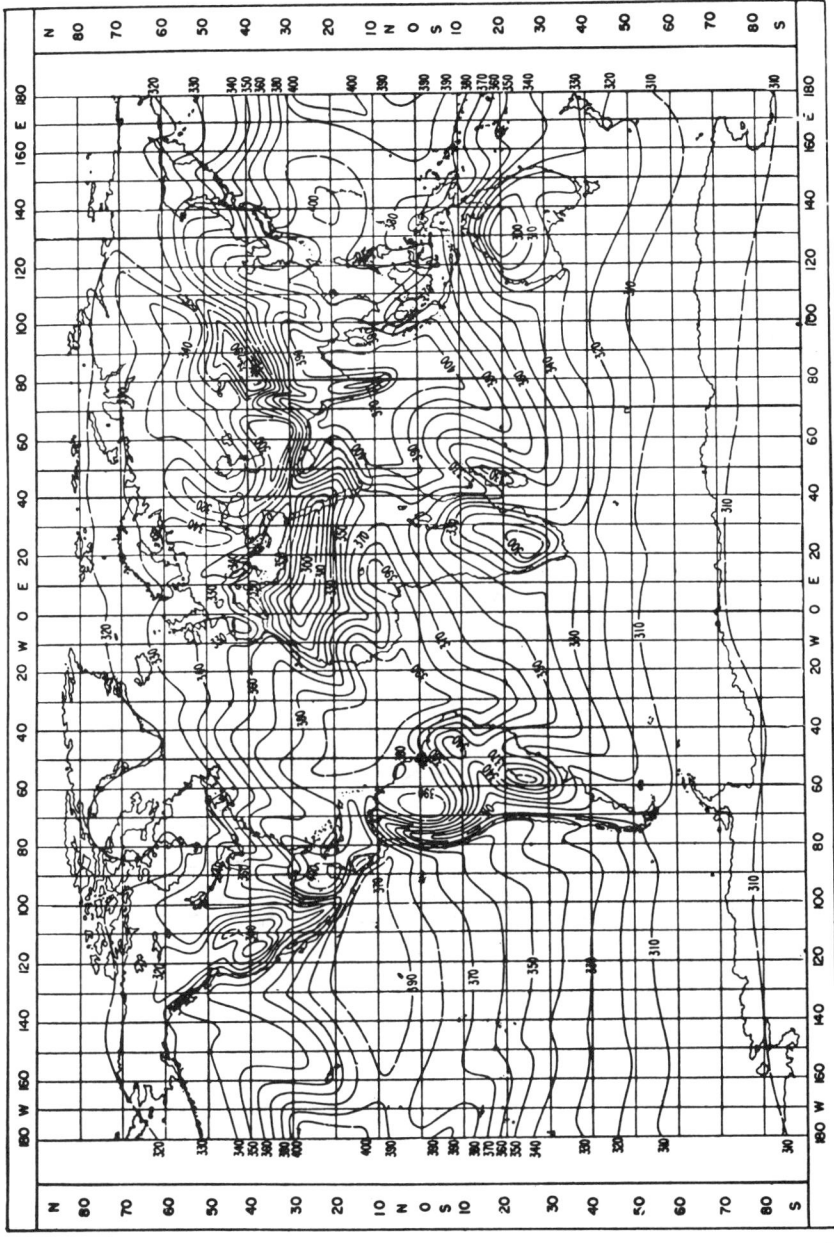

(b)

(c)

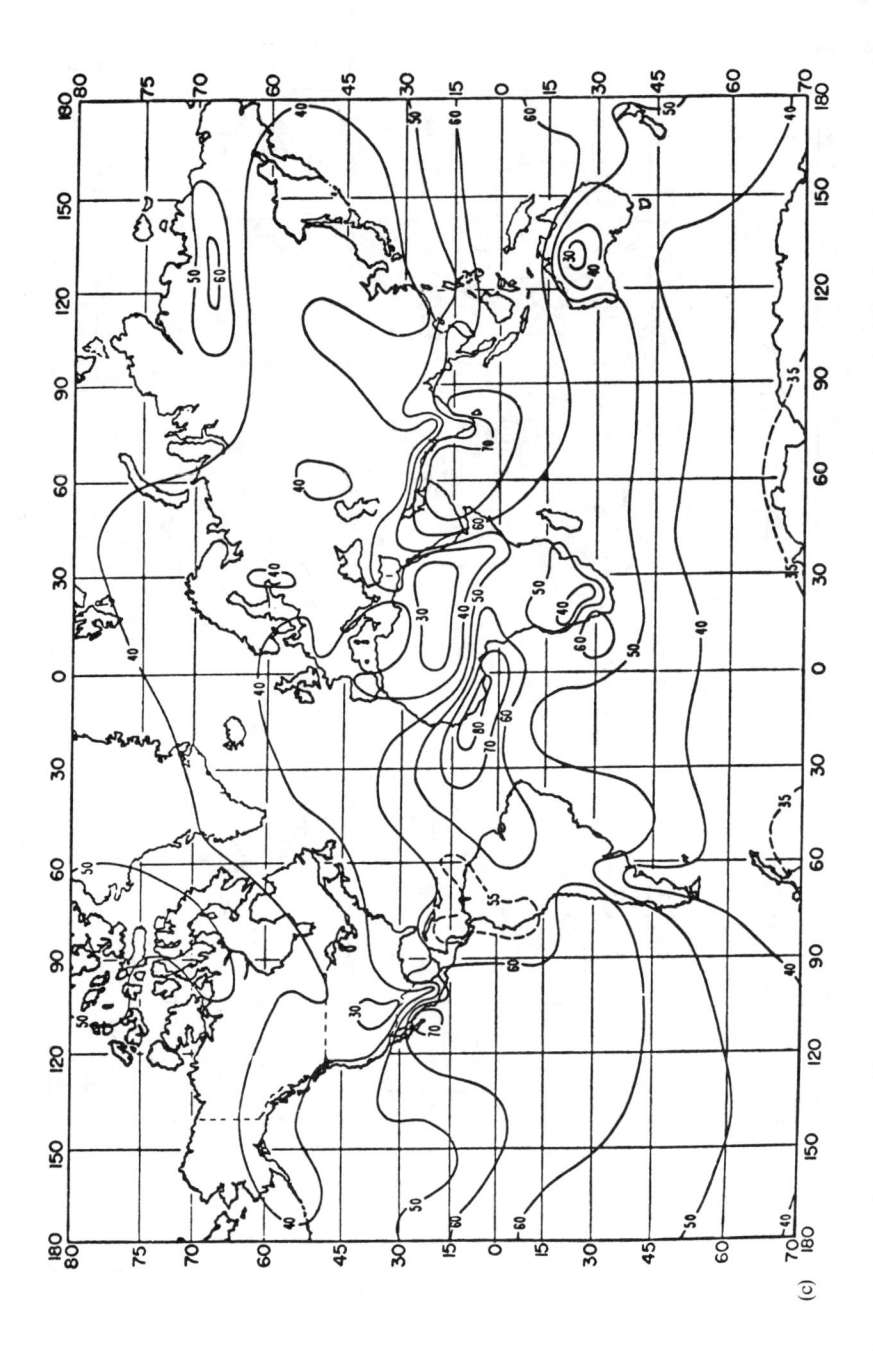

TROPOSPHERIC SCATTER MODE OF PROPAGATION 61

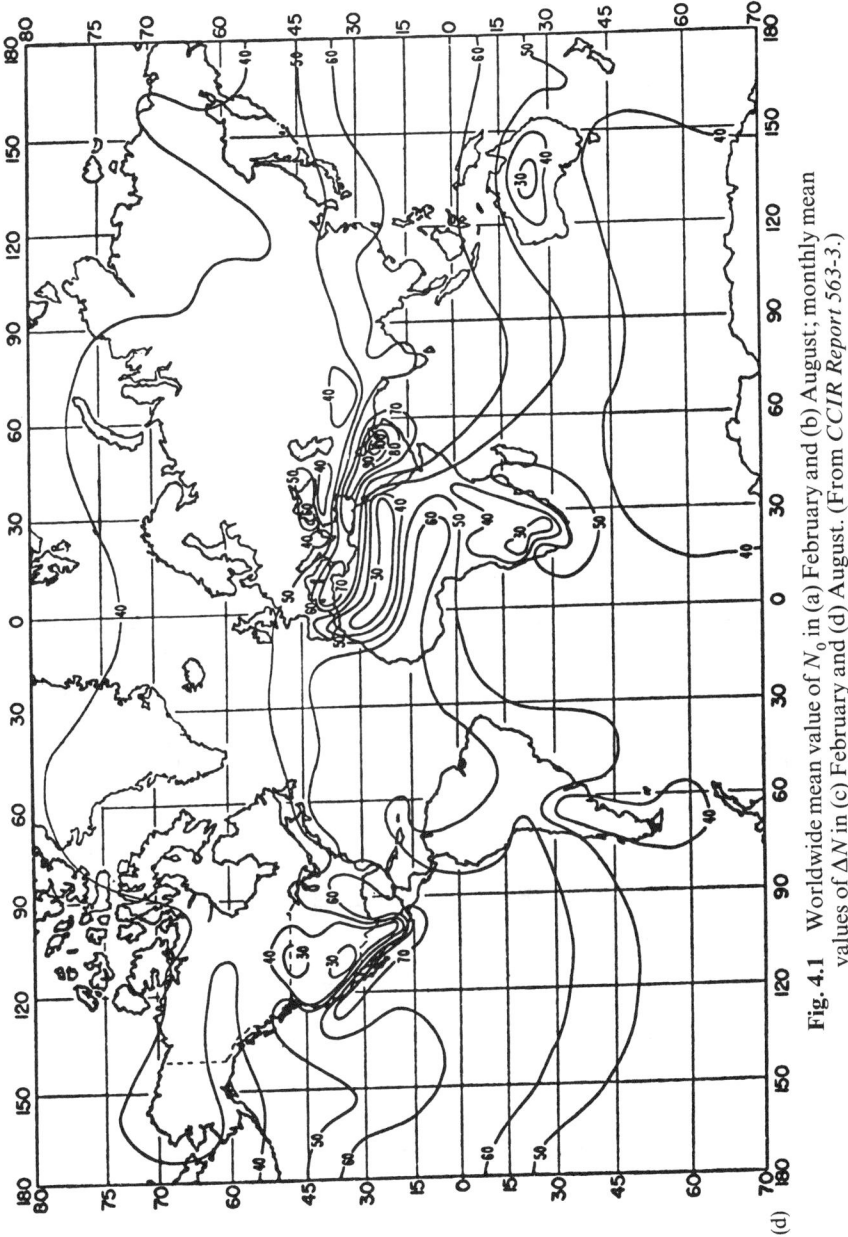

Fig. 4.1 Worldwide mean value of N_0 in (a) February and (b) August; monthly mean values of ΔN in (c) February and (d) August. (From *CCIR Report 563-3*.)

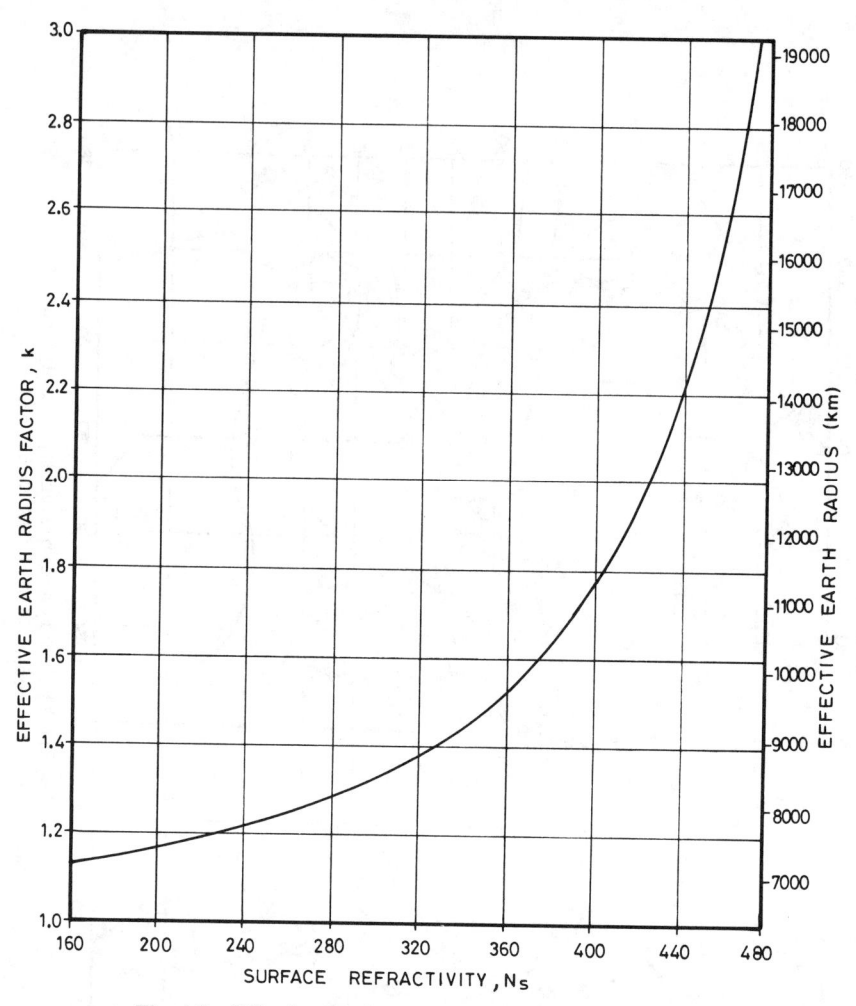

Fig. 4.2 Effective Earth radius versus surface refractivity.

soundings taken along the path. Results of measurements on LOS paths (*CCIR Report 718-2*) suggest that, for paths longer than say 100 km, the factor k is not less than unity, at least for 99.9% of the time. Furthermore, for long paths its value is much more stable.

4.1.3 Structure of the atmospheric refractivity

The radio refractivity can be calculated as a function of atmospheric pressure, temperature and water vapor pressure (*CCIR Report 563*). The spatial and temporal distribution of these variables determines the distribution of the radio refractivity. In particular, water vapor has an important role as it diffuses in various ways of greater or lesser complexity in the atmosphere. For example, in a calm atmosphere the water evaporating from the surface tends to diffuse slowly, producing a strong vertical gradient. In a turbulent atmosphere the mixing is more rapid and uniform so that the refractivity gradient is lower.

There have been many investigations of the small-scale structure of the atmosphere, which is responsible for the scattering of radio waves. This phenomenon is due to the presence of many zones with strong gradients or rapid variations in the refractive index of the air. The discontinuities can be explained using various models, particularly those based on the following:

(i) turbulent motion of the air, which generates a structure composed of vortices, eddies or blobs;

(ii) laminar motion of the air, which generates a structure composed of many small reflecting layers or feuillets.

In practice these two types of discontinuity coexist if the volume under consideration is sufficiently large. Many workers have used such models in attempts to explain the measured characteristics of the troposcatter signal and its variation with frequency, distance, scatter angle etc.

We will not analyze the physics of the troposphere nor the many theories that have been proposed but will simply consider it as being structured in sets of "scatter points" which may be individual scatterers or a continuum of scatterers or something in between. The scatter points seen by the receive antenna just above its horizon (those included in its main beam, intersecting the transmit beam in the common volume) radiate some of the received radiation to it.

An explanatory diagram of the radiation pattern is shown in Fig. 4.3. The scatterers immersed in the radiation field of the transmitting antenna act as secondary sources and scatter the incoming energy in all directions, but predominantly forward (this is why the phenomenon is also called forward

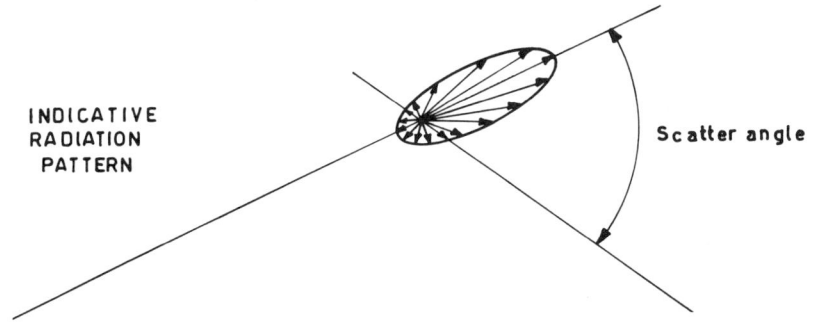

Fig. 4.3 Explanatory diagram of the radiation pattern of a scatterer.

scatter). The off-axis radiation decreases very rapidly (however, some very feeble "backward scatter" remains) and a fraction of it is scattered towards the receiving antenna, the direction of which differs slightly from the original direction (see Fig. 3.5 and the definition of the scatter angle). Of course the radiation scattered along the path of the transmitting antenna beam is a very small fraction of the total emitted radiation, most of which is dispersed in space along the beam.

In simple scattering theory it is assumed that the received power dW_s scattered by the element dV of the common volume V in a direction forming an angle θ with the direction of the incident wave is

$$dW_s = \frac{G_A G_B}{d_A^2 d_B^2 \theta^m} dV \qquad (4.5)$$

where d_A and d_B are the distances of dV from the two antennas, G_A and G_B are the gains of the antennas in the direction of dV and $m \approx 5$ according to experimental results. This formula, which will be required again in Chapter 6, Section 6.10, shows the rapid decrease in forward scattered power as the angle θ increases. The total received power is obtained by integrating eqn (4.5) over the common volume, assuming a given distribution of scatterers.

The distribution of scatterers in the troposphere appears to present a vertical variation, with the density greatest near the ground and decreasing with height in a similar manner to the refractivity gradient. This statement is justified by the results of path loss predictions (see Chapter 6, Section 6.4.3) and the decrease in antenna gain (Section 4.4). However, more detailed information is required for reliable prediction of the behavior of a troposcatter path in any region of the world. In fact there may be major differences in the structure of the troposphere from region to region, and this will affect the propagation loss (Table 4.1). At present, however, our knowledge of the worldwide characteristics of the troposphere is incomplete. The CCIR have proposed the following definitions of climate types as an aid to path loss predictions:

(1) equatorial (data from the Congo and the Ivory Coast)
(2) continental subtropical (Sudan)
(3) maritime subtropical (west coast of Africa)
(4) desert (Sahara)
(5) Mediterranean (no data available)
(6) continental temperate (Europe, USA)
(7a) maritime temperate over land (UK, France)
(7b) maritime temperate over sea (UK)
(8) polar (no data available)

These climates will be defined and their main characteristics given in Section 4.9.

Table 4.1 Climate types and maximum path loss

General climate type	Periods of maximum loss
Characterized by temperature variations (latitudes above 15°)	
Nondesert regions	Cold season
Desert regions (Sahara)	Hot season
Rainy climate (tropical and equatorial) (latitudes below 15°)	Rainy season
Monsoon climates (e.g. Dakar)	Dry season and rainy season

From ref. 4.1.

4.1.4 Radiometeorological parameters

The results presented so far can be summarized by stating that the large-scale average structure of the troposphere induces a bending of the antenna beams, which is characterized by the factor k or by the effective radius of the Earth, and that the small-scale local structure produces the scattering which is responsible for the unexpected propagation a long way beyond the horizon.

In order to simplify the path loss prediction procedures many attempts have been made to find a simple parameter that could characterize the behavior of the atmospheric structure with respect to troposcatter propagation. The only practical parameters found and used so far in the formulae (see Chapter 6) are the following:

(a) the refractivity N_s at the Earth's surface, which is currently measured in many meteorological stations worldwide;

(b) the gradient of N_s, which is also measured but only as an average over the first kilometer of height.

It has been found, and we will discuss the matter further in Chapter 6 in connection with the radiometeorological method of prediction, that the best characterization of the behavior of the common volume in all types of climate and at all times is given by the refractivity gradient in the common volume. The variations in this gradient correlate well with the path loss, and a CCIR prediction formula is based on this result. However, when a new link is being planned, it is unlikely that this parameter, together with its statistical monthly and annual variations, will be easily available. Therefore another older CCIR method based on the parameter N_s is still widely used, although it is generally only valid for temperate regions where the exponential model of the atmosphere is a good approximation and N_s is proportional to its gradient.

These parameters are used in Chapter 6 in connection with the calculation of path loss.

4.2 Troposcatter multipath transmission

Signal transmission by the troposcatter mechanism typically involves multipath transmission. The situation is shown schematically in Fig. 4.4. The signal is transmitted by the transmit antenna within the cone of its main beam in the direction of the horizon. This beam illuminates the troposphere which, as we have seen, is not perfectly physically homogeneous but contains blobs, layers and other inhomogeneities which scatter some of the illuminating energy. We can also consider that the transmitting antenna emits a beam of rays, each of which illuminates a scatterer and then proceeds to the receiving antenna. The received signal is the vector sum of the signals, which were originally identical, carried by various rays and arriving with random amplitude and phase. This is therefore a typical multipath transmission.

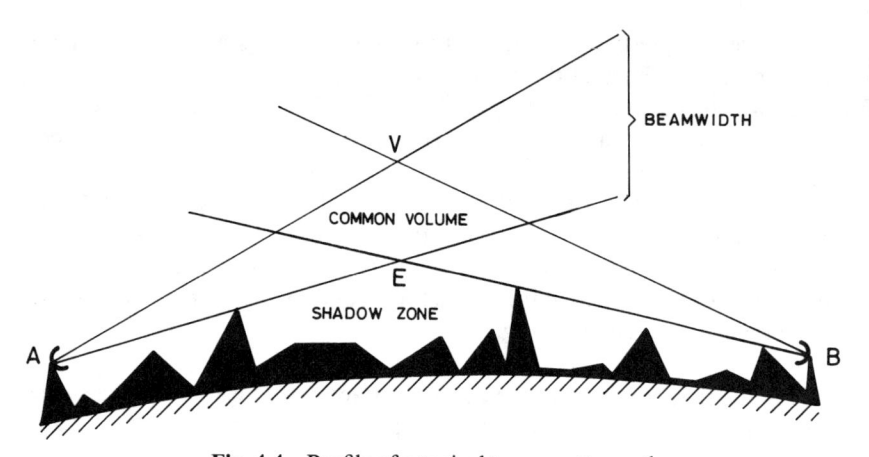

Fig. 4.4 Profile of a typical troposcatter path.

It can immediately be seen that there are long and short paths within the antenna beams, and therefore the received signal, which is the sum of many elementary signals which have undergone a variety of attenuations and delays, will be distorted with respect to the original transmitted signal. Furthermore this situation varies with time. The number, position, velocity and radiation characteristics of the scatterers vary independently and this distorts the received signal even more. Therefore the received signal must be studied on a statistical basis.

An equivalent circuit for the multipath transmission described above is shown in Fig. 4.5. The transmitted signal is split into a number of parallel paths, each with a variable delay and a variable attenuation. The delay varies because of the variation in path length. It also varies with time because of the variations in the position and velocity of the scattering center. The attenuation varies because of the changes in the position of the path in the antenna

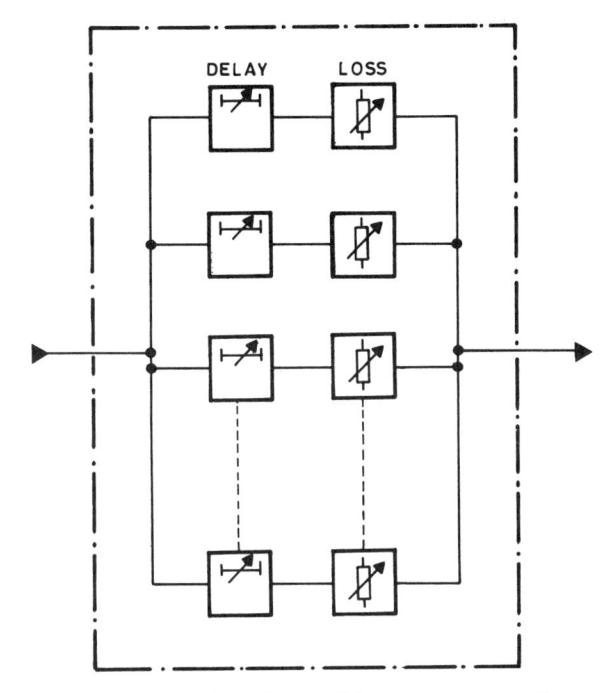

Fig. 4.5 Equivalent circuit of the troposcatter path.

radiation patterns (with consequent changes in the illumination and the reception gain) and because of the changes in the scatter characteristics of the scatterers, which also vary with time, as well as the variable number of scatter points.

The received r.f. carrier (Fig. 4.6) is the vector sum of the elementary

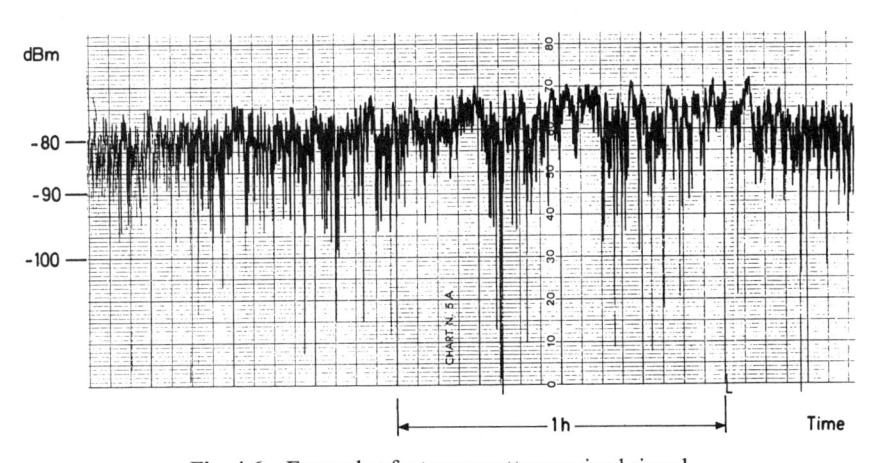

Fig. 4.6 Example of a troposcatter received signal.

signals of the constituent paths. The spread in time delays has a filtering action which starts when the average spread is no longer negligible with respect to the minimum period of the signal modulating the r.f. carrier. Therefore the troposcatter channel has a limited bandwidth (in practice a few megahertz) with a limited transmission capability (in practice 120 FDM telephone channels or 2 Mbit s^{-1}). This will be discussed in more detail later. A limitation on the transmissible bandwidth is characteristic of multipath transmission and therefore also of troposcatter.

4.3 Representation of the main propagation phenomena

Let us consider a representative path profile such as that shown in Fig. 4.4 in which the bending of the antenna beams has been removed by increasing the Earth's radius by a factor of 4/3 (valid for the standard atmosphere). The antenna beams are now represented by straight lines and the true radius of the Earth has been replaced by the effective radius. The geometrical path parameters can be calculated using the formulae given in Chapter 3 and starting from the basic data listed in Section 3.2.1 (path length, heights of the sites and the radio horizons above sea level, factor $k = 4/3$ etc.). In particular, the height of the base of the common volume above sea level can be obtained using eqn (3.11a) or eqn (3.11b).

However, this description represents only the average situation because we have chosen a fixed value of 4/3 for k with a corresponding effective radius of $a = (4/3)6400$ km = 8500 km for the Earth (eqn (3.16)). In practice we must take the following phenomena into account.

(a) Long-term changes (seasonal, monthly, hourly) occur in the structure of the refractive index and therefore the bending of the rays changes, displacing the common volume up or down. These long-term changes can be represented as variations in k and in the effective radius of the Earth. The new situations can be represented in Fig. 4.4, but with the geometrical path parameters recalculated for each new value of k. The changes in the parameters, particularly in the scatter angle, produce major variations in path attenuation.

(b) Short-term, random, rapid and statistically independent variations occur in the refractive index along the rays, particularly in the region near the surface of the Earth. These variations cause random independent bending of the antenna beams both horizontally and vertically. This phenomenon is called "scintillation": the receive antenna sees a continuous fluctuation in the angle of arrival of the rays.

Experiments confirm the above phenomena and also indicate that the two beams coupled through the random scatterers in the common volume are the free-space beams, which are not broadened by propagation in the troposphere but present random pointing angles varying rapidly and independently in both the vertical and the horizontal plane. These scintillations produce random motion of the common volume and cause variations in the time delay and the multipath spread. A long-term displacement of the common volume in the vertical plane produced by the average variation in the atmospheric refractivity and thus in the effective radius of the Earth is superimposed on this random motion.

Measurements made using swinging and scanning beams have enabled the common volume to be "seen" from the receiving antenna site. It lies just above the horizon and consists of randomly varying brilliant zones, brilliant layers, brilliant spots or similar inhomogeneities, sometimes in contact and sometimes separate, which form on the horizon a kind of irregular disk with time-variable contours and a continuous or discontinuous surface varying in shape and brilliancy.

The cone of the receiving antenna beam projects on the sky a regular disk containing the irregular disk formed by the scatterers. It can easily be seen that, if the regular disk is only partially filled by the scatterers (for example because they are concentrated in the lower part), the free-space gain of the antenna is not fully achieved and there is in practice a gain loss.

Two more phenomena which have a major effect on the transmission of signals should be noted at this point.

(c) There is a loss in the antenna gain such that an increase in the diameter D of the parabolic reflector produces a lower increase in the antenna gain than the value of $20 \log D$ predicted by the gain formula. This important effect will be discussed in more detail in Section 4.4.

(d) A spatial decorrelation occurs in the received signal so that, if two adjacent antennas are progressively separated either vertically or horizontally on a line parallel or perpendicular to the path, a volumetric cell can be defined outside which the amplitudes of the received signals are no longer correlated. This important effect is utilized in the space diversity techniques discussed in Chapter 5.

Other phenomena, which are collectively known as "anomalous propagation", will be considered in Section 4.8.

4.4 Degradation of antenna gain

The degradation of antenna gain, which was noted briefly in Section 4.3, has been studied both theoretically and experimentally, but no mathematical model describing the true situation has yet been derived. Degradations calculated using the various methods proposed in the literature can vary by up to 10–15 dB. However, very interesting experimental results have been obtained and explained qualitatively by French researchers [4.2, 4.3] and have been adopted by the CCIR in *Recommendation 617*. The basic concepts are as follows.

(a) The wavefront reaching the receive antenna is not plane but is a distorted surface whose shape varies randomly with time. This causes a random decrease in the (free-space) gain of the antenna. The average decrease is called the "gain degradation", the "drop in antenna gain" or the "antenna-to-medium coupling loss".

(b) As a result of the well-known reciprocity theorem the same phenomenon occurs at the transmit antenna, which also suffers from gain degradation.

(c) The number of "scatterers" per unit volume is greatest at the base of the common volume and, like the refractivity gradient, decreases with height.

(d) The 3 dB beamwidth of the antenna can vary from say 9° (low gain antennas) to a fraction of a degree (high gain antennas). The common volume (Fig. 4.7) for antennas with very low gain (below 25 dB) is so large that most scatterers are contained in the lower part and few are present at the top or sides. In contrast, the common volume for antennas with very high gain (50 dB or more) is so small that the scatterers can be assumed to be uniformly distributed within it.

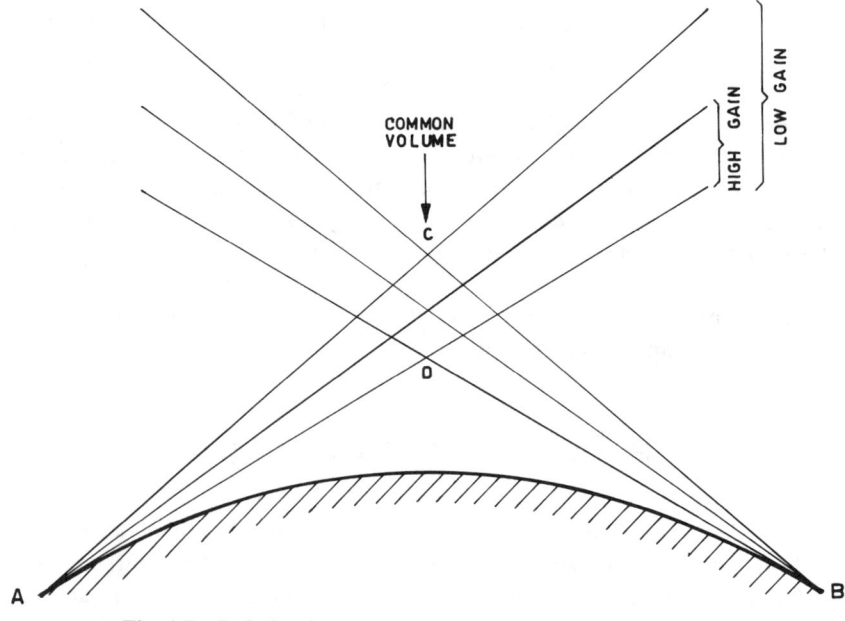

Fig. 4.7 Relation between antenna gain and common volume.

If the transmitted power is increased by n dB, the power received and scattered by each scatterer is also increased by n dB as is the power received at the distant terminal. However, the same argument does not necessarily hold if the gain of the transmitting antenna is increased by n dB. It can be seen in Fig. 4.7 that when the antenna gain is increased both the beamwidth and the common volume decrease. Therefore a smaller number of scatterers are involved in the propagation. The illumination of each scatterer is increased by n dB, but since there are fewer of them the increase in the received signal is smaller.

More precisely, with very high gain antennas (a uniform density of scatterers in the common volume) an increase in gain increases the illumination proportionally to the square of the beamwidth, but the number of illuminated scatterers decreases as the cube of the beamwidth and so there is an overall gain loss proportional to the beamwidth. With very low gain antennas a variation in gain involves only a few scatterers at the top and sides of the common volume and thus the gain degradation is negligible.

If the path length is increased the common volume is displaced upwards (where there are more uniformly distributed scatterers, producing a greater gain loss) but also becomes larger (with a smaller gain loss, as in the case of low gain antennas). These two effects compensate each other so that the gain loss is almost independent of path length up to 500 km (see below). For longer paths the gain loss decreases, and by about 1000 km it has almost disappeared because only the bottom of the common volume contains scatterers and these are seen by antennas with any gain.

Measurements of gain loss are very intricate. The signal is received on two (or more) antennas with very different diameters and gains, and the two average levels received are compared. Average values must be used because the smaller antenna receives a signal scattered by a larger number of inhomogeneities which is therefore much more variable than that of the larger antenna. The two antennas can be installed side by side or, better, superimposed on the same axis so that an identical direction is guaranteed as well as the absence of space diversity effects. Free-space gains are accurately measured, possibly on the antennas of the troposcatter terminal, using an auxiliary LOS terminal placed on the same route as the troposcatter link. The most important results of these experiments are summarized below.

(a) Gain loss becomes appreciable only for antennas with gains of at least 25–30 dB.

(b) The gain loss (statistical average) is approximately constant, independent of the level of the received signal and of its statistical distribution.

(c) The overall gain loss of the two antennas (transmit and receive) is approximately equal to the sum of the individual losses for gains below 50 dB.

(d) Gain loss is almost independent of path length from at least 150 km to 500 km, after which it decreases and vanishes at 1000 km or above.

(e) Gain loss can be assumed to depend only on the (free-space) gain of the antennas (provided that it does not exceed 55 dB and is not very different for the two antennas). This dependence can be expressed as follows:

$$\Delta g = 0.07 \exp\{0.055(G_t + G_r)\} \tag{4.6}$$

where Δg (dB) is the gain loss and G_t and G_r (dB) are the free-space gains of the transmitting and receiving antennas.

(f) Climatic influence is expected to be small except when the decrease in the refractivity gradient with height is irregular, which implies an irregular distribution of scatterers.

(g) Increasing the takeoff angles increases the gain loss.

Statement (d) is very important because it includes almost all troposcatter paths used in practice. In general path lengths do not exceed 500 km and antenna gains do not exceed 55 dB. Furthermore, the antennas at the two sites generally have the same diameter.

4.5 The troposcatter multipath signal

We have already seen in this chapter that the received troposcatter signal is the result of many randomly variable elements: the presence of scatterers in the common volume, the distribution and behavior of these scatterers, the variability of the common volume etc. We also anticipated at the end of Section 4.2 that, owing to multipath propagation, the troposcatter channel has a limited bandwidth which varies randomly in time. The randomness of the propagation characteristics also randomizes the transfer function (amplitude response versus frequency) of the troposcatter path, with the following consequences for the received signal (Fig. 4.6):

(a) a random amplitude which follows short-term Rayleigh statistics (because of the addition of many random vectors);

(b) decorrelation between the signal amplitudes at two different times exceeding a given interval (this effect, which is a consequence of (a), is utilized in the time diversity techniques described in Chapter 5);

(c) a time-variable limit on the maximum transmissible bandwidth;

(d) decorrelation between the amplitudes of two frequencies which differ by more than a given value (this effect, which is a consequence of (c), is utilized in the frequency diversity techniques described in Chapter 5);

(e) a time-variable limit on the transmissible bit rate for digital signal transmission (this is a consequence of (c));

(f) small or negligible broadening of the spectrum frequencies (this is due to the Doppler effect as explained later).

Most of these characteristics are related to the short-term behavior of the propagation medium, with the statistical analysis being performed for periods ranging from less than 1 min to several minutes.

For practical purposes the transmission capabilities of the short-term signal can be characterized in terms of the bandwidth and the associated delay spread, which are discussed in the next section, and the continuity of reception in the presence of fading is related to the random variation in the amplitude, which is examined in Section 4.7.

4.6 Spread and bandwidth parameters of the medium

The following types of test signal can be transmitted through the troposcatter channel:

(a) a sinusoidal signal;
(b) two sinusoidal signals with the same amplitude but different frequencies;
(c) a pulse.

In case (a) the sum of the multipath signals with different amplitudes and phases but the same frequency again yields a sinusoid if the time variability is neglected. However, owing to the random variability of the scatterers, the amplitude and frequency of this sinusoid vary, and thus the spectrum of the output signal is broadened and becomes a randomly variable Doppler spectrum. As is customary for statistical distributions, the width of this spectrum is expressed in terms of its standard deviation σ_d and is normally indicated in the literature as

$$B = 2\sigma_d$$

where B is known as the width of the Doppler spectrum or the Doppler spread. Fortunately the displacements of the scatterers occur slowly compared with the frequencies contained in the narrow-band troposcatter signal, so that in practice B is 0.1–20 Hz and the Doppler spread can often be ignored. However, $1/B$ is also an indication of the time after which two samples of the received signal amplitude are statistically independent (and therefore decorrelated). We can assume that correlation lasts for a time $T \ll 1/B$, and therefore, substituting the values of B given above, we have $T \ll 10$ s in the best case and $T \ll 0.05$ s in the worst case. Furthermore, B gives an indication of the frequency of occurrence of fading (in practice 0.1–20 fades per second). The Doppler spectrum is generated by the time variability of the path delays, and the fading is generated by the combined effects of attenuation and delay (phase shifts).

In case (b), if the Doppler spread is neglected, the same two frequencies can be found at the output with the same time-varying amplitude only if they are sufficiently close to each other. As their separation is increased, their amplitude variations become independent and the correlation coefficient goes from unity to zero. This enables us to evaluate the bandwidth of the troposcatter channel as well as the minimum distance necessary for adjacent channels to be independent. The troposcatter channel acts as a bandpass filter with a bandwidth of the order of a few megahertz. Two channels are statistically independent when their center frequencies differ by more than say 10 MHz.

In case (c), transmitting a pulse, or a sequence of pulses which are sufficiently well separated, is the same as transmitting a uniform spectrum. The received signal can be regarded in the frequency domain as a spectrum with the shape of the transfer function and in the time domain as a broadened pulse (Fig. 4.8). The latter spectrum is also called the "delay power spectrum" or the "multipath spectrum". The shapes of all these spectra vary randomly. As is customary for statistical distributions, the width of the broadened pulse is measured in terms of its standard deviation σ and is normally indicated in the literature as

$$L = 2\sigma$$

where L is the duration of the delay power spectrum or the multipath delay

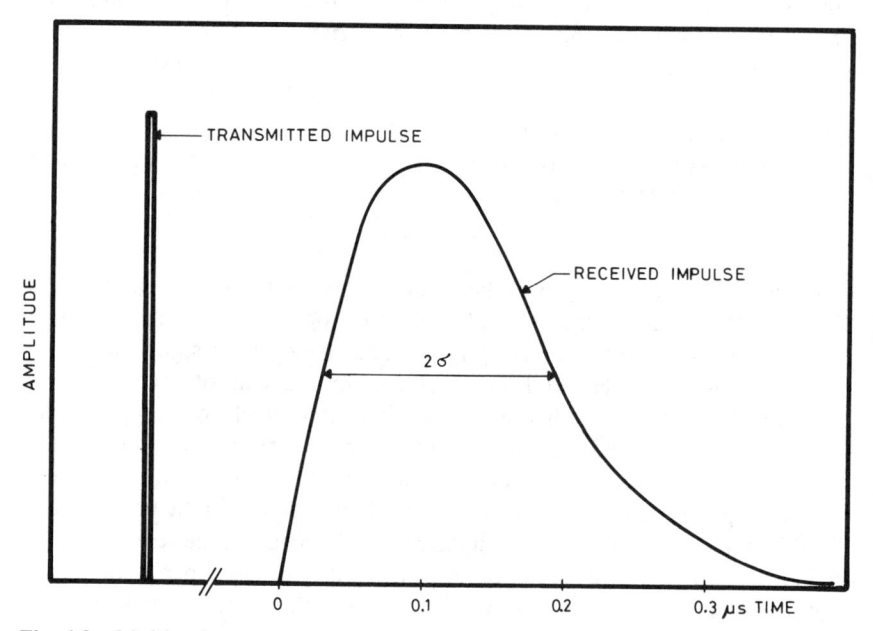

Fig. 4.8 Multipath spectrum: σ, root mean square (r.m.s.) value referred to average delay; 2σ, average duration of received impulse.

spread. In practice values of the multipath delay spread range from 0.1 to 0.5 µs. If the bandwidth of the transmitted signal is W the condition for negligible distortion is $W \ll 1/L$ and therefore, substituting the above values of L, $W \ll 10$ MHz in the best case and $W \ll 2$ MHz in the worst case. This means that the amplitudes of all frequencies in the band W are correlated and therefore fading is flat all over the spectrum.

Since bandwidth and the multipath delay spread are random variables, they can be represented by their cumulative distributions. In practice, for evaluating the transmission capabilities of the channel we consider values that are not exceeded for a given percentage of the time. Of course there are no problems if the bandwidth of the transmitted signal is well within the troposcatter channel bandwidth or if the pulse duration of a transmitted digital signal is much longer than the multipath delay spread. However, when the signal bandwidth is progressively broadened or the bit rate of the digital signal is progressively increased, the received signal becomes more and more distorted so that an increasing intermodulation noise appears in the analog signal and an increasing error rate appears in the digital signal, and both show the same time dependence as the parameters mentioned above.

The path distortions are calculated in Chapter 6.

4.7 Variability in the level of the troposcatter signal

Analysis of the experimental data obtained for the received troposcatter signal (Fig. 4.6) has suggested that the level of this randomly varying signal can be represented as the sum of three components:

$$S(\text{dBm}) = X(\text{dB}) + Y(\text{dB}) + M(\text{dBm}) \tag{4.7}$$

$S(\text{dBm})$ is the level of the r.f. carrier signal at the output of the receiving antenna. $X(\text{dB})$ is the random variation of the signal around its median value in 1 min (the statistical data were actually collected over periods of up to several minutes). This term was found to follow a Rayleigh distribution to a very good approximation, which demonstrates its multipath origin. It is responsible for the "fast fading" due to sudden dips in the signal. $Y(\text{dB})$ is the random variation of the medians of the signal obtained each minute about their median value. The statistical data were generally collected over periods of hours, days or longer. This term was found to follow an approximate Gaussian law, particularly during periods of high propagation loss, and it is the result of slow variations in the average characteristics of the troposphere. It is responsible for "slow fading". $M(\text{dBm})$ is the median value of the signal over the period of measurement (1 h, 1 day, 1 month, 1 year or some other period).

X is the short-term random variation representing the "instantaneous" variability of the signal. It is important to note that the variability of the information (baseband frequencies, bit rates) modulating the signal is much faster than that of X. For example, during a minimum correlation time of 0.05 s it is possible to transmit 12 800 bits at 256 kbit s^{-1}. This quasi-stationarity allows us to acquire information on the state of the channel before it changes, and this is exploited in diversity techniques.

Y is the long-term random variation of the medians obtained each minute, or sometimes of the hourly medians, as we shall see when we calculate path loss and performance. M is a very long term median.

Sometimes the term X is disregarded. The received signal S can be measured using an instrument with a variable time constant. If this constant is of the order of milliseconds the instrument will follow the instantaneous signal and will measure a Rayleigh distribution with a slowly varying median. If the time constant is increased to several minutes, the Rayleigh curve gradually becomes a log-normal curve.

A mathematical analysis of the characteristics of the received signal has been performed [4.3, 4.4], as a result of which the troposcatter channel has been classified as a Gaussian wide sense stationary uncorrelated scattering channel (GWSSUS). An interesting result of this analysis is that the received signal shown in Fig. 4.6 is (theoretically) identical with a white noise function. Its modulus has a Rayleigh distribution and its diagram, shown in Fig. 4.9, can be interpreted using either signal versus time, with time units $1/B$, or signal versus frequency, with frequency units $1/L$, where B and L are as defined in Section 4.6. It is important to note that the length of the units is of the order of that of the correlation intervals.

For example, if $1/B = 5$ s and $1/L = 6$ MHz, Fig. 4.9 shows that there is

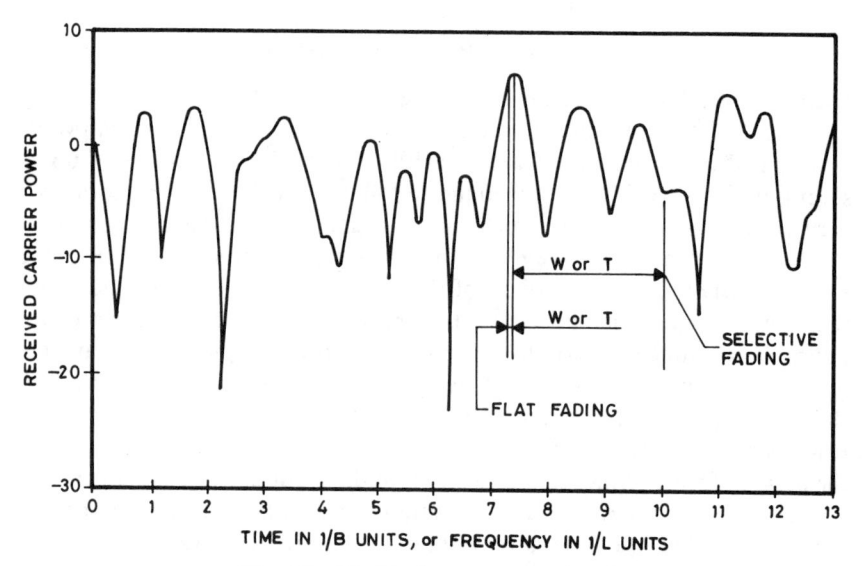

Fig. 4.9 The ideal troposcatter signal.

frequency-selective fading, in this case with two dips, within a bandwidth $W = 2.5 \times 6$ MHz $= 15$ MHz. Alternatively, there is time-selective fading in an interval $T = 2.5 \times 5 = 12.5$ s. For bandwidths much narrower than 6 MHz or time intervals much shorter than 5 s the fading is almost flat, which means that all amplitudes vary together (full correlation).

According to formulae (2.14)–(2.20) for the Rayleigh distribution, we can state that the probability of a fade deeper than X dB below the median (X negative) is given by (Fig. 2.4)

$$P(X) = 1 - \exp(-0.7 \times 10^{0.1X}) \tag{4.8}$$

This is also the outage probability if the median signal is $-X$ dB above the receiver's threshold. In practice Rayleigh fading generally occurs during the

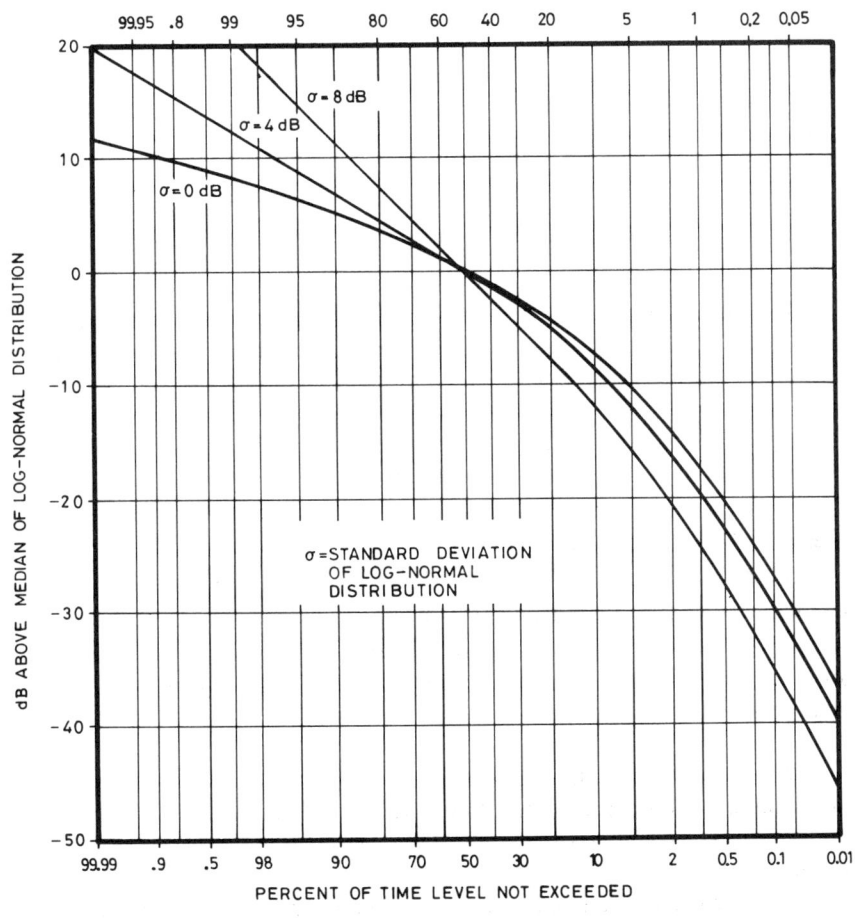

Fig. 4.10 Combined short-term–long-term signal distribution.

worst variability conditions. It gives an acceptably accurate representation of the behavior of the actual fading within the range 0.1%–99.9%. Other types of fading that may occasionally occur are generally less critical and less frequent.

The convolution of the Rayleigh distribution of X with the normal distribution Y of the medians is given by eqn (2.34). From this, over longer periods (hours, days, months), the probability of a fade deeper than Z dB below the median of the normal distribution (with standard deviation σ) is given by

$$P_c(Z) = 1 - \frac{1}{(2\pi)^{1/2}} \int_{-\infty}^{\infty} \exp(-s) \exp\left(-\frac{t^2}{2}\right) dt \qquad (4.9)$$

where $t = Y/\sigma$ and $s = 0.7 \times 10^{0.1(Z-\sigma t)}$. The diagram of this combined log-normal–Rayleigh distribution is shown in Fig. 4.10 and refers to the "instantaneous signal" over long periods.

The variability of the signal level spans tens of decibels and in practice it would cause frequent disappearance of the signal below the receiver's threshold, yielding a very poor transmission quality. The harmful effects of this variability can be substantially mitigated by using the diversity techniques described in Chapter 5.

4.8 Anomalous propagation

In all troposcatter hops there may be periods of anomalous propagation, i.e. propagation with characteristics different from those typical of troposcatter. This anomalous propagation may be due to any of the following:

(a) diffraction, when the scatter angle is originally low and occasional changes in the general refractivity of the air make it even lower;

(b) specular reflection in strong elevated and extended layers temporarily present in the troposphere;

(c) ducting due to surface or elevated layers formed under certain tropospheric conditions, particularly during fine weather and over the sea;

(d) reflections from aircraft which are frequent in some areas (e.g. Europe) and may cause nonnegligible disturbances such as Doppler effects and strong delayed signal components.

When the scatter angle is low (say less than 1°) some of the energy is propagated by diffraction. The effective radius of the Earth may increase so that the diffracted field is dominant, and the signal becomes stronger and more stable.

Ducting is essentially a fine-weather phenomenon due to a stratification of the lower troposphere into layers with very different refractivity gradients produced by a temperature inversion (temperature increasing with altitude). A normal gradient at the Earth's surface is $-50\ N$ units km^{-1}. Values greater than $-40\ N$ units km^{-1} indicate that the atmosphere has become "subrefractive": the energy is bent upwards, the effective radius of the Earth decreases and the radio horizon becomes nearer. Values below $-50\ N$ units km^{-1} and down to $-157\ N$ units km^{-1} indicate a "superrefractive" atmosphere, in which the effective radius of the Earth is increased and the radio horizon is more remote. Ducts, which behave like large waveguides with thicknesses of tens to hundreds of meters and horizontal extensions of tens to hundreds of kilometers, tend to form when the refractivity gradient falls below the critical value of $-157\ N$ units km^{-1}. Ducts are particularly common over warm seas (e.g. the Persian Gulf).

An undesirable type of propagation due to scatter by hydrometeors (rain, snow, hail, ice clouds) is occasionally present in the common volume of troposcatter antennas and of antennas of other types of radio link and can cause interference. Coupling through the sidelobes of the antennas may occur, even at large scattering angles. We shall return to these topics in Chapter 6, Section 6.9, when we attempt to evaluate path losses.

4.9 The CCIR climates

The definitions of climate types presented here are taken from *CCIR Report 238-5*.

(1) *Equatorial* corresponds to the region between latitudes 10°N and 10°S. The climate is characterized by a slightly varying high temperature and by monotonous heavy rains which sustain a permanent humidity. The annual mean value of the refractivity N_s at the surface of the Earth ($= (n-1) \times 10^6$ where n is the refractive index of the air) is about 360 N units and the annual range of variation is 0–30 N units.

(2) *Continental subtropical* corresponds to the regions between latitudes 10° and 20°. The climate is characterized by a dry winter and rainy summer. There are marked daily and annual variations in radio propagation conditions, with least attenuation in the rainy season. Where the land area is dry, radio ducts may be present for a considerable part of the year. The annual mean value of N_s is about 320 N units, and the range of variation, throughout the year, of monthly mean values of N_s is 60–100 N units.

(3) *Maritime subtropical* also corresponds to the regions between latitudes 10° and 20° and is usually found on lowlands near the sea. It is strongly influenced by the monsoon. The summer monsoon, which blows from sea to land, brings high humidity into the lower layers of the atmosphere. Although the attenuation of radio waves is relatively low at both the beginning and end

of the monsoon season, during the middle of the monsoon the atmosphere is uniformly humid to great heights and the radio attenuation increases considerably despite a very high value of N_s. There is an annual mean N_s of about 370 N units with a range of variation over the year of 30–60 N units.

(4) *Desert* corresponds to two land areas which are roughly situated between latitudes 20° and 30°. Throughout the year there are semi-arid conditions and extreme diurnal and seasonal variations of temperature. This climate is very unfavorable for forward-scatter propagation, particularly in summer. There is an annual mean value of N_s of about 280 N units, and throughout the year monthly mean values can vary over a range of 20–80 N units.

(5) *Mediterranean* corresponds to regions in both hemispheres on the fringe of desert zones, close to the sea, and lying between latitudes 30° and 40°. The climate is characterized by a fairly high temperature, which is reduced by the presence of the sea, and an almost complete absence of rain in the summer. Radio wave propagation conditions vary considerably, particularly over the sea, where radio ducts exist for a large percentage of the time in summer.

(6) *Continental temperate* corresponds to regions between latitudes 30° and 60°. Such a climate in a large land mass shows extremes of temperature, and pronounced diurnal and seasonal changes in propagation conditions may be expected to occur. The western parts of continents are influenced strongly by oceans, so that temperatures here vary more moderately and rain may fall at any time during the year. In areas progressively towards the east, temperature variations increase and winter rain decreases. Propagation conditions are most favorable in the summer and there is a fairly high annual variation in these conditions. The annual mean value of N_s is about 320 N units and monthly mean values can vary by 20–40 N units throughout the year.

(7a) *Maritime temperate, over land,* also corresponds to regions between latitudes of about 30°–60° where prevailing winds, unobstructed by mountains, carry moist maritime air inland. Typical of such regions are the UK, the west coast of North America and Europe, and the northwestern coastal areas of Africa. There is an annual mean value of N_s of about 320 N units, with a rather small variation of monthly mean values over the year of 20–30 N units. Although the islands of Japan lie within this range of latitudes, their climate is somewhat different and shows a greater annual range of monthly mean values of N_s, about 60 N units. The prevailing winds in Japan have traversed a large land mass and the terrain is rugged. Climate 6 is therefore probably more appropriate to Japan than climate 7, but duct propagation may be important in coastal and adjacent over-sea areas for as much as 5% of the time.

(7b) *Maritime temperate, over sea,* corresponds to coastal and over-sea areas in regions similar to those for climate 7a. The distinction made is that a radio propagation path having both horizons on the sea is considered to be an over-sea path (even though the terminals may be inland); otherwise climate 7a is considered to apply. Radio ducts are quite common for a small fraction of the time between the UK and the European continent and along the west coasts of the USA and Mexico.

(8) *Polar* corresponds approximately to the regions between latitudes 60° and the poles. This climate is characterized by relatively low temperatures and relatively little precipitation.

4.10 Comments and suggestions for further reading

A general description of propagation, including troposcatter, is given by Boithias [4.5], and the characteristics of troposcatter are summarized in refs 4.1 and 4.6. Scattering theories are reviewed in refs 4.7 and 4.8. Antenna gain loss is discussed in ref. 4.2. Many papers dealing with specific aspects of troposcatter are available in the literature, but they are very specialized and are not always easy to follow. Many additional references can be found in the CCIR Recommendations and Reports.

References

4.1 Boithias, L. and Battesti, J. Propagation due to tropospheric inhomogeneities, *Proc. IEEE, Part F*, **130** (7), 657–664, 1983.
4.2 Boithias, L. and Battesti, J. Etude expérimentale de la baisse du gain d'antenne dans les liaisons transhorizon, *Ann. Telecommun.*, **19** (9–10), 221–229, 1964.
4.3 Battesti, J. and Boithias, L. Propagation par les hétérogénéités de l'atmosphère et prévision des affaiblissements, *AGARD Conf. Proc.*, **70**, 43-1–43-8 *et seq.*, 1970.
4.4 Bello, P. A. A review of signal processing for scatter communications, *AGARD Conf. Proc.*, **244**, 27-1–27-23, 1977.
4.5 Boithias, L. *Propagation des Ondes Radioélectriques dans l'Environnement Terrestre*, Dunod, Paris, 1984 (English translation (revised and updated): *Radiowave Propagation*, North Oxford Academic, London, 1987).
4.6 Bello, P. A., Ehrmann, L. and Alexander, P. Signal distortion and intermodulation with tropospheric scatter, *AGARD Conf. Proc.*, **70**, Part II, 36-1–36-17, 1970.
4.7 du Castel, F. *Propagation Troposphérique et Faisceaux Hertziens Transhorizon*, Chiron, Paris, 1961.
4.8 Panter, P. F. *Communication Systems Design—Line-of-sight and Troposcatter Systems*, McGraw-Hill, New York, 1972.

Chapter 5

Diversity Techniques

In troposcatter radio links it is normal practice to use diversity techniques in order to overcome the deleterious effects of the continuous variability of the received r.f. signal. Reception by a single receiver would yield a completely unsatisfactory communication, frequently interrupted by fading and impaired by noise. The principles and applications of diversity techniques are investigated in this chapter, and some diversity combiners and special digital modems are described.

5.1 Decorrelated paths and diversity effects

We have established that the r.f. signal arriving at the receiving antenna is the vectorial sum of multipath components which are continuously varying in time, phase, angle of arrival, intensity etc. As expected, this signal has a short-term statistical distribution which has been confirmed experimentally to follow the Rayleigh law in most cases, at least in the time range 0.1%–99.9% [5.1]. Reception is therefore characterized by a number of deep fades, many of which result in a decrease in the r.f. signal intensity below the receiver's threshold so that reception is interrupted. During bad propagation conditions the number of fades per minute may vary from say 10 to 150. The fading rate is relatively low in low r.f. bands and is high in higher bands. The average duration of a fade varies from tens of milliseconds to seconds. The duration is longer in low r.f. bands, which also have lower fade rates, so that the total average fade duration per minute is roughly the same for all frequency bands. The reception would be very poor under these conditions because the r.f. signal would fall below the threshold of the FM receiver for unacceptably long percentages of time during which the circuit would be interrupted or very noisy.

"Diversity systems" have been devised to combat the effects of fast fading and to smooth out the variability of the received signal. The principle of these systems is that the r.f. signal is received at a number of receivers (the "order of diversity" is equal to this number) disposed in such a way that each receives a signal which has covered a different path and is therefore affected by a different and uncorrelated fade. In other words the signal is transmitted through two or more statistically independent "diversity branches" or "diversity channels" and is recombined at reception. The deep fades in each diversity branch will in general not be coincident, so that for most of the time at least one receiver will receive the signal even if the others are too weak. The received signals are processed using a "diversity combiner" which yields a more stable output (Fig. 5.1).

For example, as noted in Chapter 4, Section 4.3, if the r.f. signal is received at two antennas the following phenomena are observed. When the two antennas are close to each other the two received signals show the same

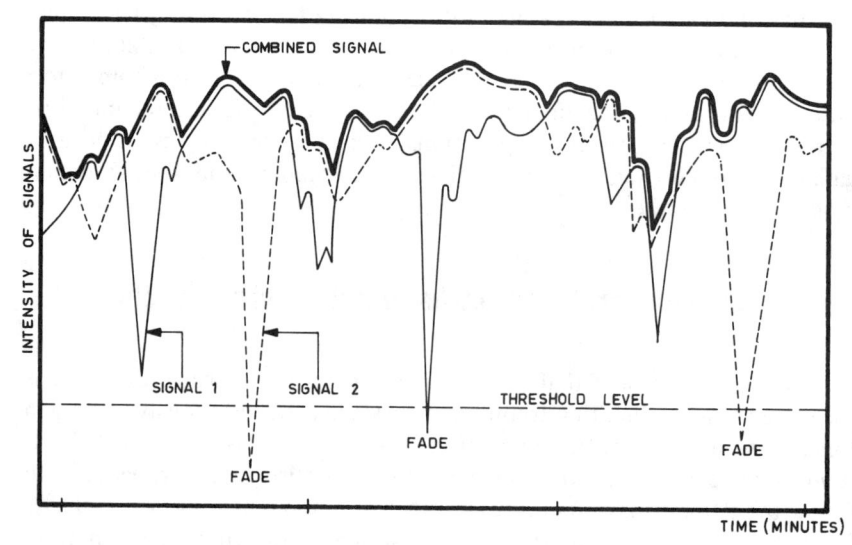

Fig. 5.1 Individual and combined diversity signals.

variations with time, i.e. they are fully correlated. This correlation decreases to zero as the distance between the two antennas is increased. Thus we obtain two independent paths. In practice it is not necessary to have a correlation coefficient of zero. An acceptable diversity effect can be obtained with correlation coefficients of 0.6 or even higher (see Section 5.6). The two (or more) independent signals are then "combined" in a simple manner (mathematically, using a "linear combination", as we will see in Section 5.3) in the diversity combiner.

More complicated devices which exploit the existence of an in-band diversity within a single branch or channel when this is wider than the correlation bandwidth of the medium have been introduced for high speed digital transmission.

5.2 Diversity systems

A diversity system is a method of obtaining statistically independent paths for the signal. These paths should have the same attenuation if possible, so that the received signals have identical median levels and the same Rayleigh statistics but uncorrelated "instantaneous" levels. This condition is generally met, although there are a few exceptions (angle diversity systems).

Independent paths can be obtained in several ways, of which the most common are the following classic diversity systems:

(a) space diversity;
(b) frequency diversity;
(c) polarization diversity;
(d) angle diversity;
(e) time diversity.

Many diversity configurations of radio stations have been proposed [5.2] in which the above methods are used to obtain solutions of greater or lesser sophistication. The simplest and most common are shown in Fig. 5.2. In the classic systems it is assumed that the diversity channels are no wider than the correlation bandwidth of the medium. Each diversity system has a direct effect on the configuration of the equipment in the stations. These systems are generally known as "explicit diversity systems" in order to distinguish them from the "implicit diversity systems" which transmit in each channel a digital signal with a bandwidth wider than that of the medium. In this case special modems are used to exploit the in-band diversity within the channel, but apart from the greater sophistication the basic principles are the same.

We will therefore begin by discussing the classic systems, which are most frequently used even in radio terminals equipped with the more recent digital adaptive modems. Implicit diversity and the special modems used in these systems will be described in Sections 5.7 and 5.8.

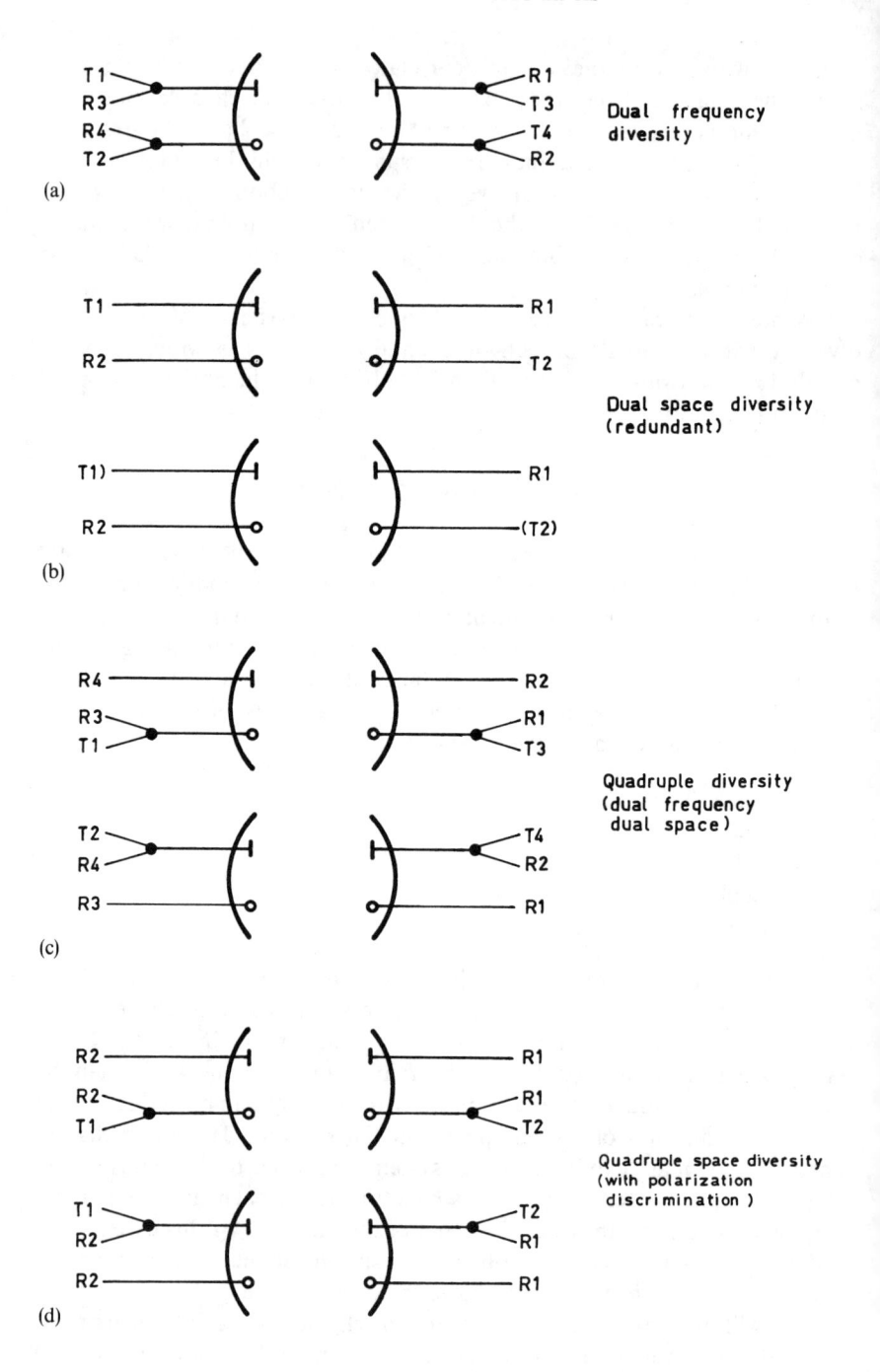

Fig. 5.2 Some common diversity configurations (T, transmitter; R, receiver; 1, frequency 1; 2, frequency 2 etc.).

5.2.1 Space diversity

It was established in Chapter 4, Section 4.3, that r.f. signal amplitudes or levels at the receiving site are not correlated outside a spatial cell with a limited time-variable width, height and depth. As these dimensions are time variable, we should refer, at least ideally, to the values that they exceed for 5% of the time for example. If the signal is received by two antennas installed at opposite sides of this cell, the levels of the two resulting signals will be sufficiently decorrelated.

It is normally more convenient to install the two antennas horizontally in a direction perpendicular to that of the path. Formulae have been proposed for calculating the minimum separation required, but often a distance of 100 wavelengths (30 m at 1 GHz) is found to give sufficient decorrelation. This criterion has been justified theoretically [5.3] and is widely adopted.

Space diversity is one of the most frequently used systems either alone or in combination with other types of system. Dual diversity systems require two antennas per site, and quadruple diversity systems require four antennas per site unless the path separation is obtained by polarization discrimination as in Fig. 5.2(d), when only two antennas are required. However, the design of this quadruple space diversity system is rather critical, because the radio paths are only partially independent and a careful choice of antenna separation is necessary to obtain acceptable decorrelation. For optimum performance this separation may be shorter or longer than 100 wavelengths and different separations may be required at each end. Design and performance prediction is not easy [5.3]. The two transmitters required at each site must be synchronized in order to avoid beats in the received signal. In dual diversity systems a second (redundant) transmitter may be used as a standby for obtaining a fully duplicated system.

5.2.2 Frequency diversity

In Chapter 4 it was shown that the medium has a limited correlation bandwidth of the order of a few megahertz. If the same information signal modulates two r.f. carriers which are separated (in megahertz) by more than the correlation distance, the two received signals are decorrelated. Formulae have been proposed for calculating the minimum separation between the two carriers, but in practice this is taken as a percentage (e.g. 2% or more) of the frequency of one carrier. The two frequencies are often much more widely separated for reasons of channelization.

This type of diversity is more economical as it requires only one antenna per site (Table 5.1), but it is not generally recommended (particularly by the CCIR) because it wastes the radio spectrum and increases the possibilities of interference.

Table 5.1 Technical comparison of dual diversity systems

Space diversity	Frequency diversity
1. Requires two transmitters per path	1. Requires four transmitters per path
2. Requires four antennas per path	2. Requires two antennas per path
3. Requires one pair of frequencies	3. Requires two pairs of frequencies
4. In each station the two antennas must be symmetrical with respect to the building and at a distance of 100 wavelengths from each other	4. In each station the only antenna can be installed very near to the building in the easiest position
5. The antenna feeders are longer (more loss)	5. The antenna feeders are shorter (lower loss)
6. The station area is greater	6. The station area is smaller
7. If the radio transmitters are also duplicated, the radio equipment is essentially the same as for frequency diversity	7. The radio equipment is essentially the same as for space diversity with duplicated transmitters
8. With some modifications it can be transformed into frequency diversity	8. With very simple modifications in the radio equipment it can be transformed into space diversity by adding a second antenna at each station
9. Without serious modifications it can be transformed into quadruple diversity	9. By addition of a second antenna at each station and without other serious modifications it can be transformed into quadruple diversity
10. Reliability and performance are the same	10. Reliability and performance are the same
11. More expensive solution	11. Less expensive solution

5.2.3 Polarization diversity

There is experimental evidence that there is no diversity effect if the radio path is the same for two polarizations because the two signals are completely correlated. This type of diversity cannot be utilized in troposcatter. However, polarization discrimination allows quadruple space diversity to be used instead of quadruple space–frequency diversity, which results in a saving of two frequencies in the path (remember Section 5.2.1).

5.2.4 Angle diversity

Large parabolic antennas can be provided with two (or more) illuminators which produce the situation shown in Fig. 5.3. In dual diversity systems the signal is transmitted on one beam and received on two separate beams at different arrival angles. In quadruple diversity systems the signal is sent at the same frequency (or at different frequencies) on two different beams, creating

Fig. 5.3 Angle diversity.

four common volumes as shown in Fig. 5.3. The various paths are decorrelated and also undergo different attenuations.

The two beams may be separated vertically, as shown in Fig. 5.3, or horizontally. It has been found that the vertical separation gives far better results, since the correlation in the horizontal separation is too high. The two beams are generated by misfocused illuminators which are optimized in size and position. The best performance is obtained when the polarization plane is aligned in the offset direction. The angular separation ("squint angle") between the beams is optimum when it is almost equal to the beamwidth. The range of the optimum is relatively broad. Since the vertical separation creates quite different common volumes with different long-term behaviors (for example the higher volume is more easily layered), it also introduces diversity in the long-term received signal. In contrast, all other diversity systems, which use virtually the same common volume, improve only the short-term signal.

This system is most effective for long-haul circuits using large antennas, and it can be used to replace quadruple space diversity (with a small amount of degradation) as it requires only one antenna per site. Because of the geometry it is necessary to optimize the design for each particular link.

The upper and lower paths have different losses; the antennas are therefore sometimes pointed to the horizon at such an elevation angle that the mean signals of the diversity branches are almost equal. However, the different path lengths require the addition of delay equalizers, particularly in digital equipment.

5.2.5 Time diversity

The information signal could be transmitted twice (or more) with a time delay longer than the autocorrelation time and the results could be combined at the receiving end. However, this method would be very inefficient and more sophisticated systems have been devised. They are employed in digital transmission using implicit diversity (see Section 5.7) and may work in real time for voice transmission with a delay not exceeding 300 ms.

5.3 Linear combination of diversity branches

We can now clarify the concepts of diversity by introducing some simple mathematics. In a diversity system of order N a signal $s(t)$ is transmitted through N different random channels, the output of which yields N copies which are similar to each other. For example the signal at the output of the ith channel is

$$f_i(t) = s_i(t) + n_i(t) = b_i s(t) + n_i(t) \tag{5.1}$$

where $s_i(t)$ is the ith copy of the original signal and is perfectly correlated with the other copies except perhaps for a different amplitude and a generally slightly different phase. The difference between the arithmetic and the vectorial sums of two equal-amplitude signals is less than 1 dB if the phase difference is less than 37.5°. Signals $s_i(t)$ are therefore assumed to be "almost in phase" and to have an amplitude $s_i(t) = b_i s(t)$. b_i is the amplitude envelope and generally follows Rayleigh statistics, and $n_i(t)$ is the noise added in the channel and is completely uncorrelated with the noise in the other channels. The output signal is normally obtained as a linear combination of the N signals. It is given by

$$f(t) = a_1 f_1(t) + a_2 f_2(t) + \ldots + a_N f_N(t) \tag{5.2}$$

$$= \sum_{i=1}^{N} a_i s_i(t) + \sum_{i=1}^{N} a_i n_i(t) = s(t) \sum_{i=1}^{N} a_i b_i + \sum_{i=1}^{N} a_i n_i(t) \tag{5.3}$$

where the coefficients a_i are proportional to the gains of the respective channels and vary randomly in time.

The signals, which are correlated and almost in phase, add in amplitude, while the noise, which is uncorrelated, adds in power, and thus the resulting signal-to-noise ratio is improved. In the combined signal $f(t)$ the ith signal is multiplied by the parameters a_i, which represents the time-varying gain of the ith diversity branch, and b_i, which represents the time-varying envelope amplitude of the ith copy of the signal $s(t)$. Therefore at the ith input of the combiner there is a time-variable signal (5.1) which is proportional to that at

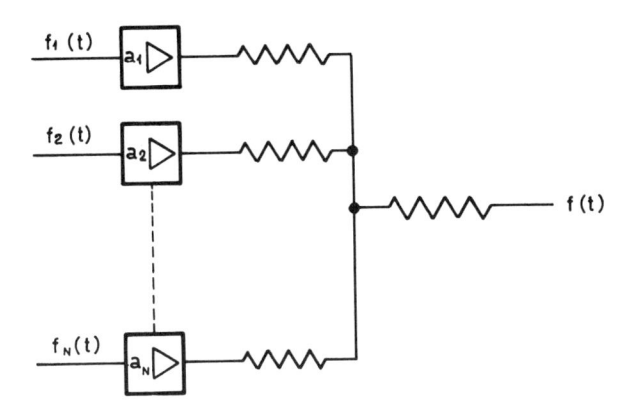

Fig. 5.4 Equivalent circuit of a diversity combiner.

the input of the corresponding receiver. The combiner multiplies this signal by a_i before adding it to the others. If the branch is provided with automatic gain control (AGC), the related variable gain can be included in either a_i or b_i.

The equivalent circuit of the combiner corresponding to (5.2) is composed of N branches, each consisting of an amplifier of gain a_i with an input signal $f_i(t)$. The output signals are simply added together (for example in a passive network or by connecting the amplifiers in parallel) to obtain the combined signal (Fig. 5.4).

In order to obtain a suitable combination at any time, the amplifier gains can be varied in time according to the amplitude of the signals and the type of combination. The control signals used to vary the gains are based on measurements of the received levels of the signal or the noise or both, depending on the type of combiner.

5.4 Predetection or postdetection combination

Each diversity branch can be considered to include a complete radio receiver from its r.f. input to its output, followed by a branch of the combiner. In FM receivers the combination can be made at either the i.f. stage (predetection combination) or the baseband level (postdetection combination). The two systems are equivalent as long as the signals remain above the threshold. Otherwise a difference arises as the threshold breaks the linearity between input and output. In predetection combination the receiver's output is at the i.f. stage, after the AGC but before the limiter, as limitation is effected after combination. In postdetection combination the receiver's output is at baseband, after demodulation.

The following signals and noise are physically or virtually present at the input of the receiver (see Chapter 7, Sections 7.3.2 and 7.4.4, and Chapter 8, Section 8.2.2):

(i) the received r.f. signal, which generally varies according to Rayleigh statistics;

(ii) constant thermal noise power $FKTB$;

(iii) background and intermodulation noise generated in the equipment and assumed to be constant;

(iv) time-variable intermodulation noise due to multipath propagation;

(v) occasional unwanted (interfering) signals.

In predetection combining, the output signal of the receiver (eqn (5.1)) represents the modulated i.f. Because of the action of the AGC this signal has compressed dynamics for large-amplitude variations, while for small variations its amplitude is proportional to the input r.f. signal. In any case its signal-to-noise ratio is the same as that of the input r.f. signal and varies according to the same Rayleigh statistics. In order to combine additional signals it is necessary to keep the $b_i s(t)$ at least approximately in phase. The combined signal $f(t)$ is sent to the limiter and the demodulator and may still be above threshold when the individual signals are just below it. If the phase relationships between branches can be maintained in these conditions (poor signals and high noise) the threshold is lowered ("threshold extension") with an obvious gain. The combination, i.e. the variation of coefficients a_i, is controlled by the level of the r.f. signals detected in the AGC circuits.

In postdetection combining, the baseband output signal (5.1) represents the modulating signal. The term $b_i s(t)$ is kept constant by the FM because it depends on a constant frequency deviation. The noise $n_i(t)$ varies with $1/b_i$. There is a characteristic threshold effect for very low r.f. signals. The output signal-to-noise ratio is higher than that at the r.f. input by a constant factor (see Chapter 7, Section 7.4.3) and varies with the same Rayleigh statistics as the input signal (in fact the true variable is the noise, as the signal is constant). The combination of additional signals is easier as they are always almost in phase since their period is much longer than the expected delays. Combination by varying the coefficients a_i could be controlled by the AGC circuits as in i.f. combination but in practice it is normally controlled by the noise level measured just above the baseband in a slot (of width say 20 kHz) simulating a high telephone channel.

Comparison of these two types of combination shows the following.

(i) Far from threshold the two systems are equivalent.

(ii) Near and below threshold the i.f. combination has a great advantage: if all input r.f. signals are just below threshold the baseband combination yields only noise, while the i.f. combined signal is still above threshold. The threshold extension produced in this way is very important, particularly in digital transmission where in practice the bit error rate (BER) may be improved by two or three orders of magnitude.

(iii) The i.f. combiner is more complex than the baseband combiner as it must keep variable signals in phase at intermediate frequencies, particularly in the presence of strong variability where the signals are masked by noise.

(iv) In normal dynamic operation postdetection combination is only slightly worse than predetection combination (average degradation around 1 dB).

(v) Postdetection combination has the great advantage of being virtually controlled by the true baseband noise, including interference, multipath intermodulation etc., from which predetection combination cannot be protected. In the control of predetection combination by AGC it is assumed that the noise is only thermal, so that this type of combination is only suitable for working near threshold where almost all noise is thermal.

(vi) Predetection combination reduces the harmful effects of multipath distortion.

5.5 Combining methods

A combining method is defined as the way in which the coefficients a_i in eqn (5.2) are chosen. Each method leads to a different type of diversity combiner in the radio receiver equipment. The classical methods of combining, which are described in this section, involve the use of the following devices.

The *selector* chooses the best signal at any instant. It can be located in the r.f. stages, in the i.f. stages or after detection. In the last case it produces troublesome switching transients. In the other cases it is rather complex and is not much used.

The *equal-gain combiner* parallels the signals from the receivers and at the same time parallels the AGC circuits so that all receivers have the same gain.

The *maximal-ratio combiner*, also known as the optimum-ratio or ratio-squarer combiner, sums the signal and the noise in such a way that the combined signal-to-noise ratio at any instant is equal to or better than the best of the single signal-to-noise ratios.

The maximal-ratio combiner is used most frequently. The principles on which these devices are based and their performances are discussed below. All of them refer to explicit diversity. More complex diversity systems specifically developed for digital transmission and exploiting implicit diversity will be studied in the final sections of this chapter.

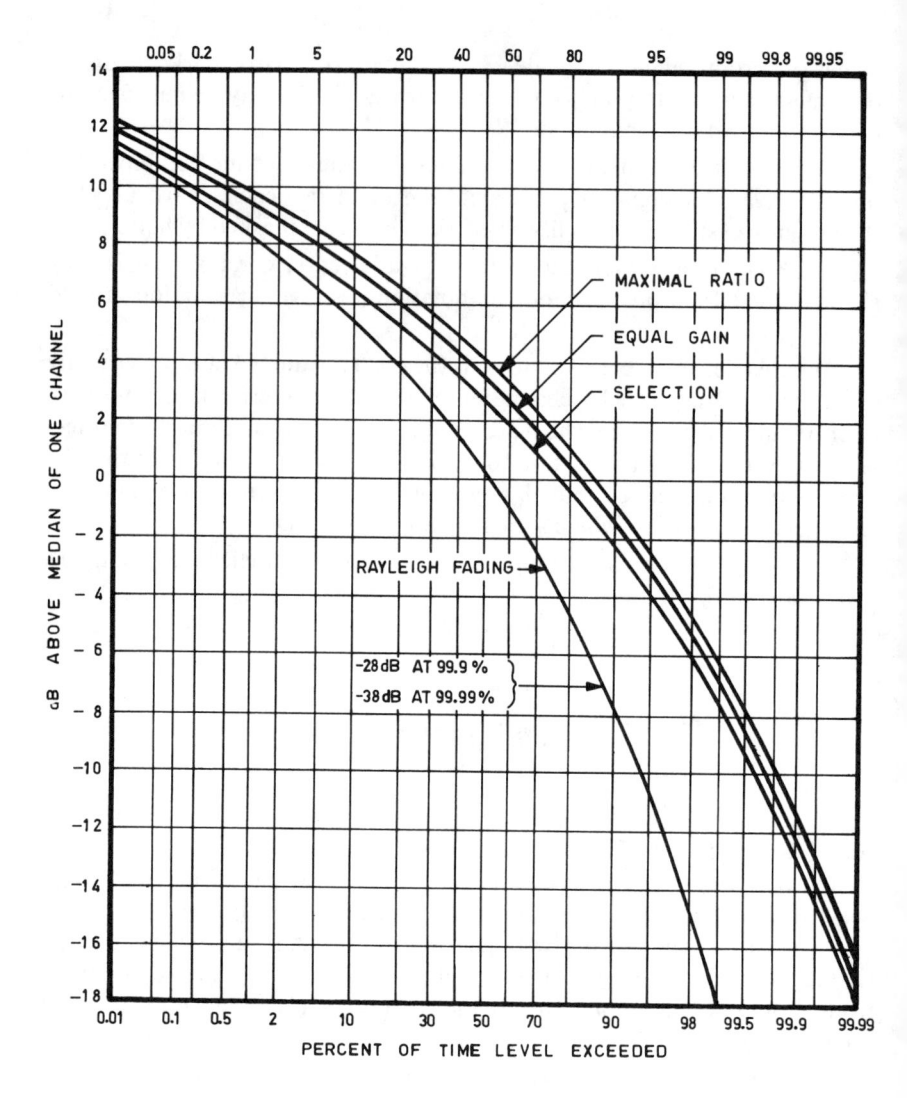

Fig. 5.5 Dual diversity distribution curves.

5.5.1 Signal selection

In selection diversity systems all coefficients a_i of eqn (5.2) except the one corresponding to the best received signal are equal to zero at any given time. In the dynamic situation the output signal will fail only if all N individual signals fall below threshold at the same time. It should be remembered that the threshold corresponds to the maximum acceptable noise in the telephone

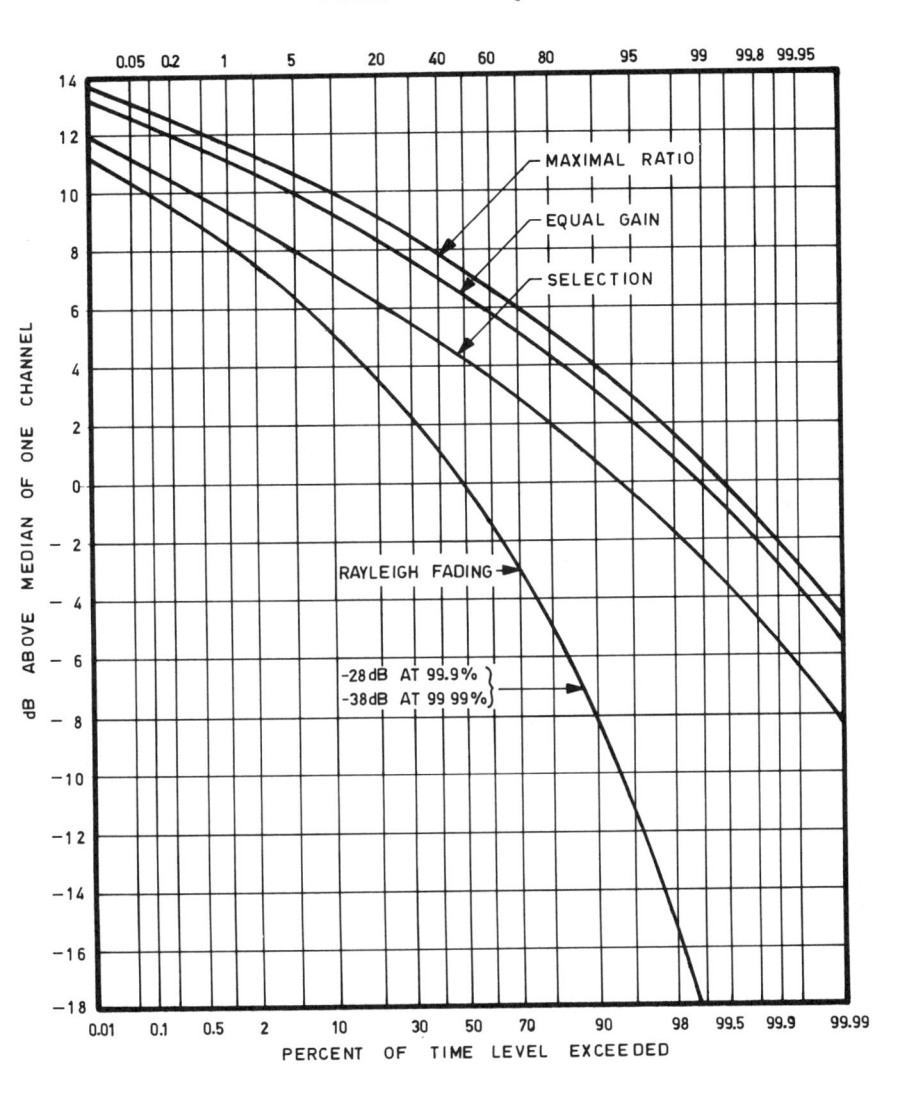

Fig. 5.6 Quadruple diversity distribution curves.

channels of an analog radio link or to the maximum acceptable BER in a digital radio link.

The individual signals follow the statistics represented by the curve labelled "Rayleigh fading" in Figs 5.5 and 5.6 (which reproduces Fig. 2.4 with a slight change in the horizontal scale). The outage probability for the output signal is therefore the product of the outage probabilities of the individual signals. If these have a Rayleigh distribution with equal medians, from eqns (2.19) and (2.20a) the resulting outage probability becomes

$$P_N(X) = \{P_1(X)\}^N = \{1-\exp(-s)\}^N \tag{5.4}$$

Alternatively, for digital systems we can write, from (2.25),

$$P_N(\gamma) = \{1-(2P_0)^{1/\gamma}\}^N \tag{5.5}$$

and similarly for (2.27), where R(dB) is the carrier-to-noise power ratio at the receiver's input and γ is its numerical value. The statistics of the combined signal are shown in Figs 5.5 and 5.6.

5.5.2 Equal-gain combination

In equal-gain combination all coefficients a_i are made identical by using a common AGC for all channels. The combined signal becomes

$$f(t) = a\{f_1(t)+f_2(t)+\ldots+f_N(t)\} \tag{5.6}$$

The static behavior yields a gain of 10 log N above the case of no diversity if the signals are all equal, but this gain decreases and becomes negative for dual diversity with signals differing by up to 8 dB and asymptotically approaches a 3 dB degradation for greater differences.

For example, in a dual diversity system the sum (combination) of the two signals, which are equal and in phase, is 6 dB greater than the single signal, while the two noises, which are equal because of the equal gain and are uncorrelated, add in power and increase by 3 dB. Thus the signal-to-noise ratio of the combined signal is 3 dB higher than that of the single signal. In the extreme case in which one of the signals disappears, the combined signal is equal to the remaining signal but the combined noise is still the power sum of the two equal noises. Therefore the combined signal-to-noise ratio is 3 dB lower than that of the single signal.

The dynamic behavior in the presence of Rayleigh fading with equal mean values is only 1 dB less than that for the maximal-ratio combination, but conditions of deep fade or disturbances may cause a marked degradation. Unless the high noise which appears when the r.f. signal is below threshold is squelched, equal-gain combiners are unsuitable for postdetection combination. The statistics of the combined signal are shown in Figs 5.5 and 5.6, but the mathematics involved is not reported here.

5.5.3 Maximal-ratio combination

In maximal-ratio combination the coefficients a_i are directly proportional to the amplitude of the envelope of signal b_i and inversely proportional to the noise power $p_i = \overline{n_i^2}$, where the bar indicates the mean value. The optimum combination is

$$f(t) = \frac{b_1}{p_1} f_1(t) + \frac{b_2}{p_2} f_2(t) + \ldots + \frac{b_N}{p_N} f_N(t)$$
$$= s(t)\left(\frac{b_1^2}{p_1} + \frac{b_2^2}{p_2} + \ldots + \frac{b_N^2}{p_N}\right) + \sum_{i=1}^{N} \frac{b_i}{p_i} n_i(t) \tag{5.7}$$

The combined numerical signal-to-noise power ratio is the sum of the numerical ratios of the individual channels. This system is also very effective when the median values of the individual channels are not equal. The degradation which takes place in equal-gain systems is absent.

If all signals in the dynamic situation follow Rayleigh statistics with identical medians, the outage probability for the combined signal is calculated to be [5.1]

$$P_N(X) = 1 - \exp(-s) \sum_{k=0}^{N-1} \frac{s^k}{k!} \tag{5.8}$$

For example, for quadruple diversity

$$P_4(X) = 1 - \exp(-s)\left(1 + s + \frac{s^2}{2} + \frac{s^3}{6}\right)$$

For digital systems, remembering (2.24), we can write

$$P_N(\gamma) = 1 - (2P_0)^{1/\gamma} \sum_{k=0}^{N-1} \frac{1}{k!} [-\ln\{(2P_0)^{1/\gamma}\}]^k \tag{5.9}$$

In predetection combiners the control signal is taken from the AGC and is a function of the r.f. received level. There is a constant thermal noise power

$$p_1 = p_2 = p_3 = \ldots = p_N = FKTB \tag{5.10}$$

at the input of each receiver (see Chapter 7, Section 7.3.2) so that eqn (5.7) can be written as

$$f(t) = \frac{1}{FKTB}(b_1 f_1 + b_2 f_2 + \ldots + b_N f_N) \tag{5.11}$$

where the envelopes of signal b_i and the "dirty" signal f_i are considered at the receiver inputs. If $1/FKTB$, which represents a fixed gain, is ignored, the amplitude of each signal f_i must be amplified by a factor b_i. Control of the variable gain b_i requires it to be extracted from the signal f_i by eliminating the noise, which is not possible. We therefore assume that $b_i = f_i$, which is approximately true except near threshold.

A maximal-ratio combination is obtained if each diversity branch has a quadratic amplifier in which the input is f_i and the output is $b_i f_i = f_i^2$. Near threshold the combination would be slightly below optimum. In practice, when the signal and noise are equal ($b_i = n_i = f_i/2$, with a signal-to-noise ratio of 0 dB), the input to the quadratic amplifier is $f_i = b_i + n_i$ and the output is $f_i^2 = 4b_i^2 = 2b_i f_i$ instead of $b_i f_i$.

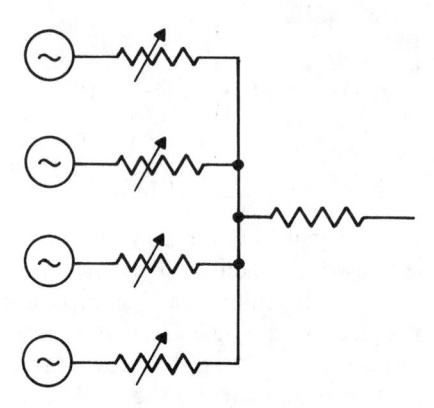

Fig. 5.7 Equivalent circuit of a maximal-ratio combiner for quadruple diversity.

In postdetection combiners the control signal is taken from the noise level existing above the baseband. It was shown in Section 5.4 that $b_1 = b_2 = b_3 = \ldots = b_N = b$ for all diversity branches. Therefore (5.7) can be written as

$$f(t) = b\left(\frac{f_1}{p_1} + \frac{f_2}{p_2} + \ldots + \frac{f_N}{p_N}\right) \tag{5.12}$$

Each diversity branch is seen from the output as an ideal voltage generator V_i in series with a resistor R_i. Equation (5.12) represents a signal current obtained by connecting the N diversity branches in parallel with $f_i = V_i$ and $p_i = R_i$ (Fig. 5.7). The maximal-ratio combination is obtained if the series resistors are proportional to the noise power in the respective branches.

In practice the performance of the maximal-ratio postdetection combination is only slightly better than that of selection. The statistics of the predetection combined signal are shown in Figs 5.5 and 5.6.

5.6 Combined signal and diversity gain

The distribution curves for the combined signal are shown in Figs 5.5 and 5.6 for dual and quadruple diversity respectively. These curves represent the following distributions:

(a) the resulting distribution of an "equivalent input r.f. level" to an ideal single receiver.

> In this case the N diversity receivers, each with a Rayleigh-distributed r.f. input signal, plus the combiner are equivalent to an ideal single receiver with an input r.f. signal with the combined distribution. The r.f. levels are referred to the median of a single branch or channel.

(b) the distribution of the signal-to-noise ratio of the combined output signal, because the level of the input r.f. signal and the signal-to-noise ratio of the output are proportional (see Chapter 7, Section 7.4.3).

> If the combination is postdetection the signals are assumed to be far from threshold, so that its effect can be ignored. Otherwise the rapid increase in noise below threshold causes the curves to bend downwards more steeply starting from the threshold level [5.4].

It can be seen from the curves that the use of diversity results in a "diversity gain" for each percentage of time and that this gain increases as the percentage of time increases. For example, with selection-type dual diversity there is a diversity gain of 15 dB since, for 99.9% of the time, a signal 28 dB below the median is raised to 13 dB below the median.

On a long-term basis, however, it is convenient to consider using a single parameter to characterize the gain obtained in a system when diversity is introduced. This parameter could be one of the following.

(1) the median diversity gain, i.e. the increase in the equivalent input r.f. level and consequently in the output signal-to-noise ratio for 50% of the time and for the same received r.f. level.

> Alternatively, it can be defined as the decrease in the received r.f. level necessary for obtaining the same median signal-to-noise ratio. For example, Fig. 5.6 shows that the introduction of quadruple diversity with maximal-ratio combination yields a median gain of 7 dB in the equivalent input r.f. level and in the output signal-to-noise ratio. Alternatively we could transmit a level 7 dB lower and keep the median received r.f. level and the median output signal-to-noise ratio unchanged.

(2) the gain in the percentage of time availability as it increases from 50% to higher values for the same received r.f. level and output signal-to-noise ratio.

> An alternative definition is the decrease in the received r.f. level necessary to maintain an availability of 50% of the time. In the example given above the time that the signal is available at 0 dB above median increases from 50% to 99.4%. As before, we could transmit at a level 7 dB lower and keep the original median values.

A relation between the parameters of the statistical distribution of the r.f. signal received by the single receiver and those of the statistical distribution of the output noise after combination is required for system design. The following important results, which are almost independent of the noise curve of the FM receiver and of the presence, if any, of a squelch device, have been obtained for maximal-ratio combining [5.4, 5.5].

(i) In dual diversity systems the mean noise power of the output is 1.6 dB below the median noise obtained in the absence of diversity, provided that the r.f. input signal to each receiver does not fall below threshold for more than 5% of the time. This corresponds to saying that the median r.f. signal level must be at least 12 dB above threshold.

(ii) In quadruple diversity systems the output mean noise power is 6.4 dB below the median noise obtained in the absence of diversity, provided that the r.f. input signal to each receiver does not fall below threshold for more than 35% of the time. This corresponds to saying that the median r.f. signal level must be at least 2 dB above threshold.

These results will prove very useful in the calculation of the performance of troposcatter analog radio links (see Chapter 8, Section 8.2.1).

The curves in Figs 5.5 and 5.6 were calculated for uncorrelated randomness of the variables, i.e. for a correlation coefficient $\rho = 0$. When ρ varies from zero to unity the diversity curves for selection move toward the Rayleigh curve and become identical with it for $\rho = 1$. However, this displacement is not uniform but starts slowly, so that in dual diversity for example the effect is negligible for $\rho < 0.3$ and the curve for $\rho = 0.6$ is less than 2 dB below the curve for $\rho = 0$. There is a similar effect in equal-gain and maximal-ratio combination, but the lower limiting curve is 3 dB above the Rayleigh curve in dual diversity and 6 dB above it in quadruple diversity. This behavior justifies the assertion that an acceptable diversity effect can be obtained with correlation coefficients as high as 0.6 or more.

5.7 Implicit diversity techniques to combat multipath distortion

In the diversity systems examined so far the distortions on signal transmission in each diversity branch are absent or moderate, as the signal bandwidth is assumed to be contained in the correlation band of the medium for at least most of the time. This technique was satisfactory until the advent of high speed digital systems which require the transmission of bandwidths wider than the correlation bandwidth of the medium. New transmission principles have been devised to combat the unavoidable distortion introduced, and in particular the possibility of exploiting an in-band diversity within a single branch has been investigated.

In the multipath propagation characteristic of troposcatter, different delays within the same branch or channel correspond to different and statistically independent paths carrying the same information. If the channel bandwidth is wider than the correlation bandwidth, this phenomenon can be exploited to establish a "multipath diversity" system within the channel itself without necessarily modifying the external configuration of the equipment in an explicit diversity system.

The diversity obtained by taking advantage of the redundancy present within the channel is called "implicit diversity" and can be viewed as implicit frequency diversity or implicit time diversity. The order of the implicit diversity is defined as follows:

(a) in the frequency domain it is the number of times the signal bandwidth exceeds the correlation bandwidth;

(b) in the time domain it is the number of times the delay spread exceeds the digital symbol length.

In fact the situation is a little more complicated, as we have to define the terms exactly. If we take the 3 dB bandwidth $W_{3\,dB}$ of the signal and the correlation bandwidth W_c corresponding to a frequency separation with $\rho = 0.707$, the order of implicit diversity is $W_{3\,dB}/W_c + 1$ and not simply $W_{3\,dB}/W_c$ as indicated above. This is because a significant diversity gain can also be provided by the small amount of decorrelation between the two signals (see Section 5.6).

When both explicit and implicit diversity are used in a system, as is customary, the "total diversity order" is the product of the explicit diversity order and the implicit diversity order. For example a quadruple (explicit) diversity system with an implicit dual diversity in each branch will act as an octuple diversity system.

Implicit diversity is used exclusively for digital transmission, and can be realized by utilizing special modulation techniques, signal coding etc. in the equipment according to the following basic principles.

(a) *Parallel processing*: the high speed data stream is split into N low speed streams transmitted in parallel over N adjacent radio channels packed in the main radio channel. The transmitted symbols become N times longer and are better able to overcome the multipath spread. This system produces an in-band frequency diversity.

(b) *Frequency hopping*: successive symbols are transmitted cyclically on a subset of frequencies contained in the radio channel so that the same frequency is repeated after a time longer than the spread of the pulses on that frequency. Adjacent symbols are spaced by at least two frequency slots, and adjacent frequencies are spaced by at least two time slots. If each symbol modulates two spaced frequencies, an in-band frequency diversity is obtained.

(c) *Adaptive equalization*: the receiver continuously tracks the variations in the transfer function of the channel by means of adaptive equalizers. These equalizers are tapped delay line filters (transverse filters) with continuously adjusted tap gains. The adjustments can be directed by the decisions of the receiver on the signal itself ("decision-directed adaptations") or by multiplexing a known bit pattern into the message stream ("reference-directed adaptation"). The bits in the stream flow along the delay line where each pulse is sampled simultaneously, at the symbol rate, at various points corresponding to the taps. During successive sampling the pulse shape varies very slowly and there is some decorrelation of the pulse amplitudes at the various taps which are conveniently displaced along the length of the distorted pulse (three taps are often sufficient). The situation is similar to that discussed in Section 5.3 with reference to Fig. 5.4: the taps distributed along the varying pulse shape on the delay line correspond to the diversity branches of an explicit diversity system. The signals from the taps can be integrated and combined as in a classical system, for example with maximal-ratio combination as shown in Fig. 5.7 where the variable resistors represent the variable tap gains. In this case the implicit diversity can be treated as a time-delay diversity. An improved system is the decision feedback equalizer which uses a second tapped line, after combination and demodulation, in a feedback scheme for further reducing intersymbol interference.

(d) *Signal recirculation with time gating*: adjacent symbols in the transmitter are shortened and separated by an off-time exceeding the multipath spread. The pulses are smeared over the symbol interval, but in order to avoid intersymbol interference they must not exceed it. In these conditions a recirculating delay line forms an average, or noiseless replica, of the received pulses. This replica is correlated with the incoming pulses, and the implicit diversity is obtained in a simpler way than with a tapped delay line. In a multiplier (correlator) the incoming signal and its replica are multiplied all along the duration of the (distorted) pulse as in a delay line with infinite taps. The quadratic output ensures maximal-ratio implicit diversity combination (see eqn (5.11) and the subsequent discussion). The principle of this technique is discussed in more detail in Section 5.8.5.

(e) *Maximum likelihood detection*: sophisticated coding and decoding techniques are used to recover the most probable transmitted sequence from the tapped line plus explicit diversity combined signal. The trellis decoding algorithm (Viterbi algorithm) introduces a complexity which increases rapidly with the delay spread. This system can overcome high intersymbol interference.

(f) *Forward error correction coding*: the digital stream is subdivided into successive sets composed of n bits plus a subset of error-correcting bits capable of correcting up to a given number of bits in error. This increases the original bit rate. To ensure that a fade does not cancel a number of successive bits, they are interleaved, i.e. transmitted with a time separation suitably related to the fade length. This implies the acceptance of some transmission delay and the necessity of storage, but if a fade cancels a number of successive bits in the rearranged stream the errors are spread among many n-bit sets, each of which is capable of self-correction, after passing through the de-interleaver. The result is a slightly delayed error-free transmission in the presence of fades which do not exceed a given maximum. This is a kind of time diversity within the channel which does not require sophisticated modems but can be added to classical radio terminals transmitting at moderate bit rates.

5.8 Combiners and adaptive modems

The first combiners to come into general use, almost exclusively for analog transmission, were classical maximal-ratio types, generally in the baseband version, and these were followed by the selection and equal-gain types. As noted in the preceding section, the advent of digital techniques and the necessity of transmitting high speed data resulted in the development of new types of modems/combiners which ensured that transmission was as immune as possible to multipath distortion. In the following sections we discuss some diversity combiners and modems described in the literature. We start with a simple baseband combiner and an example of an i.f. combiner. We then examine some of the sophisticated modems/combiners developed recently for digital transmission.

5.8.1 A simple baseband combiner

The dual diversity version of a baseband combiner is shown in Fig. 5.8. It is based on a circuit in which a constant-current generator polarizes two parallel transistors such that an increase in current flow in one of them corresponds to an identical decrease in the other. In the quadruple diversity version there are four such transistors. The baseband signals from the two diversity branches A and B are almost in phase and are of equal amplitude but have a different noise content. They are applied to the bases of the transistors together with the control signal for the combination, which is taken from the variable noise existing in a slot above the baseband and causes variable partition of the polarization currents in the transistors.

The noise, which is amplified (generally by a logarithmic amplifier for larger dynamics) and detected, becomes a very low frequency or quasi-d.c. control signal. If the two polarization currents are equal, the two signals add coherently and the two equal noises add in power at the output providing the usual 3 dB of static diversity gain. If the noise levels differ by more than 6–7 dB, the noisiest signal is cut off. In intermediate situations the noisiest signal is attenuated and contributes less.

This device acts in a similar manner to a maximal-ratio combiner. The combination takes place without variations in the level. The bandwidth of the control signal depends on the maximum variation rate in the fast fade, which can be assumed to be $20\,\text{dB}\,\text{ms}^{-1}$. A band of 0–4 kHz would allow signal variations of 0.25 ms to be followed. This band, which is often used for the

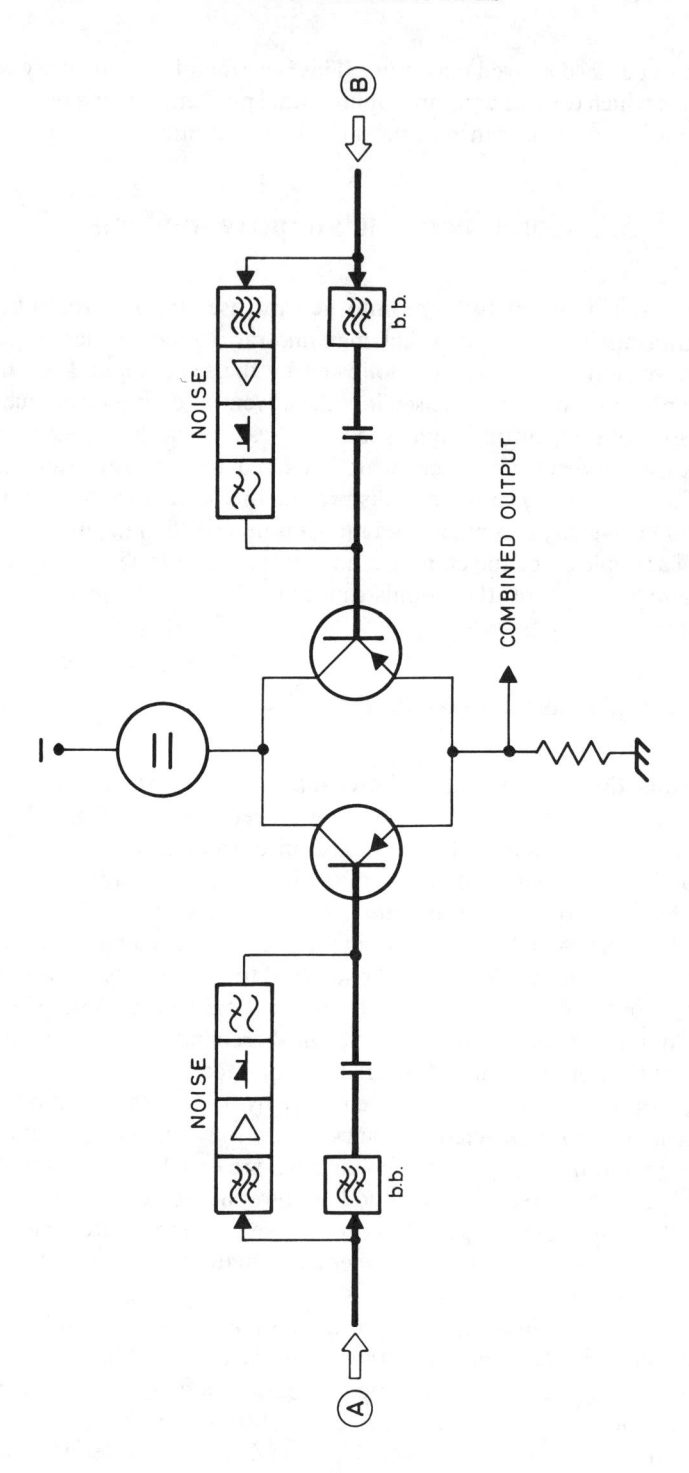

Fig. 5.8 Simple dual diversity baseband combiner.

DIVERSITY TECHNIQUES 105

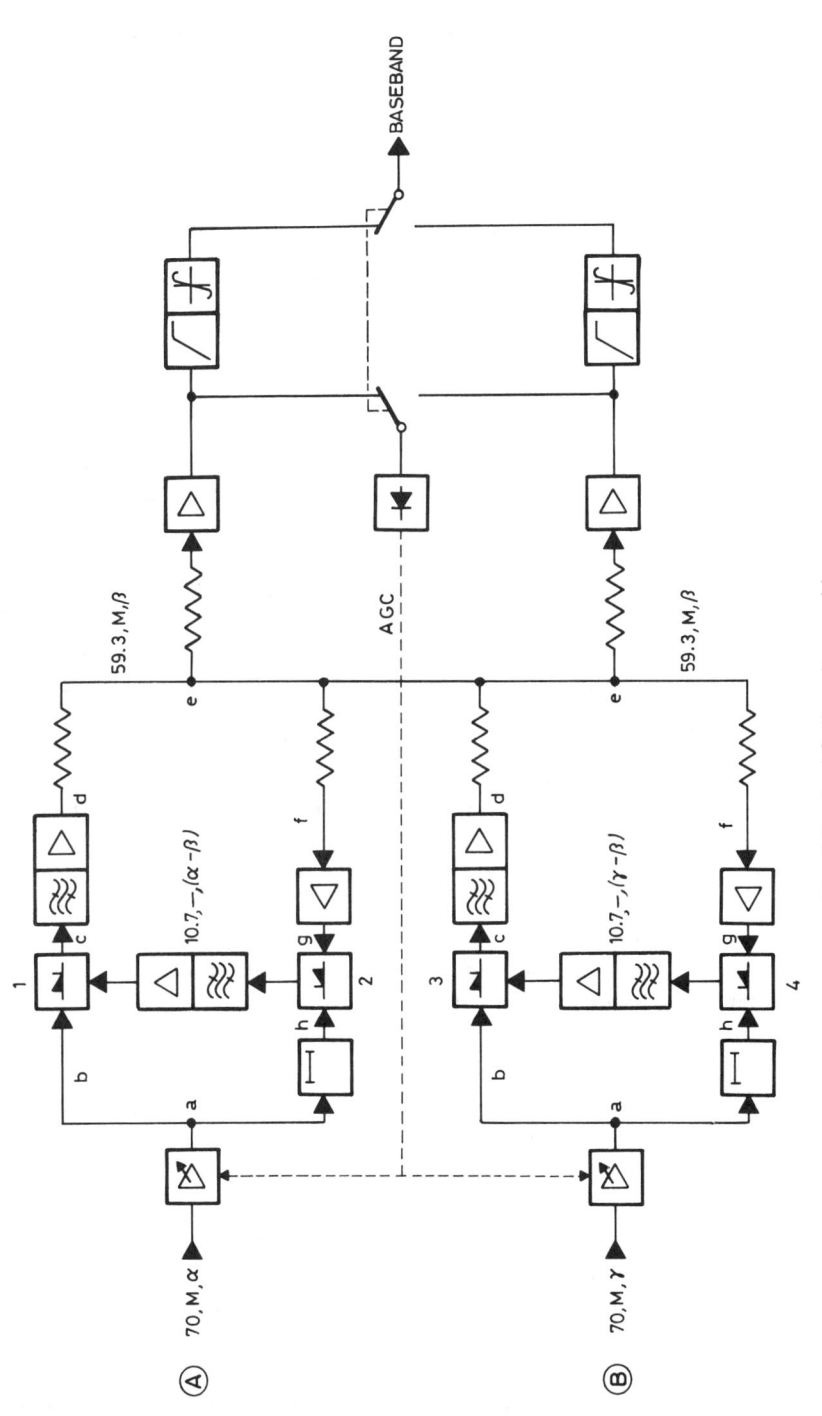

Fig. 5.9 An i.f. diversity combiner.

service channel in LOS radio links, cannot be used in troposcatter links in which the service channel must be shifted to a higher band.

5.8.2 An i.f. combiner

A dual diversity version of an i.f. combiner is shown in Fig. 5.9 [5.6]. Each of the two branches A and B includes an AGC amplifier and a phase corrector, represented by the circuitry between points a and e. The two (or four, in the case of quadruple diversity) amplifier–corrector assemblies are connected in parallel at points e, after which duplicate branches (one operating and one on standby) provide amplification, limitation and detection.

The phase correctors keep the output diversity signals at the same instantaneous phase before they are summed. These correctors work at intermediate frequency, but with different input and output frequencies. They use a double conversion system. The 70 MHz i.f. signal, with modulation M and phase α (indicated by $70,M,\alpha$), enters the phase corrector through an AGC amplifier and is split at point a and sent to mixers 1 and 2.

It enters mixer 2 with an additional delay τ and is mixed with a signal indicated by $59.3,M,\beta$ arriving from the combined output e. The output of mixer 2 is an unmodulated difference signal indicated as $10.7,-,(\alpha-\beta)$ which is filtered by a narrow filter and sent to mixer 1. The two modulations cancel each other as they have an identical frequency deviation.

Mixer 1 mixes the input signal $70,M,\alpha$ and the signal $10.7,-,(\alpha-\beta)$, yielding $59.3,M,\beta$ which is an i.f. signal with the same modulation and phase β representing the phase corrector output signal. It is also used as the reference for mixer 2.

For correct operation the following conditions must be satisfied.

(i) The delay between a and h must equal the delay abcdefg. This ensures that the inputs to mixer 2 at h and g have equal phase modulations, with a consequent cancellation of modulation.

(ii) The gain and phase variation in path cdefg must be the same for each diversity branch.

(iii) Paths de and fe must be electrically identical for all connections from the corrector to the passive network.

If the a–e transfer function is linear the combination is of the equal-gain type (see eqn (5.6)), and if it is quadratic it is of the maximal-ratio type (see eqn (5.11) and the subsequent discussion).

This combiner has the great advantage of working when the signals fall below the threshold of the FM demodulator. Other combiners with receivers using voltage-controlled local oscillators need strong signals to keep the signal in phase and cease operating near threshold, when reliable operation is essential. Furthermore, this combiner is independent of the type of modulating signal (analog, digital, TV etc.). This combiner is used by Marconi Ltd, UK.

5.8.3 Multiple-subband modem for digital transmission

The multiple-subband modem exploits the in-band frequency diversity capability of a transmitted r.f. band which is wider than the correlation band. It subdivides the original channel into a group of smaller channels and transmits them in parallel at different center frequencies.

The input digital stream with a bit rate B is subdivided into n separate streams with bit rates B/n by means of a series-to-parallel converter. Each of these streams modulates in 4PSK a subcarrier of about 70 MHz. The n subcarriers are allocated side by side as in a frequency-division multiplex and occupy an i.f. band around 70 MHz. This band is up-converted to an r.f. band and transmitted.

In the receiver the reverse operation takes place and the original bit stream is obtained by parallel-to-series conversion of the individual streams after demodulation. In this way the symbol duration in radio frequency is n times longer than with a single channel, so that the system is able to overcome path delay distortion if this is not excessive. Furthermore, each subband is within the correlation bandwidth and undergoes only flat fades instead of selective fades as is the case in the overall band.

In practice the original stream includes the service channel(s), and framing information is added to the slower streams which are then scrambled and differentially coded before modulation, as in standard digital radio equipment. The 4PSK modulation may be of the offset type, and the modulating pulses can be conveniently filtered for optimum r.f. spectrum shaping. A BER measurement derived from the framing data is provided.

The efficiency of this modem/combiner may exceed 1 bit s^{-1} Hz^{-1} with bit rates of up to 4096 kbit s^{-1}. The device has been used by Comtech Systems International, USA.

5.8.4 Independent sideband diversity modem for digital transmission

The independent sideband diversity modem is an improvement of the previous device [5.7]. The group of parallel channels (here with a lower carrier frequency), instead of being directly up-converted to radio frequency, modulate in 4PSK a 70 MHz i.f. carrier which is then converted to radio frequency. The r.f. spectrum contains several orders of sidebands but only the first-order upper and lower sidebands are considered. By optimizing the modulation index it is possible to concentrate most of the r.f. power in the first-order sidebands. These two sidebands contain the same information. Therefore if they are spaced by more than the correlation bandwidth they can be independently down-converted, i.f. combined and detected as in a dual frequency diversity system.

The spectrum of each modulated subcarrier is kept within the correlation bandwidth. In order to obtain a more compact overall spectrum the spacing between the subcarriers is made either equal to or a multiple of the corresponding bit rate (the subcarriers are then "orthogonal", i.e. the integral of their product over the bit length is zero). This permits a high degree of spectrum overlapping without interference. The spacing required between the two sidebands can be minimized by accepting a correlation coefficient of up to 0.8, which is still sufficient for acceptable operation.

When a single radio channel (no explicit diversity) is used with this in-band dual frequency diversity, the total order of diversity is 2 and the dynamic behavior is as shown in Fig. 5.5. During the infrequent periods in which the medium bandwidth is wider than that of the signal there is no diversity effect and the combined signal follows the Rayleigh curve displaced 3 dB upwards (see Section 5.6). During the long periods in which the bandwidth of the medium is narrowest the combined signal follows the maximal-ratio curve. In other periods the curve lies between these two extremes, so that the behavior of the signal can be represented by a strip limited by the two curves.

When two radio channels are used in an explicit dual diversity configuration, the total order of diversity is 4, and a similar interpretation can be made with reference to Fig. 5.6.

This modem has been used by Motorola. The upper and lower sidebands of each subcarrier were independently combined using an i.f. maximal-ratio combiner. For example, a bit stream of 12.6 Mbit s^{-1} can be transmitted in a 15 MHz r.f. band using 16 4PSK-modulated subcarriers. In each subcarrier the data rate will be 0.788 Mbit s^{-1} with a symbol duration of 2.54 µs. This system can tolerate a delay dispersion of about 1.27 µs before being seriously affected.

A further improvement of about 2 dB can be achieved by using a "gated subcarrier" technique [5.8]. This consists in gating the subcarriers coherently with the data rate so that one subcarrier is on when the other is off. It is then possible to adjust the modulation index and increase the r.f. power in each sideband.

5.8.5 Distortion adaptive modem for megabit transmission

An interesting digital modem called the distortion adaptive receiver (DAR) has been developed by Raytheon [5.9]. It employs 4PSK modulation. The principle of operation is best explained with reference to Fig. 5.10, assuming a 2PSK modulation. At the transmitter the symbols (bits) are time

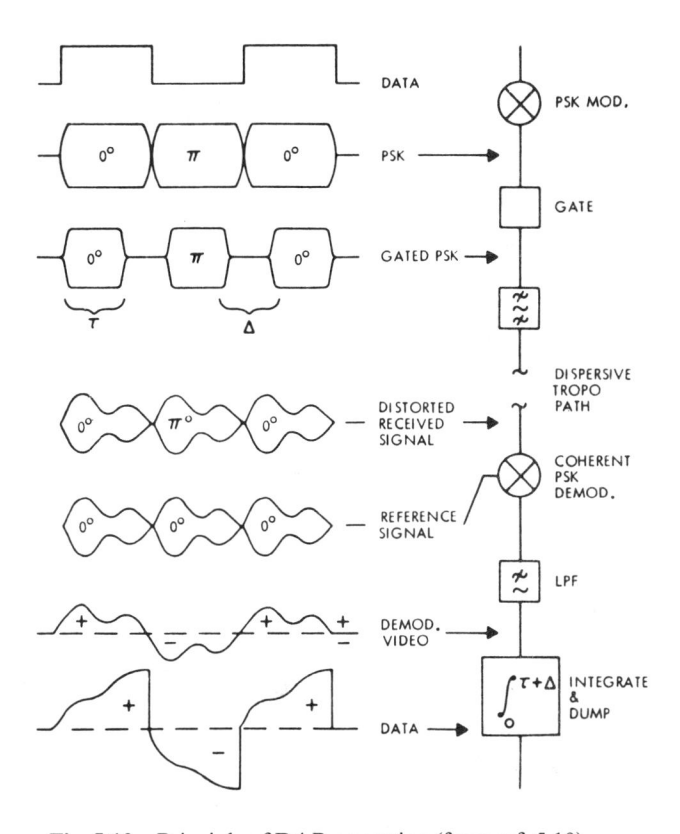

Fig. 5.10 Principle of DAR operation (from ref. 5.10).

gated after modulation, so that there is an interval between adjacent symbols during which the distortion due to multipath spread has time to vanish.

The received signal can be visualized as a stream of identically distorted nonoverlapping symbols with a distortion which varies very slowly compared with the symbol duration. In the demodulator (Fig. 5.11) each symbol is multiplied by an unmodulated replica of itself, and thus a d.c. signal is obtained which is integrated over the symbol duration and dumped at the end. The unmodulated replica is obtained from the incoming symbol by delaying it by 1 bit and inverse modulating it by the output d.c. signal corresponding to the same bit. This adds 0° to a 0° modulation and 180° to a 180° modulation. This replica could be used directly as a reference for demodulation, but its signal-to-noise ratio can be improved by a "recirculating filter". This is shown in Fig. 5.11 as a delay line of 1 bit and an amplifier with an overall loop gain g slightly less than unity. This recirculator loop coherently adds amplitudes of previous pulses and the present pulse as

$$1 + g + g^2 + \ldots = \frac{1}{1-g}$$

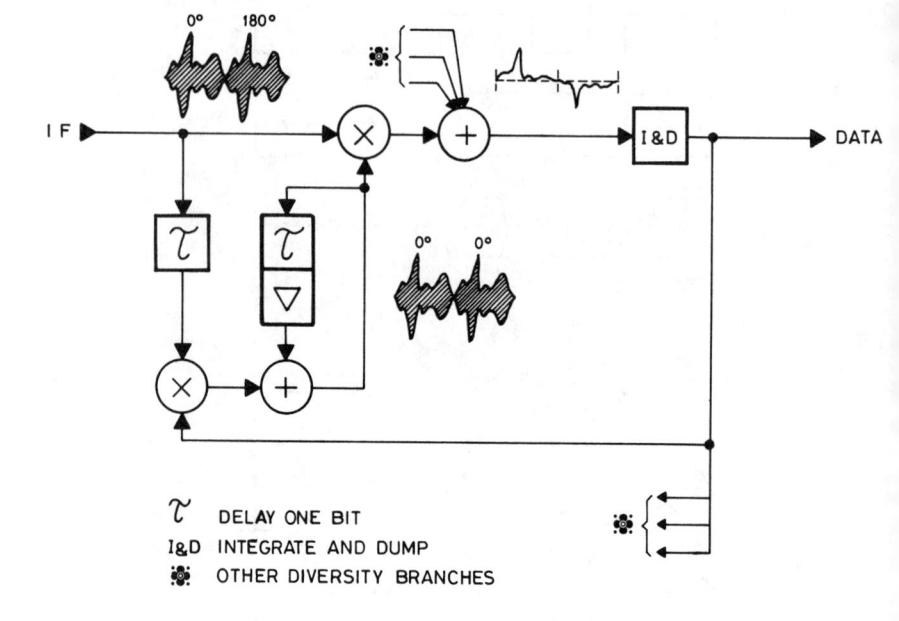

Fig. 5.11 Simplified binary model of a DAR demodulator (after ref. 5.10).

and adds noise powers incoherently as

$$1+g^2+g^4+\ldots = \frac{1}{1-g^2}$$

so that the output signal-to-noise power ratio is improved by

$$\frac{1-g^2}{(1-g)^2} = \frac{1+g}{1-g}$$

When $g = 0.9$ an improvement of 13 dB, corresponding to an almost noiseless reference, is obtained if the channel is quasi-time-invariant for about 20 pulses. This condition is easily satisfied at the bit rates considered. The effect of an output bit in error gives a degradation of less than 1 dB.

The other explicit diversity branches are connected as shown in Fig. 5.11. In each diversity receiver the signal is variable and the noise is identical. Each signal is multiplied by its unmodulated replica, which in turn is proportional to the signal, so that in the combination each branch gives a contribution proportional to the square of the signal amplitude. The combination is therefore of the maximal-ratio type (see eqn (5.11) and the subsequent discussion) and, owing to the linearity of the detection process, it behaves as a predetection combination.

The modem is "adaptive" as it adapts to the distorted pulse shape. As the multipath spread and consequently the distortion increase, the performance improves owing to the implicit diversity effects. The r.f. level for a BER of 10^{-5} may decrease by as much as 10 dB. However, as the distortion is increased, a point is reached at which the BER curve (see Section 7.5.1) no longer improves but starts producing an irreducible error floor (see Chapter 9, Section 9.1.3) because the tail of the pulses overlaps the adjacent pulses.

An improvement in the technique described above can be obtained by transmitting a second gated channel on a different radio frequency interleaved with the first, thus transmitting a continuous r.f. signal. This would permit a more efficient transmission, closer to constant amplitude, with a virtual 3 dB gain, and it would also be possible to split the modulating data stream into two streams of half bit rate, with symbols of double duration, which are capable of overcoming a double path delay distortion.

5.9 Comments and suggestions for further reading

Brennan's classic paper [5.1] on the theory of diversity is still valid and can be read with profit. The book by Schwartz *et al.* [5.11] contains a long chapter and many other sections, mainly mathematical, on diversity with special application to digital techniques. The paper by Larsen [5.3] gives a good description of space diversity. There are several interesting publications on angle diversity, mostly based on experimental results [5.12–5.16]. The theory is discussed in ref. 5.17. Time diversity and its applications are dealt with theoretically in ref. 5.11 and experimentally in refs 5.18–5.21 and a number of other papers. There are a number of papers dealing with problems of implicit diversity, adaptive modems etc., but they are not easily accessible to the nonspecialist or the engineer interested in actual applications. A good synthesis is given in ref. 5.22.

References

5.1 Brennan, D. G. Linear diversity combining techniques, *Proc. IRE*, **47**, 1075–1102, June 1959.
5.2 Altman, F. Configurations for beyond-the-horizon diversity systems, *Electr. Commun.*, 161–164, June 1956.
5.3 Larsen, R. Quadruple space diversity in troposcatter systems, *Marconi Rev.*, 28–55, First Quarter 1980.
5.4 Boithias, L. and Battesti, J. Puissance moyenne de bruit dans les faisceaux Hertziens transhorizon à modulation de fréquence, *Ann. Telecommun.*, **18** (5–6), 88–93, 1963.
5.5 *Recommendations and Reports of the CCIR, 16th Plenary Assembly, Dubrovnik, 1986*, Vol. 9, *CCIR Report 376-5*, Section 3.3, ITU, Geneva, 1986.
5.6 Skingley, B. S. *Advances in Tropospheric Scatter Techniques*, Brighton, 1974.

5.7 Carlton, B. F. Digital transmission over troposcatter links using independent sideband diversity, *Int. Conf. on Communications (ICC 75), San Francisco, 16–18 June 1975*, Vol. 1, pp. 5-20–5-23, IEEE, New York, 1975.

5.8 Jost, R. and Gohlke, K. Gated subcarrier sideband diversity modem for digital transmission, *Natl Telecommunications Conf., New Orleans, 1–3 December 1975*, Vol. 2, pp. 28-6–28-9, IEEE, New York, 1975.

5.9 Unkauf, M. G., Connor, W. J. et al. *Technical Publications on Digital Troposcatter*, Raytheon Company, Sudbury, MA, June 1980.

5.10 Unkauf, M., Liskov, N., Curtis, R. and Boak, S. Advanced digital troposcatter modem technology, *Eastcon Conference Record*, September 1977.

5.11 Schwartz, M., Bennett, W. R. and Stein, S. *Communication Systems and Techniques*, McGraw-Hill, New York, 1966.

5.12 Gough, M. W. et al. Troposcatter angle diversity in practice, *Marconi Rev.*, 199–217, Fourth Quarter 1978.

5.13 Krause, G. and Monsen, P. Results of an angle diversity field test experiment, *Natl Telecommunications Conf., Birmingham, AL, 3–6 December 1978*, Vol. 2, pp. 17.2.1–17.2.6, IEEE, New York, 1978.

5.14 Morita, S, Tachibana, H., Hoshino, T. and Kawasaki, H. Effect of angle diversity in troposcatter communication systems, *NEC (Nippon Electr. Co.) Res. Dev.*, **45**, 83–93, 1977.

5.15 Gough, M. W. and Rider, G. C. Angle diversity in troposcatter communications, *Proc. IRE*, **122** (7), 713–719, 1975.

5.16 Crisholm, J. H., Rainville, L. P., Roche, J. F. and Root, H. G. Angular diversity reception at 2290 MHz over a 188 mile path, *IRE Trans. Commun. Syst.*, 195–201, September 1959.

5.17 Koono, T., Hirai, M., Inone R. and Ishizawa, Y. Antenna beam deflection loss and signal amplitude correlation in angle diversity reception in UHF beyond the horizon communications, *J. Radio Res. Lab., Tokyo*, **9** (41), 21–49, January 1962.

5.18 Chase, D. The application of error-correction coding for troposcatter links, *Int. Conf. on Communications (ICC 75), San Francisco,16–18 June 1975*, pp. 5-11–5-14, IEEE, New York, 1975.

5.19 Osterholz, J. L. Megabit digital communications over a dispersive channel, *Natl Telecommunications Conf., Birmingham, AL, 3–6 December 1978*, Vol. 2, pp. 17.1.1–17.1.7, IEEE, New York, 1978.

5.20 Rogers, J. D. Introduction to digital tropo for military tactical communication, *Commun. Broadcast.*, **6** (3), 3–9, 1981.

5.21 Rogers, J. D. et al. Radio equipment for digital tropo military tactical communication, *Commun. Broadcast.*, **7**, 61–71, 1982.

5.22 Monsen, P. Fading channel communications—adaptive processing can reduce the effect of fading on beyond-the-horizon digital radio links, *IEEE Commun. Mag.*, 16–25, January 1980.

Chapter 6

Path Loss and Path Distortion

Methods for calculating the loss and distortion of a troposcatter path are given in this chapter. The loss between isotropic antennas must always be known for dimensioning the system, and a knowledge of the path distortion is necessary for wideband digital systems but is less essential for analog systems.

No method that can provide accurate predictions of path loss in all circumstances has yet been developed. All methods are approximate and semi-empirical and may give significantly different values of loss for the same path. We shall pay particular attention to the methods provisionally accepted by the CCIR (see *Recommendation 617*). It should be noted that the CCIR admit that the data available at present do not allow reliable prediction methods to be developed which would provide adequate accuracy worldwide, but methods that are acceptably accurate for some regions have been devised. It is even more difficult to predict the multipath delay spread for digital transmission. Methods that are available at present are described in this chapter.

Beyond-the-horizon transmission includes the special case of diffraction, which is dominant when the scatter angle is very small. Methods for calculating the path loss under these conditions are also given.

6.1 Path loss between isotropic antennas

We define the path loss L(dB) as the attenuation between isotropic antennas placed at the two ends of the path. More exactly we can write

$$L(\text{dB}) = P_t(\text{dBm}) + G_t + G_r - S(\text{dBm}) \tag{6.1}$$

where P_t(dBm) is the transmitted signal level (constant) at the input of the transmitting antenna, $G_t = G_r = 0$ dB are the (isotropic) gains of the transmitting and receiving antennas, and S(dBm) is the level of the received signal at the output of the receiving antenna (given by eqn (4.7)). This loss is the same as that defined by *CCIR Recommendation 341* as the "basic transmission loss". Remembering (4.7) we can write

$$L(\text{dB}) = P_t(\text{dBm}) - X(\text{dB}) - Y(\text{dB}) - M(\text{dBm}) \tag{6.2}$$

Since M(dBm) is the median value of the signal over a long period, the median loss over the same period can be written as

$$L(50\%) = P(\text{dBm}) - M(\text{dBm}) \tag{6.3}$$

and the path loss becomes

$$L(\text{dB}) = L(50\%) - X(\text{dB}) - Y(\text{dB}) \tag{6.4}$$

The variability of the path loss, which is represented by the terms X(dB) and Y(dB), is the same and is given by the same diagrams (e.g. Figs 2.2, 2.4 and 4.10) as for the received signal except that the sign is opposite. In fact, an increase of 1 dB in the path loss involves a decrease of 1 dB in the level of the received signal. The combined Rayleigh–normal path loss distribution (and other similar curves) is still given by Fig. 4.10, but with the ordinate values taking opposite signs. However, if the variability of the loss is considered as a sequence of fades of varying depths and if we indicate these fades by negative decibel values, the same diagrams apply both to signals and to fades.

It should be noted that, with reference to the long-term variability Y(dB) of the path loss, it has become customary to use hourly median values, even if the statistical data obtained during the various tests were normally taken over periods of say 5–30 min. However, the characterization of link performance is based on minute-by-minute values. Fortunately there is experimental evidence that the distributions of hourly and 1 min values are almost identical.

In path loss calculations the terms in (6.4) are evaluated either separately or together depending on the data available and the type of approach adopted. We shall see that the path loss used by the CCIR is the distribution of hourly medians

$$L(\text{dB}) = L(50\%) - Y(\text{dB}) \tag{6.5}$$

which can be represented by a diagram similar to that of Fig. 2.2. The value of the term $Y(dB)$ for a given percentage of time q can be written $Y(q\%)$, and the loss referred to that percentage is given by

$$L(q\%) = L(50\%) - Y(q\%) \tag{6.6}$$

which is another way of writing the distribution (6.5). Therefore we have for example (see Chapter 2, Section 2.2.1)

$$L(84.13\%) = L(50\%) + \sigma(dB)$$

The CCIR assume that the missing term $X(dB)$ is smoothed out by the use of diversity and can be ignored.

In other approaches the path loss is calculated as the distribution of the instantaneous values

$$L(dB) = L(50\%) - Z(dB) \tag{6.7a}$$

or

$$L(q\%) = L(50\%) - Z(q\%) \tag{6.7b}$$

where $Z(dB) = Y(dB) + X(dB)$ is the convolution of the long-term and short-term variations (see eqn (4.9) and Fig. 4.10).

The random variability of the path loss throughout the year implies that there is a period, which is taken as 1 month for practical reasons, during which the loss reaches a maximum and consequently the radio link exhibits its worst performance. This month is usually February in Europe and the USA. To ensure appropriate dimensioning of the link the path loss calculations are normally referred to the worst propagation conditions expected in a given period (year, month). Although some prediction methods give the annual median path loss, the CCIR normally refers to the worst month for the evaluation of performance, which is related to the path loss, and gives some diagrams for the necessary data transformations.

The concept of the "worst month" for use in performance evaluation is introduced and discussed in *CCIR Recommendation 581*. It can be defined in various ways according to the type of problem, but the general requirement is to estimate the percentage of time within a calendar month that radio, electrical or meteorological parameters (fading, rain rate etc.) exceed (or do not exceed) a given threshold. In our case we define the worst month as the calendar month in which the path loss exceeds a given value, related to the threshold of the radio equipment or to the maximum allowable noise, for the longest percentage of time.

For the correct design of a radio link it should be possible to calculate the following parameters:

(a) the maximum path loss which is not exceeded for specified high percentages of time, e.g. for 99.9% of the worst month (this affects the dimensioning of the troposcatter system);

(b) the minimum path loss that can be expected for small percentages of time, which is related to both the possibility of receiver saturation, with consequent distortion, and the risk of interference beyond the area covered by the troposcatter system. However, this figure is difficult to evaluate as it is generally related to anomalous propagation conditions.

If possible these data should be completed by a predicted distribution of path loss. The results of the calculations are used to plan the system and, if necessary, to provide protection against receiver saturation and interference with other links. For digital links in particular it is also necessary to estimate the path distortion in terms of the delay spread or the bandwidth of the medium expected for high percentages of the time.

6.2 Method of calculation

Path loss and path distortion are calculated using the following procedure.

(1) The type of climate in the area that the link is required to traverse is established (see Chapter 4, Section 4.9), and the values of the atmospheric refractivity at the Earth's surface, the refractivity gradient (if required) and the effective radius of the Earth are determined for the worst month.

(2) The basic data for calculating the geometrical parameters (Chapter 3, Section 3.2.1) are taken from maps of the area. These parameters are the path length, the heights of the sites and horizons above sea level, and the relative separations of the sites and horizons. Alternatively, the elevation angles of the horizons from the sites can be measured directly.

(3) The geometrical parameters of interest are calculated using the formulae given in Chapter 3. The most important parameters are the scatter angle and, when required, the height of the common volume and the 3 dB path differences for various antenna diameters.

(4) The long-term median path loss $L(50\%)$ in the worst month (or in any other period of interest) is calculated for the center frequency of the band using the methods described below.

(5) The statistical distribution of hourly medians is evaluated and the values of interest, e.g. $L(80\%)$ and $L(99.9\%)$, are extracted.

(6) Rayleigh fading and improvements due to the application of diversity techniques are taken into account if warranted by the approach adopted (see below).

(7) The multipath delay is evaluated and the delay power spectrum is calculated if required.

Sometimes the scattering angle is very small (very much less than 1°) and the diffraction mode of propagation may have to be taken into account. If this is the case the diffraction loss for the same path must be calculated and compared with the scatter loss. For completeness we shall also consider calculations of the path loss of the LOS tails often provided at troposcatter end-stations for connecting them to main communications centers in towns or at more accessible sites.

The calculations listed above are discussed in more detail in the remainder of this chapter.

6.3 Path geometry calculations

Before the troposcatter path loss can be calculated, the geometrical parameters must be determined. The three types of profile normally found are shown in Table 6.1, which also includes calculations of the relevant geometrical parameters. The most important result is the value of the scatter angle, which is used in subsequent path loss calculations.

A correction factor of $k = 4/3$ for the effective radius of the Earth is assumed by most designers in calculations of troposcatter path loss. In some cases, however, the effective radius is calculated using the assumed value of the refractivity of air at the Earth's surface (Fig. 4.2). The latter procedure was used by the National Bureau of Standards in their classic study [6.1]. The angles involved are generally very small and are measured in radians (rad) or milliradians (mrad). Calculations are made using the formulae given in Chapter 3.

6.4 CCIR methods of calculating path loss

In *Recommendation 617* (which includes *Report 238*) the CCIR propose two main empirical methods for calculating the path loss and its statistical distribution throughout an average year. These methods are based on measured data and are more reliable for losses which are not exceeded for more than 50% of the time. The CCIR also provide curves for the change from annual values to worst-month values. The two methods are in reasonable agreement for temperate climates, although there are some differences between them for other climates. Two other methods have also been reported by the CCIR. All four CCIR methods are described below.

Calculations of path loss using CCIR methods go only as far as the distribution of the hourly medians during the year or during the worst month. The fast fading represented by the term $X(dB)$ in eqn (6.4) is not taken into account. This is because the CCIR assume that diversity (and particularly quadruple diversity), which smoothes out fast fading, will be used.

TROPOSCATTER RADIO LINKS

Table 6.1 Examples of path geometry calculations

	Station 1: Site A	Station 2: Site B
Ground level a.s.l.	2450 m	400 m
Tower height	10 m	10 m
Antenna height a.s.l.	$h_1 = 2460$ m	$h_2 = 410$ m
Radio horizon distance	$d_1 = 115$ km	$d_2 = 141$ km
Radio horizon height a.s.l.	$h_1' = 1725$ m	$h_2' = 910$ m
Radio horizon elevation	$\theta_1 = -13.1$ mrad	$\theta_2 = -4.75$ mrad
$\theta_{1,2} = (h_{1,2}' - h_{1,2})/d_{1,2} - d_{1,2}/2a$	$-0.75°$	$-0.27°$
Elevation above chord	$\alpha = 17.3$ mrad	$\beta = 16.3$ mrad
Crosspoint distance	$r_0 = \beta d/\theta = 212$ km	$s_0 = \alpha d/\theta = 225$ km
Path length $d = 437$ km		
Angular length $\theta_0 = d/a = 51.4$ mrad $= 2.94°$		
Scatter angle (angular distance) $\theta = \theta_0 + \theta_1 + \theta_2 = 33.6$ mrad $= 1.92°$		
Equivalent distance $\theta a = 286$ km		
Crosspoint height h_0 a.s.l. $= 2316$ m		
Symmetry factor $s = 0.94$		

PATH LOSS AND PATH DISTORTION

	Station 1: Island	Station 2: Mainland
Ground level a.s.l.	474 m	30 m
Tower height	10 m	10 m
Antenna height a.s.l.	$h_1 = 484$ m	$h_2 = 40$ m
Radio horizon distance	$d_1 = (2ah_1)^{1/2}$	$d_2 = 35$ km
	$= (17\,000 h_1)^{1/2}$	
	$= 91$ km	
Radio horizon height a.s.l.	$h_1' = 0$	$h_2' = 229$ m
Radio horizon elevation	$\theta_1 = -d_1/a$	$\theta_2 = (h_2' - h_2)/d_2 - d_2/2a$
	$= -10.7$ mrad	$= 3.34$ mrad
	$= -0.61°$	$= 0.19°$

Elevation above chord $\alpha = 5.8$ mrad $\beta = 16.3$ mrad
Crosspoint distance $r_0 = \beta d/\theta = 184.4$ km $s_0 = \alpha d/\theta = 65.6$ km
Path length $d = 250$ km
Angular length $\theta_0 = d/a = 29.4$ mrad $= 1.7°$
Scatter angle (angular distance) $\theta = \theta_0 + \theta_1 + \theta_2 = 22.1$ mrad $= 1.3°$
Equivalent distance $\theta a = 188$ km
Crosspoint height h_0 a.s.l. $= 514$ m
Symmetry factor $s = 0.36$

	Station 1: Seashore	Station 2: Island
Ground level a.s.l.	20 m	410 m
Tower height	10 m	10 m
Antenna height a.s.l.	$h_1 = 30$ m	$h_2 = 420$ m
Radio horizon distance	$d_1 = 22.8$ km	$d_2 = 84.5$ km
$d_{1,2} = (2ah_{1,2})^{1/2}$		
$= (17000 h_{1,2})^{1/2}$		
Radio horizon elevation	$\theta_1 = -2.66$ mrad	$\theta_2 = -9.94$ mrad
$\theta_{1,2} = -d_{1,2}/a$	$= -0.15°$	$= -0.57°$
$= d_{1,2}/8500$		
Elevation above chord	$\alpha = 12.4$ mrad	$\beta = 7.9$ mrad
Crosspoint distance	$r_0 = \beta d/\theta = 109.5$ km	$s_0 = \alpha d/\theta = 170.7$ km
Path length $d = 280.2$ km		
Angular length $\theta_0 = d/a = 32.96$ mrad $= 1.86°$		
Scatter angle (angular distance) $\theta = \theta_0 + \theta_1 + \theta_2 = 20.4$ mrad $= 1.16°$		
Equivalent distance $\theta a = 173$ km		
Crosspoint height h_0 a.s.l. $= 441$ m		
Symmetry factor $s = 0.64$		

$a = (4/3) \times 6400 = 8500$ km

6.4.1 Method I

Method I is a slightly simplified version of the National Bureau of Standards procedure described in ref. 6.1. It has been well known to designers for many years, and it has been implemented in a number of computer systems.

The long-term (annual) median transmission loss in decibels due to forward scatter is given approximately by the following formula:

$$L(50\%) = 30 \log f - 20 \log d + F(\theta d) - V(d_e) \qquad (6.8)$$

where f is the radio frequency in megahertz, d is the path length in kilometers, θ is the scatter angle in radians, $F(\theta d)$ is the function shown in Fig. 6.1, d_e is the effective distance (defined below) in kilometers and $V(d_e)$ is the function shown in Fig. 6.2 which corrects for the various types of CCIR climate. This formula can be used for most land-based paths, particularly in temperate regions, and is most accurate for frequencies of 200–4000 MHz. The attenuation due to rain and atmospheric absorption is almost negligible at these frequencies. For example, the atmospheric absorption in a 1000 km path adds a loss of 2 dB at 1 GHz.

The function $F(\theta d)$ for a radio refractivity N_s at the Earth's surface of 301 is given by the following formulae [6.1]:

$$F(\theta d) = 30 \log(\theta d) + 0.332 \theta d + 135.82 \qquad 0.01 \leqslant \theta d \leqslant 10$$
$$F(\theta d) = 37.5 \log(\theta d) + 0.212 \theta d + 129.5 \qquad 10 \leqslant \theta d \leqslant 70 \qquad (6.9a)$$
$$F(\theta d) = 45 \log(\theta d) + 0.157 \theta d + 119.2 \qquad \theta d > 70$$

For other values of N_s the formulae become

$$F(\theta d, N_s) = F(\theta d, 301) - 0.1(N_s - 301) \exp\left(-\frac{\theta d}{40}\right) \qquad (6.9b)$$

These formulae are valid for a symmetry factor $0.7 \leqslant s \leqslant 1$ (see eqn (3.7)), which holds for most troposcatter paths. The path loss for more asymmetrical paths with the same value of θd is significantly less than that given by (6.9), particularly for $s < 0.5$. The appropriate value of N_s is determined by substituting the annual mean value of N_0 (averaged from Fig. 4.1) in eqn (4.3). If the path connects points with different values of N_s, an average value is taken.

The effective distance in kilometers is defined as

$$d_e = \frac{130 d}{d_1 + d_{s1}} \qquad d < d_1 + d_{s1}$$
$$\qquad\qquad\qquad\qquad\qquad\qquad\qquad\qquad (6.10)$$
$$d_e = 130 + d - (d_1 + d_{s1}) \qquad d > d_1 + d_{s1}$$

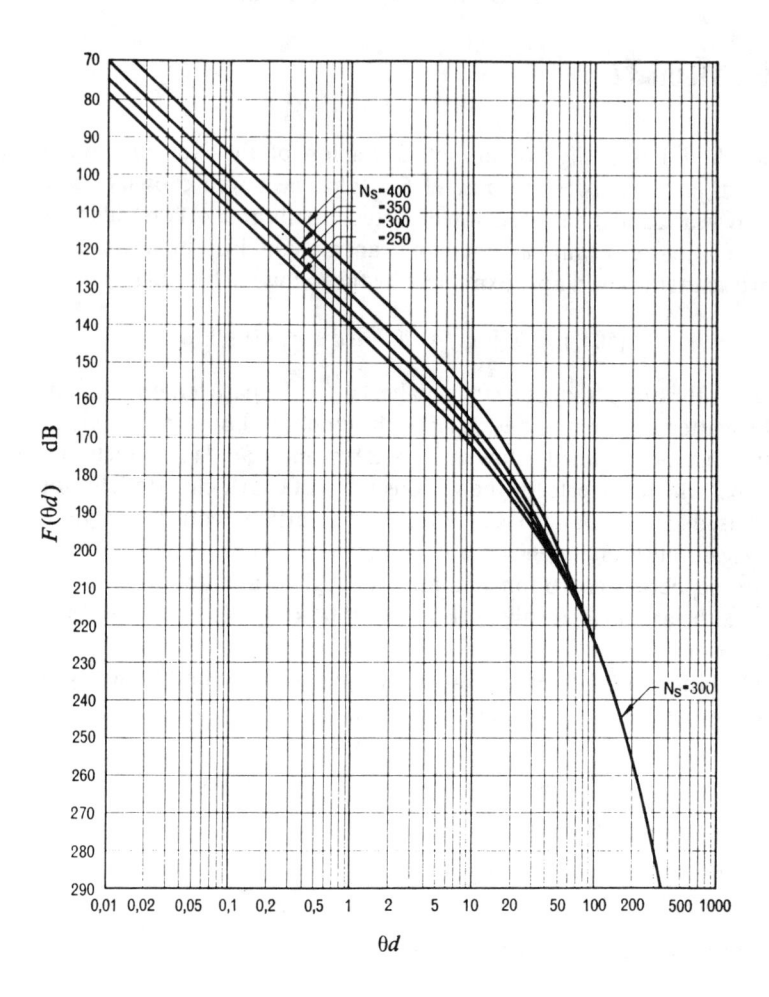

Fig. 6.1 Function $F(\theta d)$, where d is in kilometers and θ is in radians. (From *CCIR Report 238-5*.)

where

$$d_1 = 4.1(h_t^{1/2} + h_r^{1/2})$$

$$d_{sl} = 302 f^{-1/3}$$

h_t and h_r are the effective antenna heights in meters (i.e. the heights of the antennas above an ideal smooth Earth with a height above sea level equal to the average height of the central 80% of the terrain between the antenna and its horizon), d_1 is the sum of the distances to the radio horizon calculated for the ideal smooth Earth and d_{sl} is related to the angle at which troposcatter loss and diffraction loss are approximately equal. Two examples of this calculation are

PATH LOSS AND PATH DISTORTION

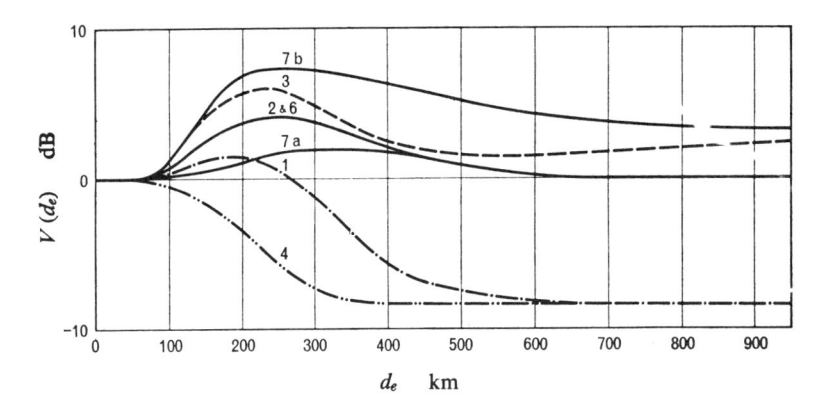

Fig. 6.2 Function $V(d_e)$ which corrects for the climatic types indicated on the curves. (From *CCIR Report 238-5*.)

shown in Table 6.2; the path profile of the first example is the first profile in Table 6.1. The calculation is self-explanatory. It can be seen that when the frequency is increased by a factor of 3 (from 900 MHz to 2700 MHz) the loss increases by $30 \log 3 = 14.3$ dB.

The distribution of the path loss during the year can be found by subtracting from $L(50\%)$ the term $Y(q\%)$ corresponding to the desired percentage of time q. *CCIR Recommendation 617* (completed by *Report 238*)

Table 6.2 Calculation of annual path loss using method I (*CCIR Report 238-5*)

Climate	7a	5
Path frequency f (MHz)	900	2700
Path length d (km)	437	366.8
Scatter angle θ (mrad)	33.6	28.2
Equivalent distance $d_q = 8.5\theta$ (km)	286	240
Hop parameter θd	14.7	10.3
Distance $d_1 = 4.1(h_1^{1/2} + h_2^{1/2})$ (km)	129	118
Distance $d_{s1} = 302 f^{-1/3}$ (km)	31	22
Effective distance d_e (km)	407	357
Annual mean refractivity N_s	325	320
Function $F(\theta d)$ (dB)	174.7	168.3
Function $V(d_e)$ (dB)	1.6	5.5
Annual median path loss		
$+ 30 \log f$ (dB)	$+88.6$	$+102.9$
$- 20 \log d$ (dB)	-52.8	-51.3
$+ F(\theta d)$ (dB)	$+174.7$	$+168.3$
$- V(d_e)$ (dB)	-1.6	-5.5
$= L(50\%)$ (dB)	208.9	214.4

gives diagrams of hourly median values of $Y(q\%)$ for 90%, 99%, 99.9% and 99.99% of time in various climates (Fig. 6.3). With some loss of accuracy we can assume that these curves are symmetrical about the line $Y = 0$ and also give the minimum path loss at the various percentage times. The distribution of $Y(q\%)$ for other climates that can be considered as a mixture of two CCIR climates can be obtained by averaging the percentage times of the two curves for each loss value.

It is possible to plot a point on probability paper for each percentage time and loss. The line joining these points gives the statistical distribution of the hourly median path loss during the year. This distribution is approximately normal and its standard deviation σ can easily be obtained from the diagram as described in Chapter 2, Section 2.2.1 and Table 2.1 (σ(dB) = $Y(84.13\%)$). Annual values can be transformed to worst-month values by using the curves of Fig. 6.4, also taken from *CCIR Report 238-5*, in which the equivalent distance is that defined by eqn (6.15).

The standard error of prediction for any specified percentage time is estimated as

$$e(q\%) = \{13 + 0.12 Y^2(q\%)\}^{1/2} \text{ dB} \qquad (6.11)$$

It is therefore possible to draw on probability paper the entire distribution including the prediction error:

$$L(q\%) \pm e(q\%) = L(50\%) - Y(q\%) \pm \{13 + 0.12 Y^2(q\%)\}^{1/2} \qquad (6.12)$$

An example of a distribution of this type is shown in Fig. 6.5. The curve broadens into a strip called the "uncertainty zone" within which the true loss should be found. Figure 6.5 also shows the free-space loss for an LOS path of the same length, the maximum path loss that the equipment is able to overcome in this example, and the received r.f. levels corresponding to the various path losses and calculated using the methods described in Chapter 8. It can be seen that there is a risk of outage for a small percentage of time.

We can continue the calculations of the path loss throughout the year for the first example in Table 6.2 as follows:

$$
\begin{aligned}
L(q\%) \quad &= L(50\%) - Y(q\%) \pm e(q\%) \\
L(99\%) \quad &= 208.9 \quad +16.5 \quad \pm \; 6.8 \quad = 225.4 \pm \; 6.8 \text{ dB} \\
L(99.9\%) \quad &= 208.9 \quad +22.0 \quad \pm \; 8.4 \quad = 230.9 \pm \; 8.4 \text{ dB} \\
L(99.99\%) &= 208.9 \quad +27.0 \quad \pm 10.0 \quad = 235.9 \pm 10.0 \text{ dB}
\end{aligned}
$$

The values for $Y(q\%)$ were taken from Fig. 6.3(f). The annual data can be transformed into those for the worst month by adding the corrections taken from Fig. 6.4(d) to the values calculated above.

The prediction error is assumed to have a Gaussian distribution. The standard error $e(q\%)$ of eqn (6.11) (taken with the plus sign) has a probability of

84.13%, and the other errors have different probabilities (see Chapter 2, Section 2.2.1). A "degree of confidence" (or "service probability" with respect to the behavior of the final link) of 84.13% is conventionally attributed to the loss $L(q\%) + e(q\%)$. If a different degree of confidence is required, the standard error of prediction should be multiplied by the factor corresponding to the required probability in the normal distribution (Table 2.1). For example a loss $L(99\%)$ estimated with a degree of confidence of 95% (the factor for 95% is 1.7) is given by

$$L(99\%) = L(50\%) - Y(99\%) + 1.7e(99\%) \qquad (6.13)$$

Before concluding this section it should be noted that in some cases the antennas can be placed in such a way that there is a reflection of energy in the foreground. According to the particular geometry of this situation, if the antennas are high enough the scatterers can receive more incident power and thus the path loss is decreased. Conversely, with low antennas and low frequency the reflected energy tends to cancel the rays directed to the lower part of the common volume (where scattering efficiency is greatest), thus increasing the path loss. However, these effects can be neglected for most paths.

6.4.2 Method II

It should be possible to obtain reliable predictions by using sets of data measured in various climatic regions and collected in suitable form for prediction purposes. This is the philosophy underlying method II, which is based on the curves given in Fig. 6.6. These curves give the loss $l(q\%)$ at 1000 MHz for $q = 50\%, 90\%, 99\%$ and 99.9% of the year for various climates and for horizon elevation angles of 0° as a function of an equivalent distance which is defined below.

For frequencies below 5 GHz the path loss in decibels can be written as

$$L(q\%) = l(q\%) + 30 \log f + 20 \log\left(\frac{d}{d_q}\right) \qquad (6.14)$$

where $L(q\%)$ is the long-term (annual) path loss in decibels for $q\%$ of hours, $l(q\%)$ is the corresponding path loss found on Fig. 6.6, f is the frequency in gigahertz, d is the path length in kilometers and d_q is the equivalent distance in kilometers given by

$$d_q = d + 8.5(\theta_1 + \theta_2) = 8.5\theta \qquad (6.15)$$

where θ_1 and θ_2 are the elevation angles of the horizon in milliradians and θ is the scatter angle in milliradians. By applying the above formula for the

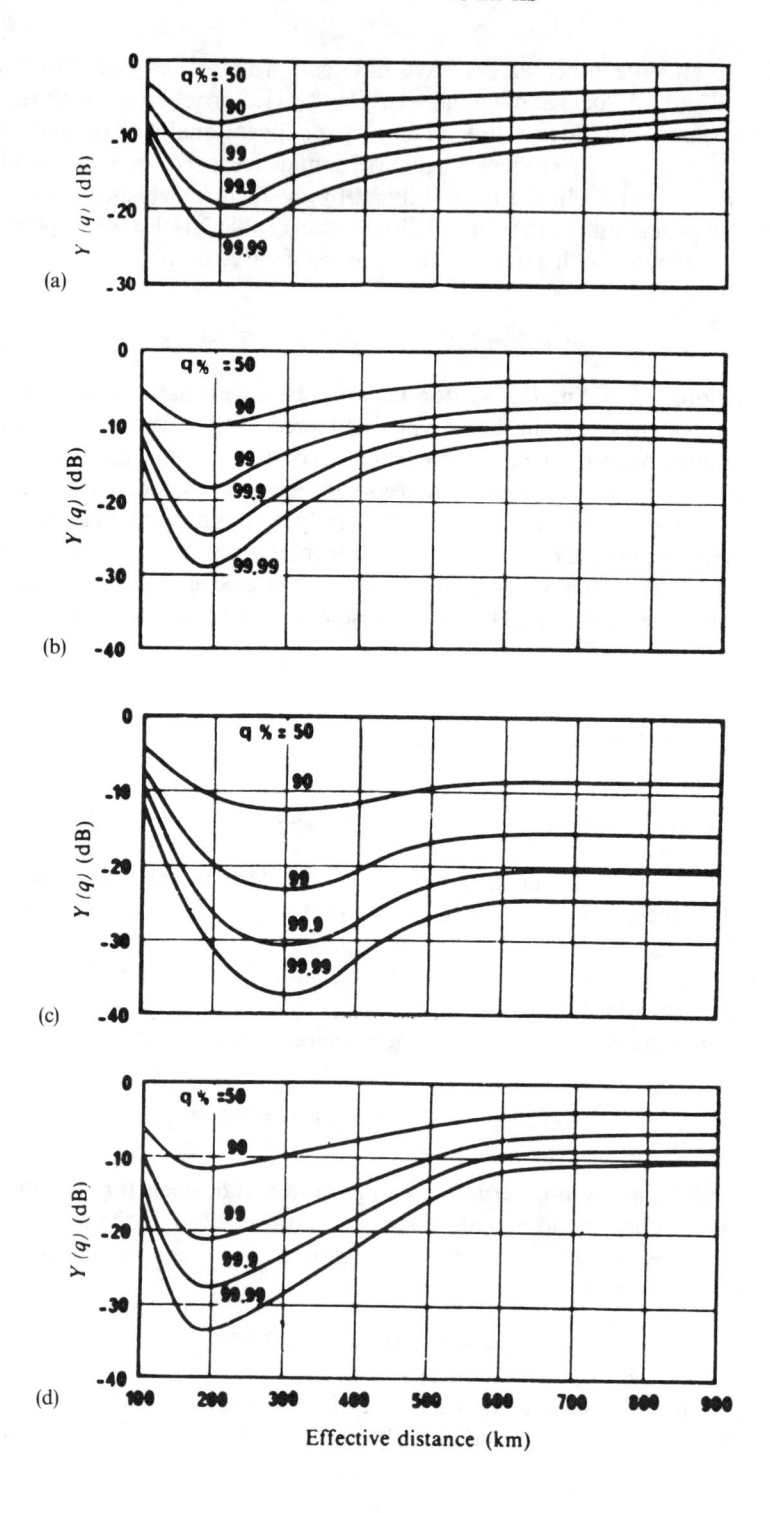

(a)

(b)

(c)

(d)

Effective distance (km)

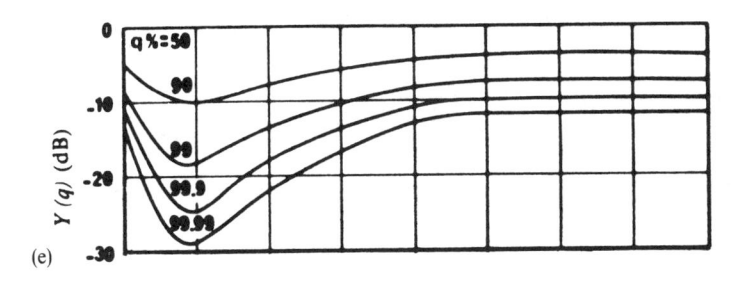

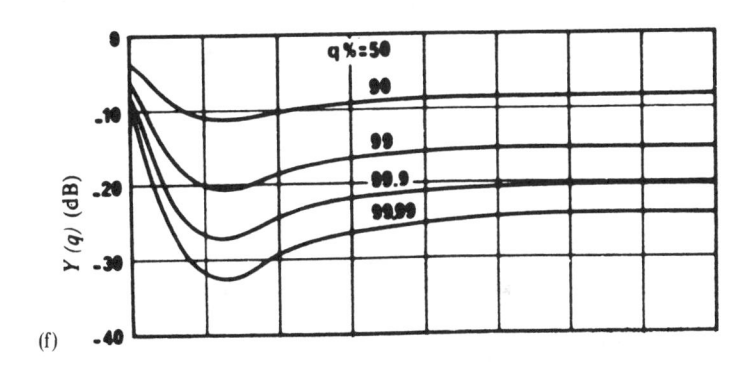

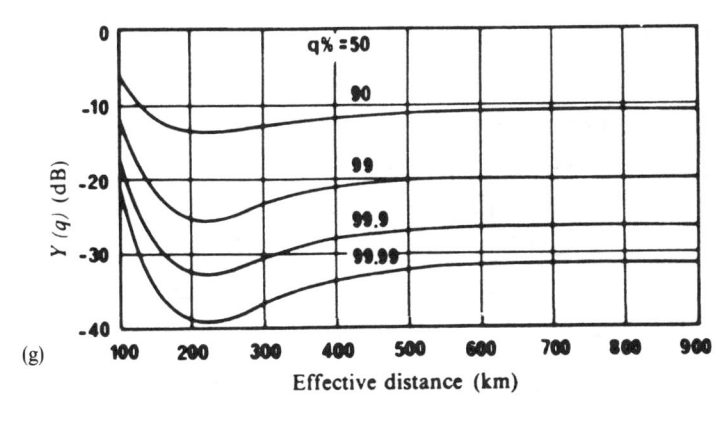

Fig. 6.3 Function $Y(q\%)$ for various climates: (a) equatorial climate (type 1); (b) continental subtropical climate (Sudan) (type 2); (c) maritime subtropical climate (type 3); (d) desert climate (Sahara) (type 4); (e) continental temperate climate (type 6); (f) maritime temperate climate over land (type 7a); (g) maritime temperate climate over sea (type 7b). (From *CCIR Report 238-5*.)

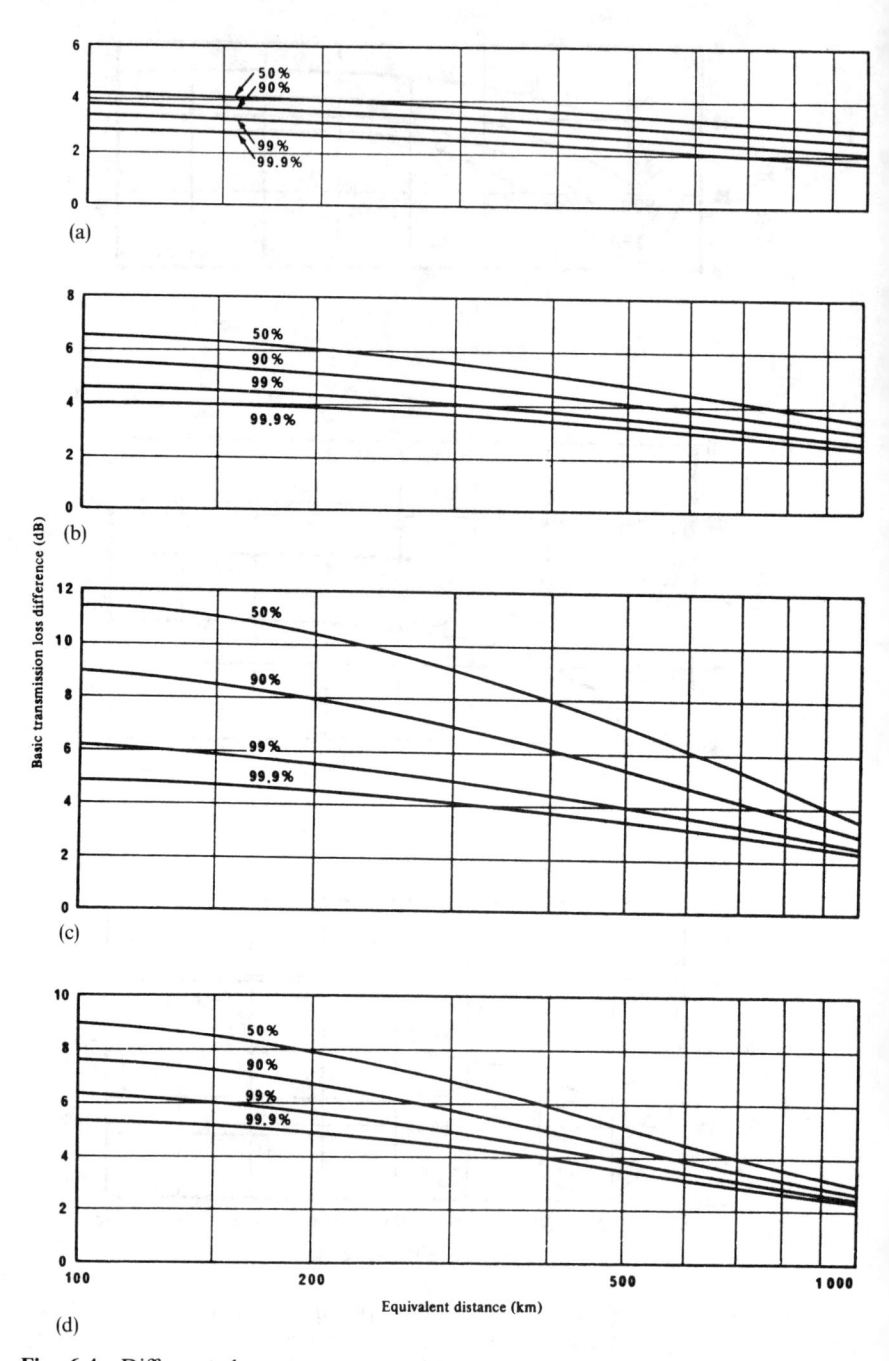

Fig. 6.4 Difference between worst-month and annual path losses: (a) equatorial climate; (b) humid tropical climate; (c) desert climate; (d) temperate climate. (From *CCIR Report 238-5*.)

PATH LOSS AND PATH DISTORTION   129

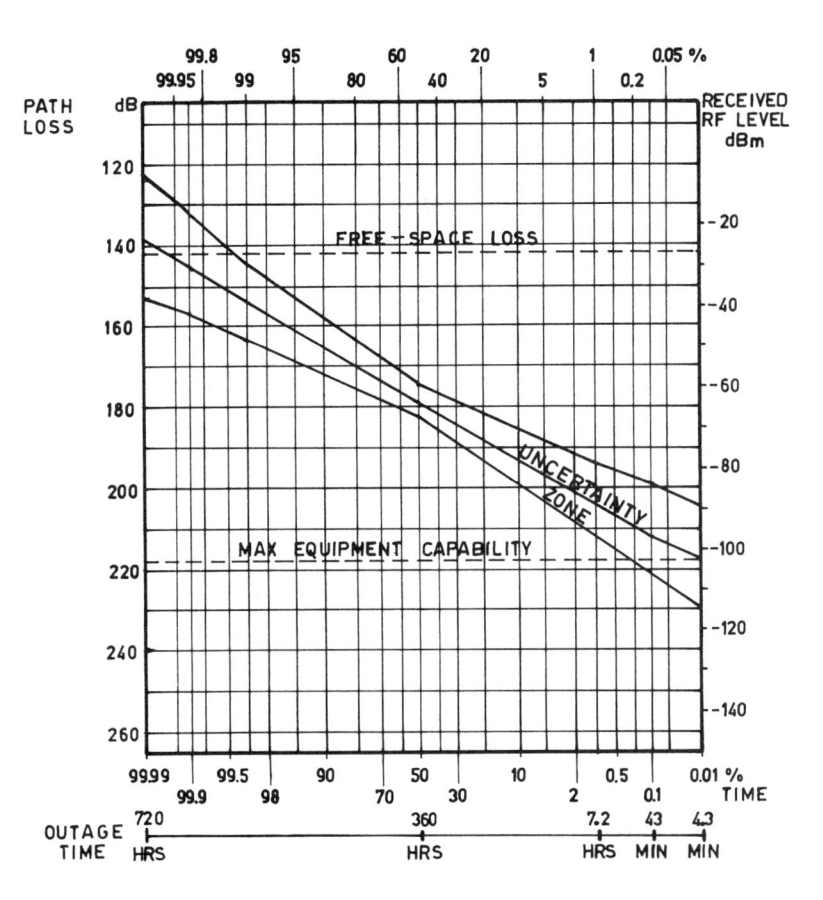

Fig. 6.5 Example of the distribution of hourly medians during the worst month.

appropriate climate and the various percentages of time it is possible to determine the distribution of hourly median values of path loss throughout the year, which can then be drawn on probability paper.

For comparison we can use this method to recalculate the path losses of the first example in Table 6.2. From Fig. 6.6(d) we obtain the values to which we must add the two final terms of (6.14) (-1.4 dB and 3.7 dB respectively):

$l(50\%) = 210.3$ dB   from which $L(50\%) = 212.6$ dB
$l(99\%) = 225.5$ dB   from which $L(99\%) = 227.8$ dB
$l(99.9\%) = 229.9$ dB   from which $L(99.9\%) = 232.2$ dB

We see that there is a difference of only a few decibels between the results of the two methods.

The values for the worst month can be obtained in the same way from eqn

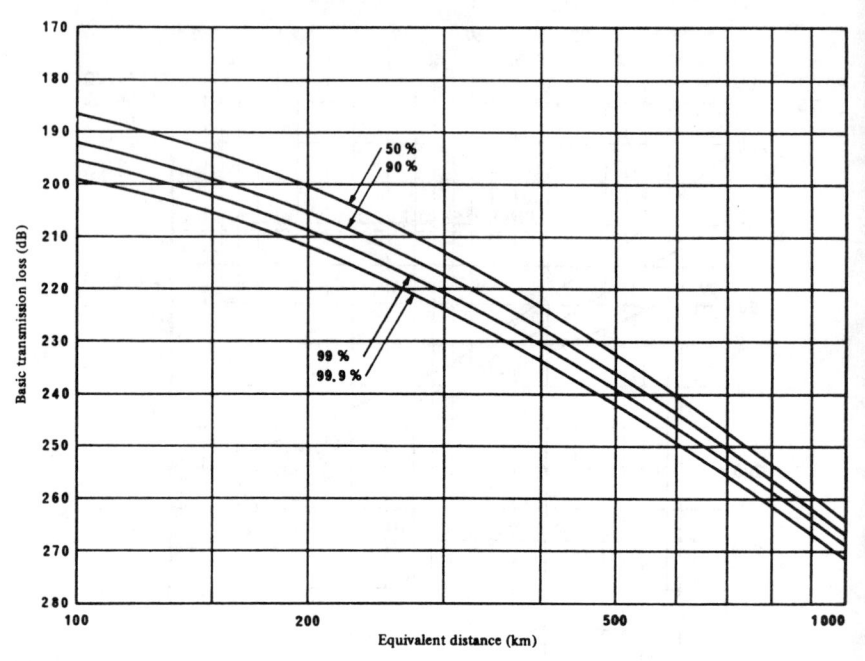

TROPOSCATTER RADIO LINKS

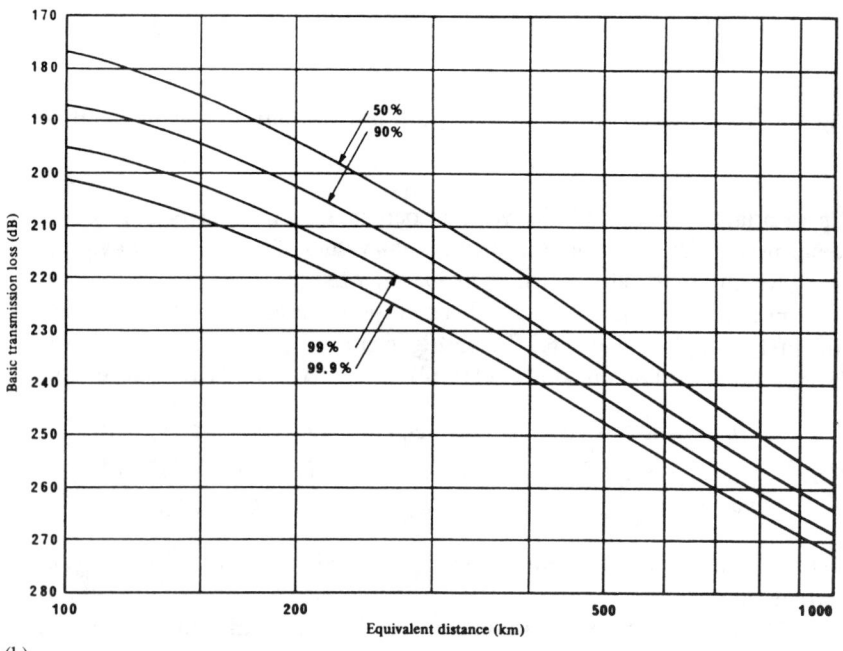

(a)

(b)

PATH LOSS AND PATH DISTORTION 131

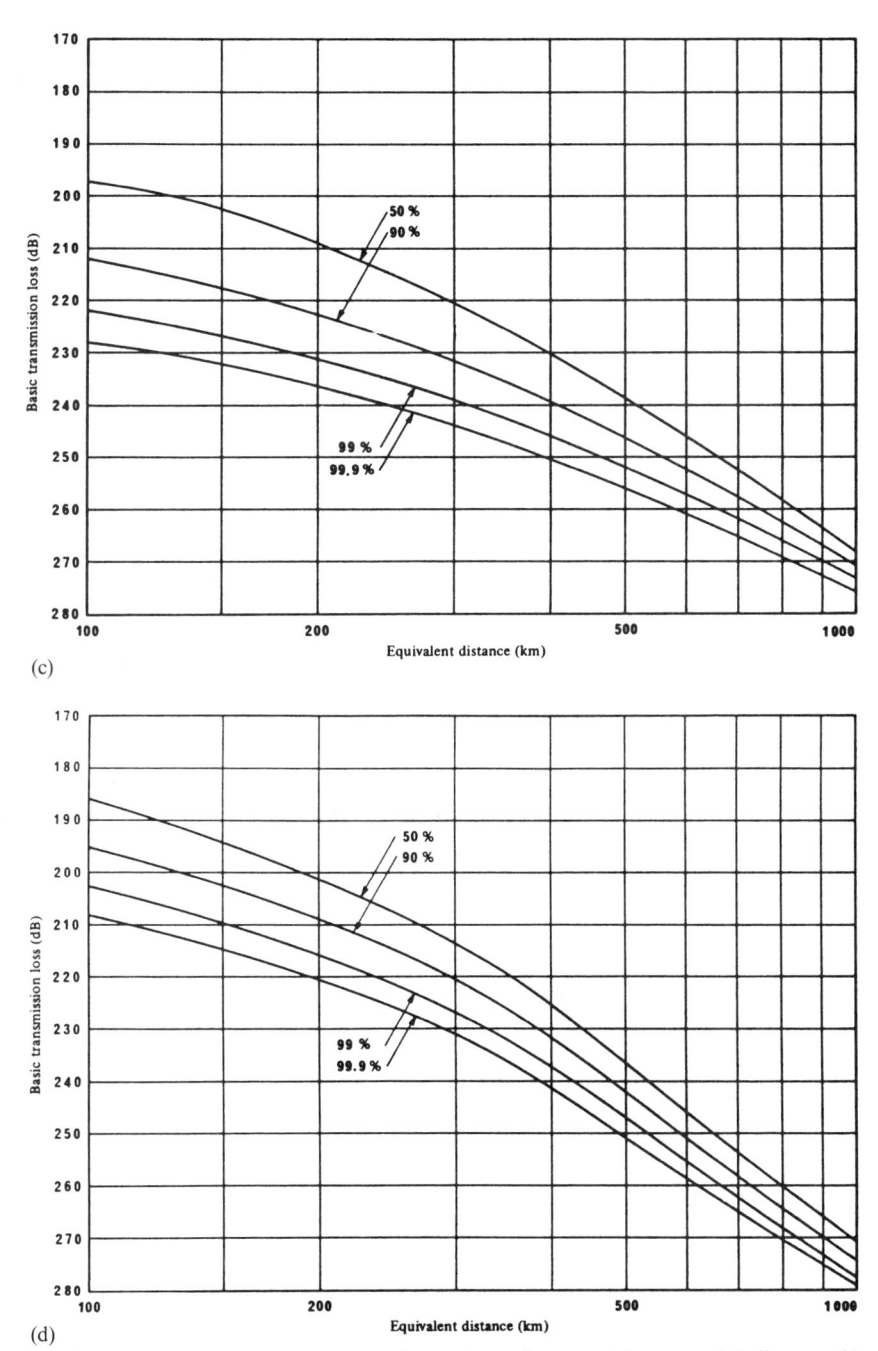

(c)

(d)

Fig. 6.6 Annual path loss at 1 GHz for various climates: (a) equatorial climate; (b) humid tropical climate; (c) desert climate; (d) temperate climate. (From *CCIR Report 238-5*.)

(6.14) and the curves shown in Fig. 6.7. The curves in Fig. 6.4 represent the differences between the curves in Figs 6.6 and 6.7, which were obtained from measurements made primarily in Europe and Africa.

6.4.3 Radiometeorological method

It was shown in Chapter 4, Section 4.1.4, that there is a correlation between the refractivity gradient in the common volume and the path loss. When the gradient is greater the loss due to the scatterers is lower and the increased bending of the rays further reduces the height of the common volume and the value of the scatter angle, thus reducing the path loss.

This radiometeorological parameter is the basis of the semi-empirical formula for path loss proposed in *CCIR Report 718-2*, which is valid for all climates and all percentage times:

$$L(\text{dB}) = 30 \log f + 30 \log d + 1.5 G_c + 102 \qquad (6.16)$$

where f is the frequency in megahertz, d is the path length in kilometers and G_c is the refractivity index gradient in the common volume in N units per kilometer (note that this gradient is normally negative). The only random element is G_c, which appears to have an approximately normal statistical distribution. This implies that the path loss is also normally distributed. Distance appears in the term $30 \log d$ and is also included in G_c, as the height of the common volume changes with distance and this affects the value of G_c. The effect of distance is closely related to the decreasing density of scatterers with increasing altitude. The values of G_c should be measured or obtained from meteorological stations in the vicinity of the path. Then we can immediately calculate the path loss distribution from the known distribution of G_c.

For comparison we can recalculate the annual median path loss of the first example in Table 6.2. For an assumed annual median value of $-40 N$ units km^{-1} for G_c, we obtain from eqn (6.16) $L(50\%) = 209.8$ dB, which is very near the value calculated in the table.

6.4.4 Chinese method

Measurements made in China over a long period of time suggested the following empirical formula for the long-term (annual) path loss in decibels:

$$L(50\%) = 30 \log f + 30 \log \theta + 10 \log d + N(H) - 0.08(N_s - 320) + 124.6 \qquad (6.17)$$

where f is the frequency in megahertz, d is the path length in kilometers, θ is the scatter angle in radians, N_s is the annual mean value of refractivity at the Earth's surface,

$$N(H) = 20\log(5+0.3H)+0.65H$$
$$H = \theta d_1 d_2/d$$

and d_1 and d_2 are the distances from the site to the crosspoint of the radio horizon rays.

The variability of the hourly median path losses appears to be normally distributed with a standard deviation given by

$\sigma = 7.0+0.09\Delta N \exp(-0.003 d_s)$ dB   over flat terrain

$\sigma = 4.6+0.06\Delta N \exp(-0.003 d_s)$ dB   over mountainous terrain (6.18)

$\sigma = 8.8+0.11\Delta N \exp(-0.003 d_s)$ dB   over the sea

where ΔN is the difference between the mean values of the refractivity at the Earth's surface in winter and summer, and d_s is the distance in kilometers between radio horizons. It is therefore possible to draw a diagram of the distribution like that shown in Fig. 6.5 by plotting on probability paper the two points with coordinates $L(50\%)$, 50% and $L(50\%)+\sigma$, 84.13% (or more precisely $L(50\%)+3.1\sigma$, 99.9%).

Comparisons with tests made in temperate climates elsewhere in the world suggest that the Chinese method is slightly less accurate than method I in such regions. Table 6.3 gives the results of a recalculation of the path losses of the first example in Table 6.2 using the Chinese method. The values differ by a few decibels from those calculated using the other methods.

6.5 Calculation of instantaneous path loss with fast fading and diversity

The approach adopted by the CCIR is to calculate the distribution of hourly median path losses. However, the "instantaneous" path loss is determined in other approaches, and thus any diversity effects are taken into account and the "fade distributions" are calculated. Within a short time of transmission the signal amplitude is Rayleigh distributed around its median value according to eqns (2.19) and (2.20) and presents frequent fades which can be smoothed using diversity techniques. The signal distribution, and consequently the fade distribution, is given by eqn (5.4) for selection and eqn (5.8) for maximal-ratio combination. Figure 6.8 shows the distribution curves for the maximal-ratio combination. As established in Chapter 5, Section 5.6, there is a diversity gain as if reception takes place through an ideal single receiver and the attenuation is lower with a narrower variability range.

This applies to the fast fading term $X(\text{dB})$, but we should consider the total variability $Z(\text{dB})$ of eqn (6.7) in calculating the distributions over long periods. For a single signal (no diversity or a single branch of a diversity

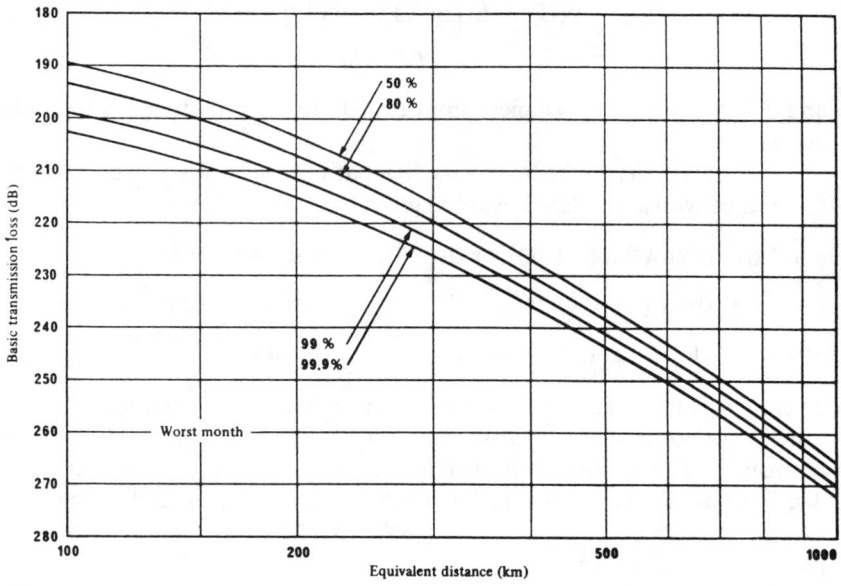

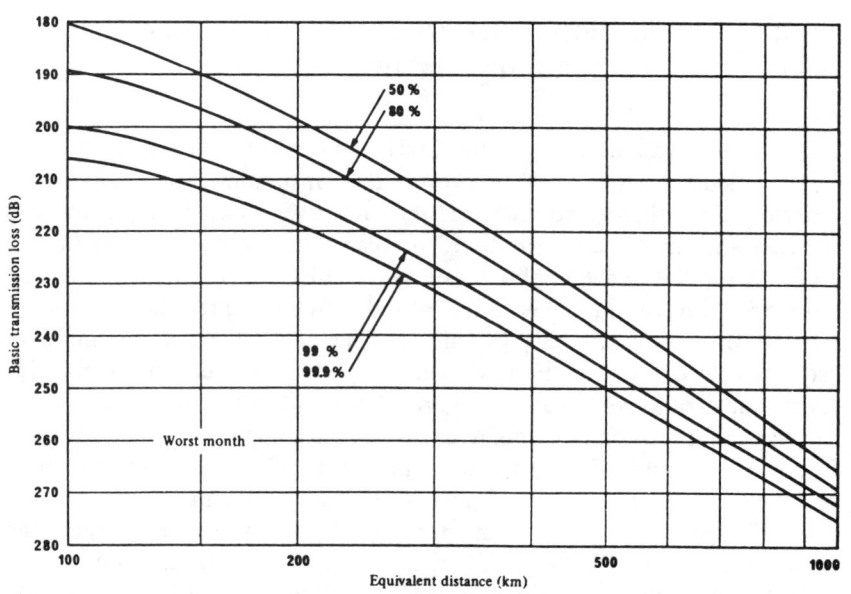

(a)

(b)

PATH LOSS AND PATH DISTORTION 135

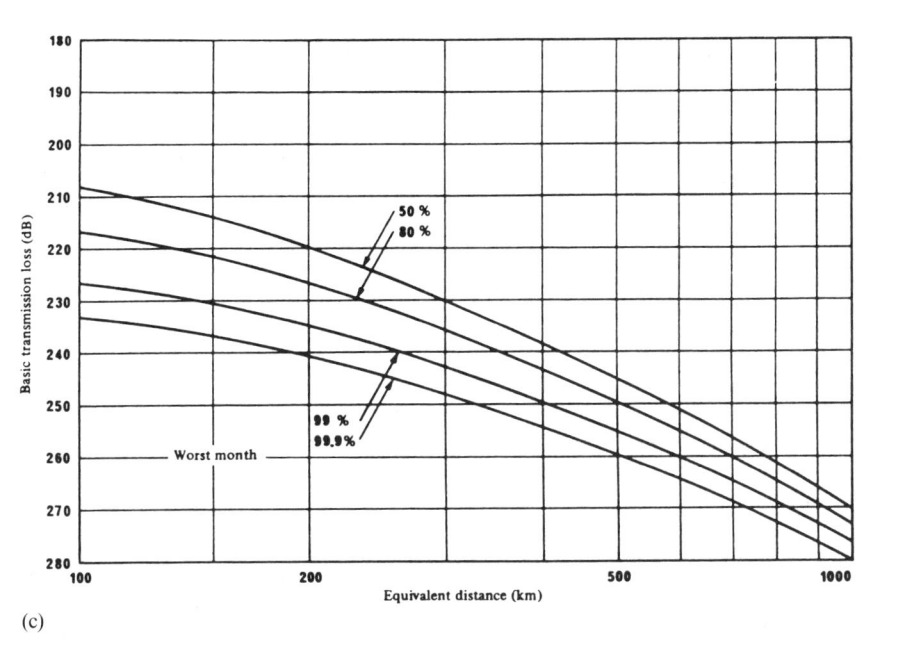

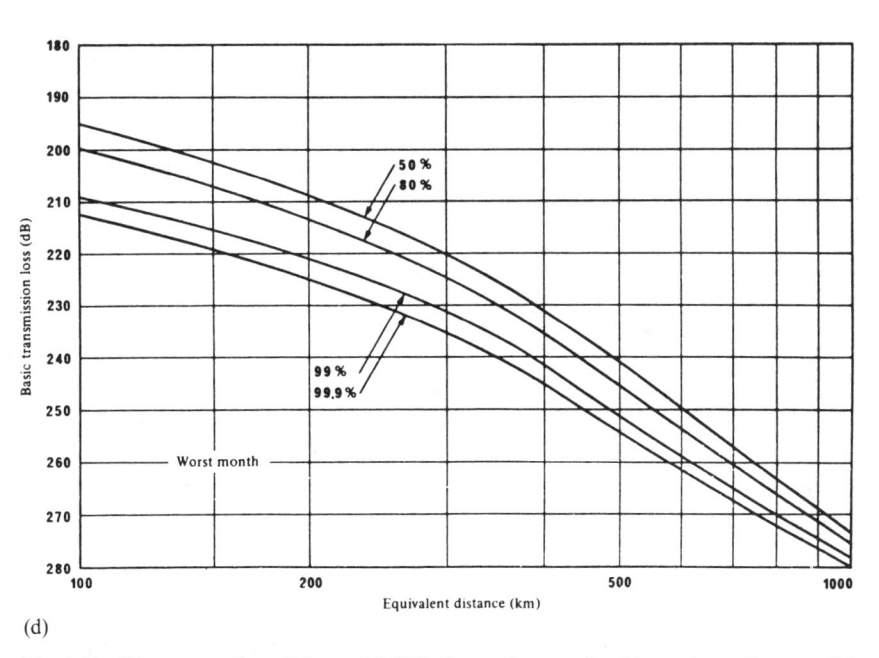

(c)

(d)

Fig. 6.7 Worst-month path loss at 1 GHz for paths over land in various climates: (a) equatorial climate (type 1); (b) maritime subtropical climate (type 3); (c) desert climate (type 4); (d) temperate climate (types 6 and 7a). (From *CCIR Report 238-5*.)

Table 6.3 Calculation of annual transmission loss using the Chinese method
(*CCIR Report 238-5*)

Path frequency f (MHz)	450	900	2700
Path length d (km)	437	437	437
Distance d_s between radio horizons (km)	181	181	181
Scatter angle θ (rad)	0.0336	0.0336	0.0336
Distance d_1 from site 1 to crosspoint (km)	212	212	212
Distance d_2 from site 2 to crosspoint (km)	225	225	225
Parameter $H_s = \theta d_1 d_2/d$ (km)	3.7	3.7	3.7
Function $N(H_s)$ (dB)	18.1	18.1	18.1
Annual mean refractivity N_s at Earth's surface	325	325	325
Variation ΔN_s in refractivity	30	30	30
Standard deviation δ of hourly medians (dB)	5.6	5.6	5.6
Annual median path loss			
$+30 \log f$ (dB)	79.6	88.6	102.9
$+30 \log \theta$ (dB)	−44.2	−44.2	−44.2
$+10 \log d$ (dB)	26.4	26.4	26.4
$+N(H_s)$ (dB)	18.1	18.1	18.1
$-0.08(N_s - 320)$ (dB)	−0.4	−0.4	−0.4
+constant (dB)	124.6	124.6	124.6
$= L(50\%)$ (dB)	204.1	213.1	227.4
Hourly median path loss not exceeded for 99.9% of the year:			
$L(99.9\%) = L(50\%) + 3.1\delta$ (dB)	221.5	230.5	244.8

system) this distribution is given by eqn (4.9) and shown in Fig. 4.10. The figure shows that if, for example, the standard deviation of the log-normal distribution is 4 dB an overall combined fast–slow fade deeper than 30 dB will take place for only 0.1% of the time or, alternatively, for 99.9% of time fading will not exceed 30 dB. It is unclear whether this 0.1% of time is made up of many short fades or a few long fades or whether they are concentrated or randomly distributed.

In the case of diversity reception the N signals (4.9) with the same log-normal component and decorrelated short-term components are combined in a distribution which can be obtained from (4.9) by substituting the simple Rayleigh term $\exp(-s)$ by a combination of N such terms. After some rearrangement the expression becomes

$$P(Z_c) = 1 - \frac{1}{(2\pi)^{1/2}} \int_{-\infty}^{\infty} F(Z_c, t) \exp\left(-\frac{t^2}{2}\right) dt \qquad (6.19)$$

where $P(Z_c)$ is the probability of a fade deeper than Z_c (dB) (taking Z_c negative),

$$F(Z_c, t) = 1 - \{1 - \exp(-S)\}^N$$

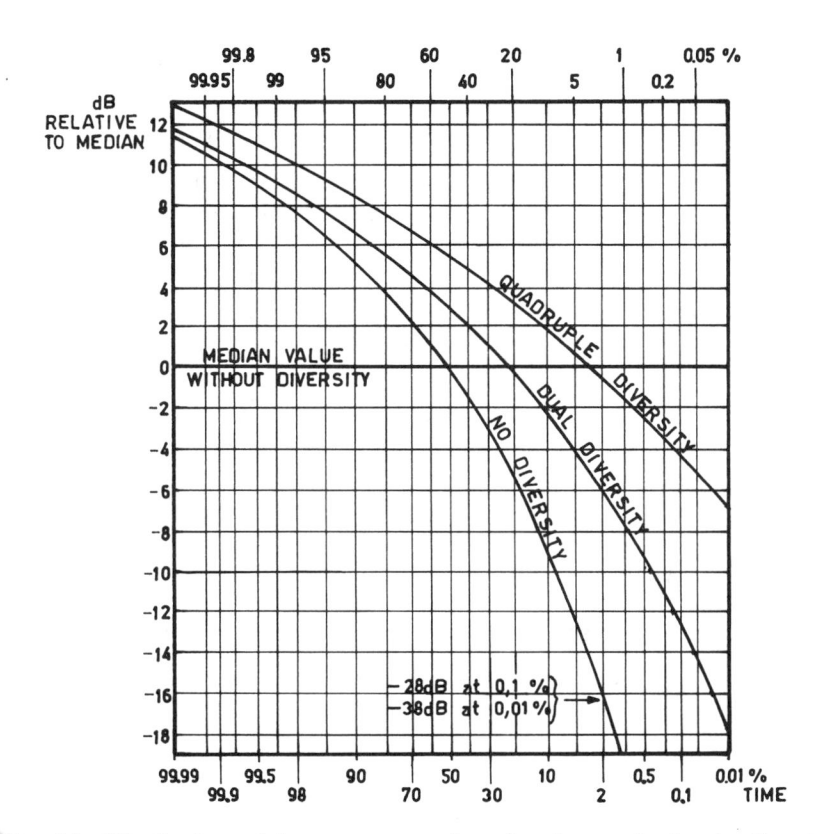

Fig. 6.8 Distribution of instantaneous values for the maximal-ratio diversity combination (Rayleigh fading).

for selection,

$$F(Z_c, t) = \exp(-S) \sum_{k=0}^{N-1} \frac{S^k}{k!}$$

for maximal-ratio combination,

$$S = (\ln 2) \times 10^{0.1(Z_c - \sigma t)}$$

and σ is the standard deviation in decibels of the log-normal distribution.

Distribution curves for this combined fast–slow fade Z_c(dB) are shown in Fig. 6.9. If we consider the fade values (in decibels) to be negative, these values also represent the signal referred to its median and the same distributions apply to both fades and signal variations. The distribution of the equivalent "instantaneous loss" can now be calculated using eqn (6.7b):

$$L(q\%) = L(50\%) - Z_c(q\%)$$

This approach is adopted, for example, in Yeh's method [6.2]. $L(50\%)$ is calculated using an empirical approximate formula. Then a fading margin

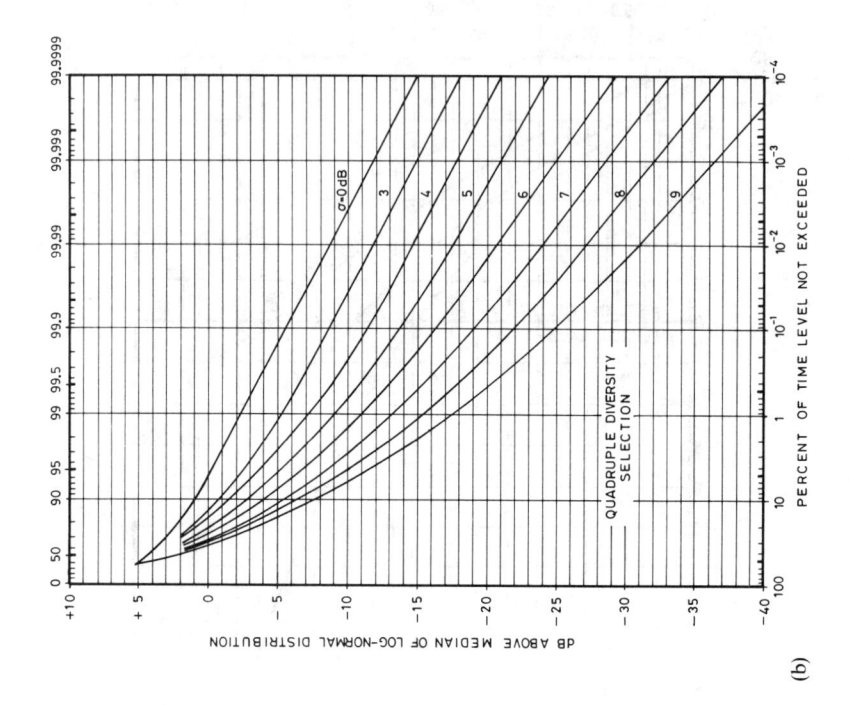

138 TROPOSCATTER RADIO LINKS

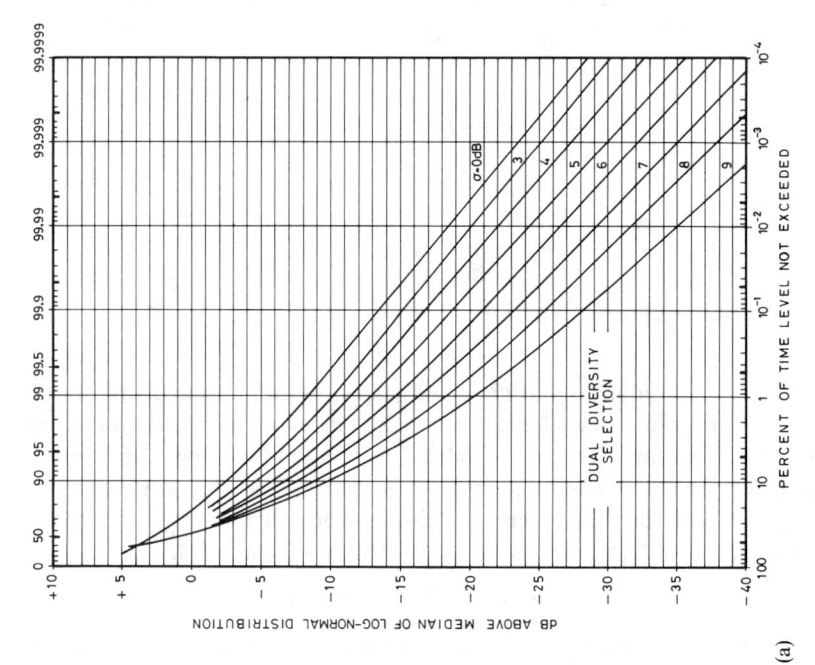

(b)

(a)

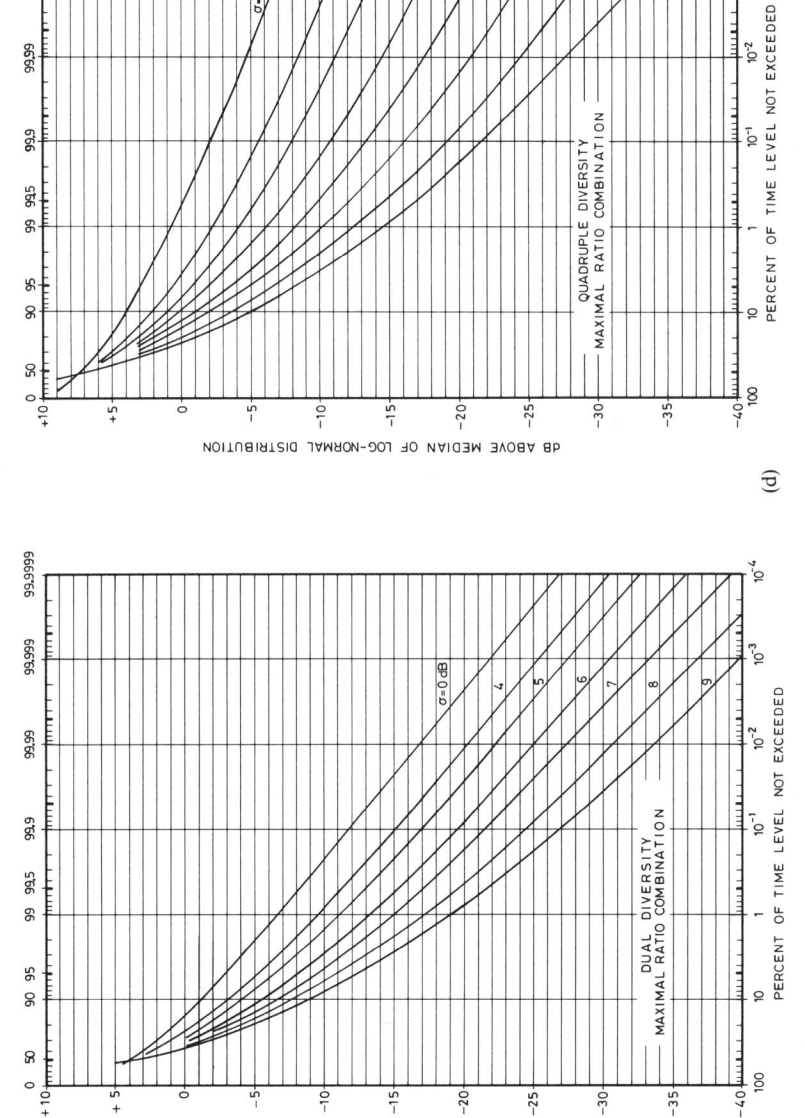

Fig. 6.9 Distribution of instantaneous values with combined fast–slow fading: (a) dual diversity with selection; (b) quadruple diversity with selection; (c) dual diversity with maximal-ratio combination; (d) quadruple diversity with maximal-ratio combination.

which depends on the order of diversity and the type of combination is added for each given percentage time and the maximum loss which is not exceeded for those time percentages is obtained. An example of this procedure is shown below:

Order of diversity	4
Type of combination	Selection
Standard deviation σ of normal distribution	7 dB
Required reliability $q\%$	99.9%
Calculated monthly path loss $L(50\%)$	209 dB
Depth of fade not exceeded for $q\%$ of the time (from Fig. 6.9(b))	19 dB
Maximum path loss $L(q\%)$ not exceeded for $q\%$ of the time	228 dB

6.6 Dependence of path loss on the main parameters and its actual variability

It has been found experimentally and it is generally agreed that up to about 3 GHz or more the frequency f makes a contribution of $30 \log f$ to the long-term median path loss. The CCIR formulae given above are based on this statement. It should therefore be possible to test a troposcatter path at one frequency and to transform the results to a different frequency. However, this simple relation does not always hold exactly. It has been shown experimentally that the frequency term may have a statistical distribution with values varying from $40 \log f$ to $16 \log f$. The CCIR formulae for $L(50\%)$ appear to be valid up to 12 GHz, but there are suggestions that the frequency term is approximately $20 \log f$ for f between 2 and 6 GHz. Therefore it is suggested in *CCIR Report 569* that the term $30 \log f$ is replaced by the term $20 \log f + 96$, where f is in gigahertz, for interference calculations.

Distance and scatter angle are interdependent. With zero antenna elevation angles the only parameter would be the distance d, which would appear in the term $30 \log d$. However, in practice a more complex term $F(\theta d)$ is used in method I, and in the radiometeorological method there is a dependence on the refractivity gradient.

If all other parameters are kept constant, an increase of 1° in the scatter angle or in the antenna elevation angle produces an increase of 9–12 dB in the path loss. Therefore the scatter angle should be kept to a minimum and the elevation angles should be negative if possible in order to minimize the path loss.

We are now able to improve on the information given in Table 4.1 about the changes in the path loss in the various climatic regions. In temperate climates the monthly median loss may vary during the year by 10–15 dB on overland paths 150–250 km long and by a few decibels on very long paths, with higher values in winter than in summer. On overland paths 100–200 km long the diurnal variations are greater in summer with a range of 5–10 dB. The greatest path loss takes place in the afternoon and the smallest losses are observed in the early morning. Larger variations occur in paths over the sea where super-refraction and elevated layers may be present.

Dry hot desert climates present the maximum loss in summer; the annual variation in the monthly median exceeds 20 dB for medium-length paths and the diurnal variations are very large. In monsoon climates the path loss is a minimum between the wet and the dry season, with maximum values of N_s in the wet season. In equatorial climates there are small variations in path loss during the day and the year.

This behavior can be explained with reference to eqn (6.16). The lower path losses are associated with more negative values of the refractive index gradient G_c in the common volume, e.g. in hot humid regions. Conversely the path loss is higher in regions with low absolute values of G_c, e.g. in desert areas. However, in humid regions with a monsoon climate the atmospheric conditions are sometimes such that the temperature and humidity change very little with height, so that G_c has small absolute values and the path loss is large even for very large values of N_s.

Since the path loss is associated with the vertical gradient G_c, the use of N_s in the prediction formulae (e.g. in CCIR method I) is valid only for certain climates (e.g. temperate climates) which can be approximated by an "exponential atmosphere" where the gradient of the refractivity is proportional to the refractivity at the Earth's surface. These two parameters are often decorrelated in other types of climate, and the gradient may deviate substantially from proportionality, particularly in the lower part of the atmosphere.

6.7 Path loss for line-of-sight links

Sometimes a troposcatter link is terminated with an LOS tail which connects the radio station, which is normally in an elevated position, to the town or the main communications center, which is generally located in a lower position unsuitable for the troposcatter station. The formula for free-space loss is well known:

$$L(\mathrm{dB}) = 20 \log f(\mathrm{MHz}) + 20 \log d(\mathrm{km}) + 32.45 \qquad (6.20)$$

The maximum attenuation in the presence of fading can be obtained by adding the semi-empirical term $10 \log(fd) - 30$ dB to the above loss or by evaluating the fade allowance from the characteristics of the profile and the climate.

6.8 Diffraction path loss

When the scatter angle is very small (less than 1° say) the dominant propagation mechanism in a beyond-the-horizon path may become diffraction, at least for a certain percentage of the time. In such cases it is necessary to calculate the median path losses due to both troposcatter and diffraction and compare them.

For short periods (minutes) the mechanism with the lower loss is dominant if the two losses differ by 6 dB or more. For longer periods (days, months), in view of the variability of the signals, the dominant mechanism is definitely that with the lower loss if the median path losses differ by more than 20 dB. The diffracted signal is much more stable than the troposcatter signal, and in a "mixed path" the deep fades of the latter may be limited by diffraction as shown schematically in the distribution diagram of Fig. 6.10.

The diffraction path loss is normally considered as the sum of the free-space loss which exists in the absence of obstacles and the "diffraction loss" introduced by the obstacles. The latter can be reliably calculated for the two

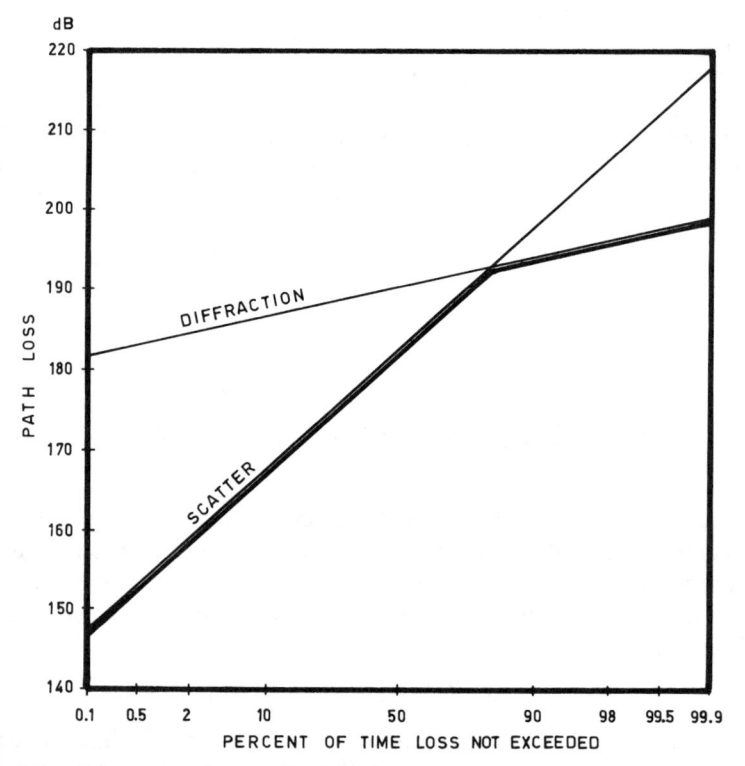

Fig. 6.10 Schematic diagram of the path loss distribution for a combined scatter–diffraction path.

ideal cases described below. The formulae, which are taken from *CCIR Report 715*, were derived using sophisticated calculations which are not reproduced here. It will be noted that the loss in a given diffraction path is smaller at lower radio frequencies. This means that a diffraction path operates best at lower frequencies.

The diffracted signal is normally very stable and its variations are mainly due to variations in the refractivity of the air. The path geometry calculations are based on the appropriate effective radius of the Earth derived from N_s. As this varies throughout the year there will be corresponding variations in the path loss, which should be calculated at least for the extreme values of N_s.

6.8.1 Knife-edge diffraction

Knife-edge diffraction occurs when the two terminal stations have the same radio horizon formed by a sharp transverse obstacle. The diffraction loss is given by

$$L_k(\text{dB}) = 6.4 + 20 \log\{(v^2+1)^{1/2} + v\} \tag{6.21}$$

where

$$v = 2\left(\frac{\Delta d}{\lambda}\right)^{1/2} = 2.58\theta\left(\frac{f d_A d_B}{d}\right)^{1/2} \tag{6.22}$$

f is the frequency in megahertz corresponding to a wavelength λ, d_A and d_B are the distances of the obstacle from the sites in kilometers, d is the path length in kilometers, θ is the diffraction angle (instead of scatter angle) in radians and Δd is the path difference (eqn (3.22)).

Equation (6.21) is accurate to within 0.5 dB and is valid for $v > -1$. A negative v means that the obstacle is slightly below the line joining the two terminal antennas, the first Fresnel zone is partially obstructed and the diffraction angle is negative. If the obstacle is tangential to the line joining the antennas, only half the energy will pass the obstacle and the diffraction loss will be 6 dB. A larger obstruction will increase the loss.

It should be noted that the conditions on the mountain top may affect the propagation positively or negatively. For example, a layer of snow may produce refraction that significantly decreases the path loss. Conversely, as mountains are not generally good knife edges, the actual diffraction loss has been found to be 10–20 dB higher than that calculated using eqn (6.21).

In a profile with two knife-edge radio horizons there will be double diffraction over the two horizons if the angles are sufficiently small. If A and D are the terminal sites and B and C are the obstacles, the diffraction loss for each obstacle can be calculated using eqn (6.21) by considering first a path ABC and then a path BCD and summing the two diffraction losses with the free-space loss AD. The result is an acceptable approximation.

An example of path loss calculations made using *CCIR Report 715* is shown in Table 6.4. The calculations were performed for two values of the effective radius of the Earth, and an allowance degradation due to bad knife-edge conditions was estimated. A further margin for shorter-term fades should be added. It is evaluated by calculating the variability of hourly medians as a convolution of the fading distributions on two cascaded LOS links (site A to obstacle and obstacle to site B) [6.1].

Table 6.4 Calculation of path attenuation by knife-edge diffraction (*CCIR Report 715-2*)

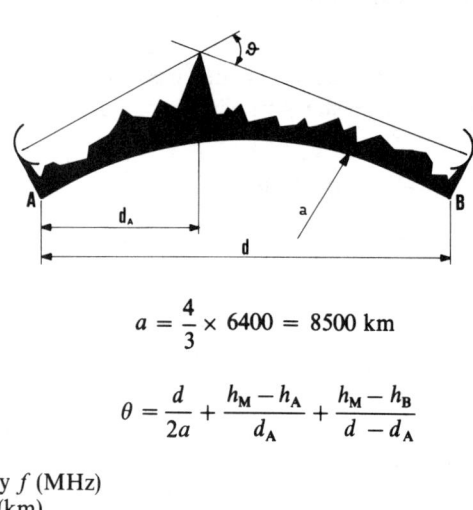

$$a = \frac{4}{3} \times 6400 = 8500 \text{ km}$$

$$\theta = \frac{d}{2a} + \frac{h_M - h_A}{d_A} + \frac{h_M - h_B}{d - d_A}$$

Path frequency f (MHz)	450	450
Path length d (km)	61	61
Effective radius a of the Earth (km)	8500	6400
Height h_A of antenna A (km)	0.020	0.020
Height h_B of antenna B (km)	0.065	0.065
Height h_M of obstacle (km)	0.190	0.190
Distance d_A of obstacle from antenna A (km)	25	25
Distance d_B of obstacle from antenna B (km)	36	36
Diffraction angle θ (mrad)	13.9	15.0
Parameter v	2.9	3.2
Diffraction loss L_k (dB)	21.9	22.7
Allowance l_k for bad knife edge (dB)	12.0	12.0
Free-space loss		
$+20 \log f$ (dB)	+53.1	
$+20 \log d$ (dB)	+35.7	
$+$ constant (dB)	+32.4	
$= L_s$ (dB)	= 121.2	121.2
Path attenuation $L_p = L_s + L_k + l_k$ (dB)	155.1	155.9

6.8.2 Smooth-earth diffraction

Smooth-earth diffraction occurs when the path profile is smooth, e.g. over the sea or a flat desert. The formulae given in *CCIR Report 715*, which are valid for any type of transmission and for various polarization conditions (horizontal or vertical) and surfaces (water, terrains with a variety of conductivities and permittivities) can be simplified for troposcatter links. The diffraction loss, which is generally negative, is given by the following expression:

$$L_d(\text{dB}) = F(D_n) + G(H_A) + G(H_B) \qquad (6.23)$$

where

$$F(D_n) = 10 \log D_n - 17.6 D_n + 11$$

$$G(H) \approx 17.6(H-1.1)^{1/2} - 5\log(H-1.1) - 8 \qquad \text{for } H > 2$$

$$G(H) \approx 20 \log(H + 0.1 H^3) \qquad \text{for } 10 K_n < H < 2$$

$$G(H) \approx 2 + 20 \log K_n + 9 \log\left(\frac{H}{K_n}\right)\left\{\log\left(\frac{H}{K_n}\right) + 1\right\} \qquad \text{for } H < 10 K_n$$

$$D_n = 2.2\, d\left(\frac{f}{a^2}\right)^{1/3}$$

is the normalized path length and

$$H = 9.6\, h\left(\frac{f^2}{a}\right)^{1/3} \times 10^{-3}$$

is the normalized height. In the above equations d is the path length in kilometers, a is the effective radius of the Earth in kilometers, h is the antenna height in meters, f is the frequency in megahertz and K_n is the normalized factor for surface admittance. The maximum values for K_n at frequencies above 300 MHz are as follows: for vertical polarization

$$K_n = 4.5 \times 10^{-2} \qquad \text{on water}$$
$$K_n = 1.5 \times 10^{-2} \qquad \text{on land}$$

and for horizontal polarization

$$K_n = 2.0 \times 10^{-3} \qquad \text{always}$$

The first two expressions for $G(H)$ are normally used for any polarization and type of surface for frequencies above 300 MHz and antennas installed at heights greater than 20 m. The third expression is used in particular cases, for example when an antenna is installed on an island at a height of less than 20 m.

An example of such a calculation is shown in Table 6.5, in which we have varied first the height of antenna A and then the effective radius of the Earth.

Table 6.5 Calculation of path attenuation by smooth-earth diffraction (*CCIR Report 715-2*)

$$a = \frac{4}{3} \times 6400 = 8500 \text{ km}$$

$$\theta = \frac{d}{a} - \left(\frac{2h_A}{a}\right)^{1/2} - \left(\frac{2h_B}{a}\right)^{1/2}$$

Path frequency f (MHz)	450	450	450	450
Path length d (km)	135	135	135	135
Effective radius a of the Earth (km)	8500	8500	8500	7000
Height h_A of antenna A (m)	83	47	15	83
Height h_B of antenna B (m)	128	128	128	128
Diffraction angle θ (mrad)	6.0	7.1	8.5	8.4
Polarization (V, H)	V	V	V	V
Normalized factor K_m	—	—	0.045	—
Normalized path length D_n	5.46	5.46	5.46	6.22
Normalized height H_A of antenna A	2.29	1.30	0.41	2.45
Normalized height H_B of antenna B	3.53	3.53	3.53	3.77
Function $F(D_m)$ (dB)	−77.8	−77.8	−77.8	−90.5
Function $G(H_A)$ (dB)	10.8	3.6	−7.9	11.8
Function $G(H_B)$ (dB)	17.5	17.5	17.5	18.6
Diffraction loss L_d (dB)	−49.5	−56.7	−68.2	−60.1
Free-space loss				
$+20 \log f$ (dB)	+53.1			
$+20 \log d$ (dB)	+42.6			
+ constant (dB)	+32.4			
$= L_s$ (dB)	= 128.1	128.1	128.1	128.1
Path attenuation $L_p = L_s - L_d$ (dB)	177.6	184.8	196.3	188.2

6.9 Path loss in particular propagation conditions

Various anomalous propagation phenomena that may occur in a troposcatter path were mentioned in Section 4.8. In fact there may be periods in the year when propagation takes place by mechanisms different from pure troposcatter, for example diffraction (which may be dominant or may remain mixed with troposcatter propagation), specular reflection from elevated layers or ducting.

In the case of mixed troposcatter and diffraction propagation, which is not infrequent, the received signal may become severely distorted. The signal with the minimum delay, propagated by diffraction, will be received first, followed by the signal propagated by troposcatter with a longer delay and multipath dispersion. This mixed propagation mechanism may cause serious difficulties, particularly in digital transmission. The statistics of the "instantaneous" signal follow a Nakagami–Rice distribution (Nakagami n distribution) in which the transmitted energy is divided between the two mechanisms. In extreme cases the signal becomes either essentially constant, when diffraction propagation is dominant, or Rayleigh distributed, when troposcatter propagation prevails.

In other cases, for example paths over the sea, ducts may be formed for large percentages of time. The ducts behave as large waveguides in which the propagation loss tends to vary as $10 \log d$ instead of $20 \log d$. If both antennas are included in a duct, they will receive extremely strong and stable signals and will not experience any gain degradation. Ducts can trap energy at f MHz if they have thicknesses exceeding $5062 f^{-2/3}$ m [6.1]. Frequencies below 300 MHz can be trapped by ducts but higher frequencies are more easily trapped. However, if the antenna beam elevation is greater than a minimum value ($0.5°$ say), the trapped energy can escape.

Such anomalous propagation conditions may last for several hours. Radio receivers may be overloaded and severe distortion may occur. In addition, there may be overshoot problems causing interference with other radio links. There are no definite rules for predicting the minimum expected path loss and its frequency of occurrence in these cases, and each situation should be evaluated on the basis of experience with similar links in similar regions. For example, it has been observed that ducts with thicknesses of more than 100 m occur over the North Sea for about 0.7% of the year. Ducts are very common over warm seas (e.g. the Persian Gulf) and may be present continuously on summer days.

Hydrometeors may cause unwanted propagation of signals between a troposcatter antenna (normally radiating a high power) and antennas associated with other links in the region when their beams have a common volume which is located in the hydrometeor. Because of the high power radiated, coupling could also occur through sidelobes. Specular reflections and ducting may also cause interference with other radio links working at the same frequencies as the troposcatter link and located either in the region crossed by it or at great distances from it in the same direction or sometimes even laterally.

It is very difficult to evaluate the attenuation encountered by these unwanted signals, although the CCIR suggest some methods. However, when there is an effective risk of interference due to anomalous propagation conditions, precautions should be taken in the design of the link: repetition of frequencies within a sufficiently large area should be avoided, antennas with narrow beams and low sidelobes should be used, high transmission powers should be decreased during anomalous propagation etc.

6.10 Evaluation of path distortion and the multipath delay spectrum

After the path loss and its behavior have been evaluated, the second basic problem for the path design is to determine the distortion generated by multipath delays, or at least its effects. If it is difficult to obtain reliable predictions of path loss, it is even more difficult to predict the expected path distortion and its effects reliably. The predictions obtained using the current methods are generally approximate and semi-empirical.

Path distortion produces the following effects:

analog links—additional nonlinear noise due to multipath propagation;

digital links—additional intersymbol interference with a consequent increase in the BER, which could be more negatively affected by path distortion than by path loss.

The nonlinear noise power generated by multipath propagation in analog links is calculated in Chapter 8, Section 8.3.2, together with other nonlinear noise contributions.

For digital links it is important to evaluate, at least approximately, the multipath delay spectrum discussed in Chapter 4, Section 4.6, from which the multipath delay spread and the coherence bandwidth can be derived. We start by recalling that the energy following the lines tangential to the horizons is propagated with the minimum delay (in practice 1 ms per 300 km) and take this as the reference. The other rays exhibit a differential delay with respect to the reference. If we now refer to Fig. 3.2 and consider two identical antennas with a path beamwidth ω which is 0.6 times the 3 dB beamwidth, the differential delay between rays AFB and AEB is given by expression (3.14) divided by the velocity of light. It is shown in Fig. 6.11 as a function of the path length and the 3 dB beamwidth of the antenna for zero elevation angles at the terminal sites.

If a rectangular delay power spectrum (Sunde model) is assumed, the multipath delay spread (in seconds) is given by

$$L = 2\sigma = \frac{d}{2c}(\omega^2 + \omega\theta) \qquad (6.24)$$

where σ (in seconds) is the parameter defined in Chapter 4, Section 4.6, d is the path length in kilometers, c is the velocity of light (300 000 km s^{-1}), ω is the path beamwidth (in radians) of the two equal antennas and θ is the scattering angle in radians. This is a good approximation for small delay spreads with

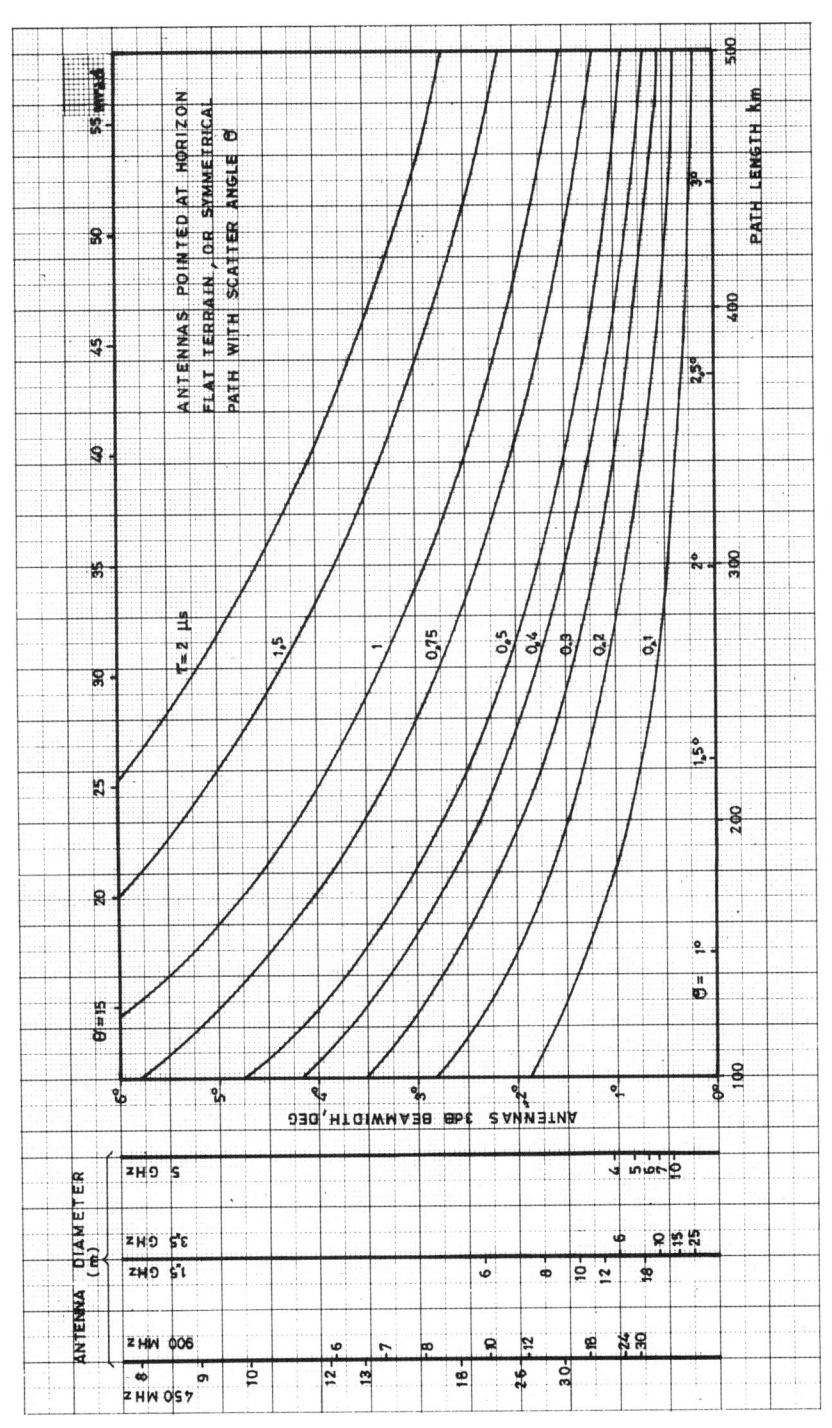

Fig. 6.11 Multipath differential delay in troposcatter paths.

large antenna gains. The corresponding coherence bandwidth can be taken as

$$W = \frac{0.7}{L} = \frac{1.4c}{d(\omega^2 + \omega\theta)} \tag{6.25}$$

If we assume a power spectrum due to a scattering contribution inside the common volume which decreases from the center of this volume according to a three-dimensional Gaussian law (Rice model), the delay spread and the correlation bandwidth become

$$L = 2\sigma = \frac{d}{c}\frac{\omega\theta}{3^{1/2}} \tag{6.26}$$

and

$$W = \frac{2c \times 3^{1/2}}{\pi\omega\theta d} \tag{6.27}$$

respectively.

A more realistic delay power spectrum can be calculated using an integral which takes the multipath geometry and the antenna patterns into account (Bello model) [6.3]. The calculation is made for all possible rays whose intensity depends on their position in the antenna patterns. The spectrum of the received power is obtained as a function of the differential delays. This spectrum is also the shape of the envelope of the received signal corresponding to an extremely short transmitted pulse (a sufficiently wide band is assumed for the system).

The computation starts from eqn (4.5) which is the expression of the power dW_s scattered by the volume element dV. The volume elements dV providing the same path delay form an ellipsoidal shell with foci at the two antennas. The approximate integral of dW_s over the fragment of shell contained in the common volume provides the following expression for the delay power spectrum:

$$Q(\delta) = \frac{1}{\delta^u} \int_r^s \frac{G(\varphi) G(\psi)}{x(x+1/x)^v} dx \tag{6.28}$$

In eqn (6.28) $Q(\delta)$ is the power density of the spectrum as a function of δ, $\delta = \Delta d/d = \Delta t/t = \frac{1}{2}\alpha\beta$ (see eqn (3.12)), t is the delay, δ_0 is the minimum value of δ, corresponding to horizon tangent rays, and is given by $d^2/8a^2$, where d is the path length (in kilometers) and a is the effective radius of the Earth in kilometers,

$$r = \left(\frac{\delta_0}{\delta}\right)^{1/2} \qquad s = \left(\frac{\delta}{\delta_0}\right)^{1/2}$$

$$u = 1 + \frac{m}{2} \qquad v = m - 2$$

where $m = 5$, according to measurements reported in ref. 6.3, or $m = 31.211 - 0.225d$ km, according to an improved Bello model [6.4],

$$G(\omega) = \exp\left\{-\left(\frac{\omega}{0.6\omega_0}\right)^2\right\}$$

is the approximate antenna pattern, ω_0 is the 3 dB beamwidths of the antenna,

$$\varphi = x(2\delta)^{1/2} - \frac{d}{2a}$$

and

$$\psi = \frac{(2\delta)^{1/2}}{x} - \frac{d}{2a}$$

Integral (6.28) can be evaluated numerically using a microcomputer. If the origin is taken at the start of the received impulse and the variable is the differential time delay τ, we can write $\delta = \delta_0 + \tau/t$ and substitute τ for δ in (6.28).

Curves calculated in this way and normalized to unit area are shown in Fig. 6.12. The multipath delay spread 2σ is the width of the received pulse. The Fourier transforms of these curves are shown in Fig. 6.13 and represent the

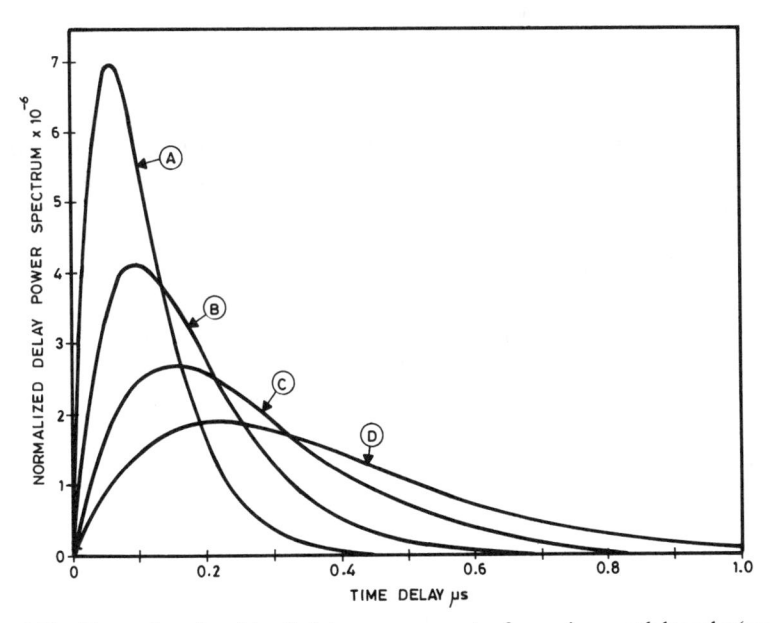

Fig. 6.12 Examples of multipath delay power spectra for various path lengths (carrier frequency, 850 MHz; antenna takeoff angle, 0°; parabolic dishes of diameter 18 m; effective radius, four-thirds Earth radius): curve A, 320 km; curve B, 400 km; curve C, 480 km; curve D, 560 km.

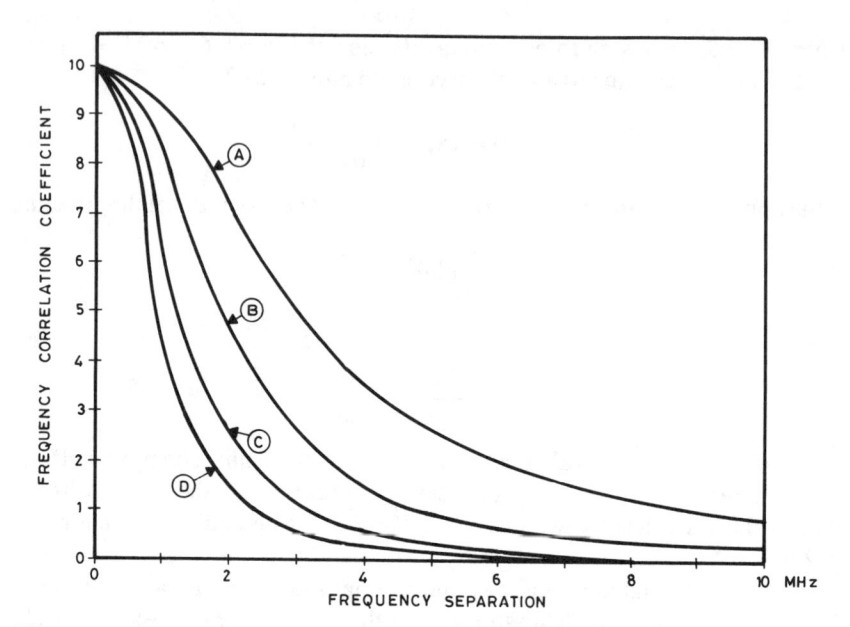

Fig. 6.13 Examples of frequency correlation functions (× 10) for various path lengths (carrier frequency, 850 MHz; antenna takeoff angle, 0°; parabolic dishes of diameter 18 m; effective radius, four-thirds Earth radius): curve A, 320 km; curve B, 400 km; curve C, 480 km; curve D, 560 km.

frequency correlation function of the path. It can be seen that the longer the path is, the larger is the spread and the narrower is the correlation bandwidth.

The above results are derived from the theoretical Bello model of the troposcatter path. Field measurements of the delay power spectrum are only in partial agreement with the theoretical predictions. The experimental curves are not always as regular as the theoretical curves, and the values of the multipath delay spread are statistically variable and may be double the theoretical value [6.5]. In general the delay spread is broader than the calculated value, and for 99.9% of the time it does not exceed twice its median value. It increases with increasing fade rate and short-term path loss, but there is no correlation with the median path loss. In contrast, the delay spread decreases in conditions of high atmospheric pressure. An improved Bello model, which uses a different value of m, appears to yield results which are in better agreement with experiment [6.4].

As longer paths use larger antennas with smaller beamwidths, simple geometrical considerations justify the fact that the measured spreads change very little with path length. For prediction purposes and in the absence of adequate data it can tentatively be assumed that the multipath delay spread is normally distributed and its 99.9% value is four times the value calculated using (6.28).

6.11 Comments and suggestions for further reading

The classic NBS note [6.1] can always be read with profit. The CCIR documents cited in the text, which include other references, should be read. The radiometeorological method is discussed in ref. 6.6. Diffraction theory is discussed in ref. 6.7. Reference 6.3, which is often cited in the literature, gives the approximate derivation of the formula for the delay power spectrum (which is not particularly difficult to follow) and can be read with profit.

References

6.1 Transmission loss predictions for tropospheric communication circuits, *NBS Tech. Note 101* (National Bureau of Standards, U.S. Department of Commerce), Vols 1 and 2, January 1967.
6.2 Yeh, L. P. Simple methods for designing troposcatter circuits, *IRE Trans. Commun. Syst.*, **8** (3), 1960.
6.3 Bello, P. A. A troposcatter channel model, *IEEE Trans. Commun. Technol.*, **17** (2), 130–137, April 1969.
6.4 Daniel, L. D. and Reinman, R. A. Modification of the Bello model for performance prediction of troposcatter links, *Int. Conf. on Communications (ICC 76), Philadelphia, PA, 14–16 June 1976*, pp. 46-24–46-26, IEEE, New York, 1976.
6.5 Cairns, J. B. S. A review of digital troposcatter links, *Conf. Publ. Telecom 79*, ITU, Geneva, 1979.
6.6 Battesti, J., Boithias, L. and Misme, P. Calcul des affaiblissements en propagation transhorizon à partir des paramètres radiométéorologiques, *Ann. Telecommun.*, **23** (5–6), 129–140, 1968.
6.7 Boithias, L. *Propagation des Ondes Radioélectriques dans l'Environnement Terrestre*, Dunod, Paris, 1984 (English translation (revised and updated): *Radiowave Propagation*, North Oxford Academic, London, 1987).

Chapter 7
Tropospheric Scatter Equipment

The various types of equipment used in troposcatter systems are dealt with in this chapter. Their general descriptions and characteristics are given. The most important equipment parameters used in system design are discussed in detail. The reader is assumed to have some familiarity with conventional LOS equipment.

7.1 Composition of a troposcatter terminal

All tropospheric scatter radio stations, both analog and digital, are arranged in an (explicit) diversity configuration. The order of diversity is generally 2 for the more economical systems and 4 for high quality systems. Block diagrams of typical dual diversity and quadruple diversity terminals are shown in Figs 7.1 and 7.2.

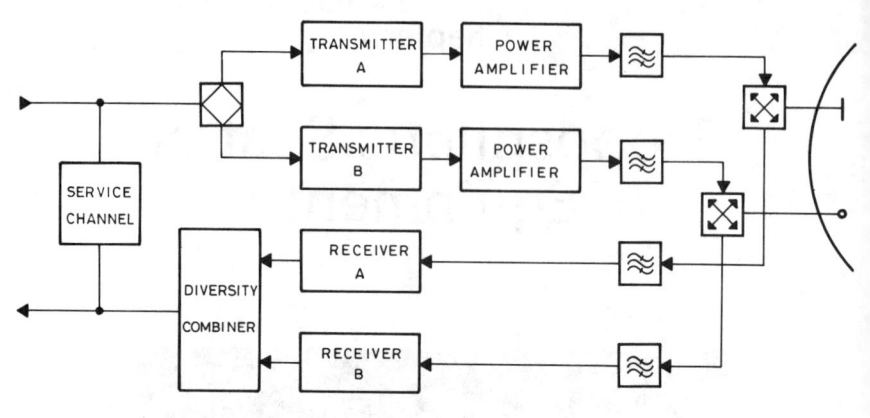

Fig. 7.1 Dual frequency diversity terminal.

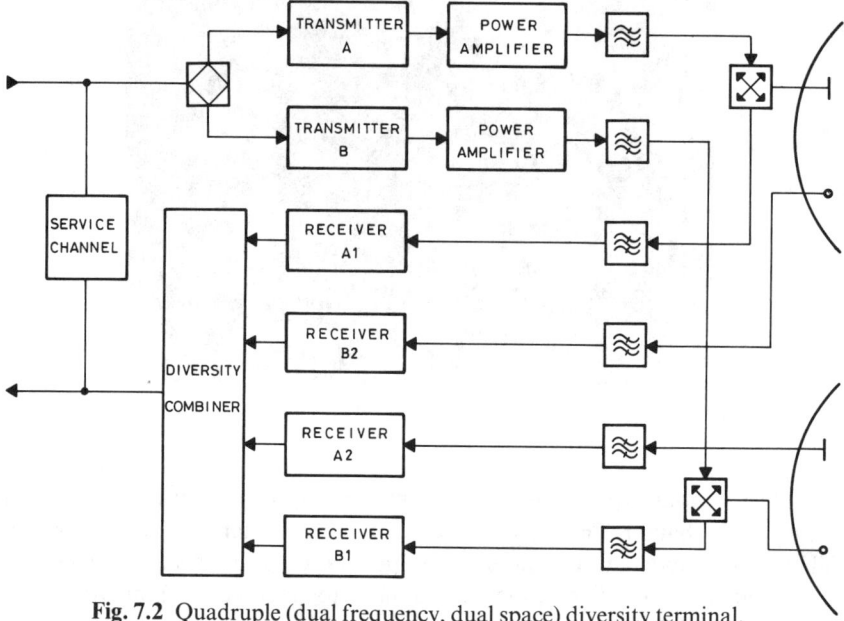

Fig. 7.2 Quadruple (dual frequency, dual space) diversity terminal.

Only one transmitter is necessary for dual space diversity, in which two antennas per terminal are required, but a second transmitter is often added to enhance reliability because it completes the duplication of the equipment and ensures that communications are not interrupted if the first transmitter fails. The second transmitter is either on hot standby or radiating. In this and other cases in which two transmitters are operating simultaneously at the same frequency, the two carriers must be perfectly synchronized in order to avoid beats in the received signal.

In an analog system a pilot tone (generally the CCIR pilot) is added to the baseband signal for monitoring the connection in all diversity branches and for the operation of the baseband combiner. One or more service channels, as well as optional supervisory channels, are added to both analog and digital systems for use by the operators. The composite signal is sent to both transmitters A and B. It is always possible to exclude one transmitter for alignment and maintenance purposes. Each transmission path includes a high power amplifier which raises the signal power to the required output value. In military systems it may be necessary to change the operating frequencies in the field in as short a time as possible. In this case frequency synthesizers, equipped with easily tunable power amplifiers, r.f. filters, receivers etc., are provided.

Transmitters and receivers are connected to the antennas through filters and duplexers which enable economic use of antennas and/or polarizations by connecting one transmitter and one or more receivers to the same feedhorn polarization. On the receive side the antennas are connected to the receivers through appropriate bandpass r.f. filters which eliminate unwanted out-of-band signals. There is a receiver for each diversity branch and the number of branches is equal to the order of (explicit) diversity. The feeder lines between the radio equipment and the antennas should be approximately of the same length so that the attenuation and the delay in the transmission are the same for all diversity branches. Otherwise the troposcatter receivers could need a delay equalizer in order to avoid intermodulation in the combined signal. The receivers differ from standard LOS receivers mainly with respect to the attention paid to minimizing noise, lowering the threshold and optimizing the carrier-to-noise ratio.

In digital reception with coherent demodulation provision is made for avoiding loss of synchronization during fading with complete loss of r.f. signal. The clock should also be maintained after interruptions of several seconds. The squelch circuit can be deactivated or set just below threshold, as its function is usually taken over by the diversity combiner. The outputs of the receivers are connected to the diversity combiner, which may be either the i.f. type or the baseband type. A switch or other method for progressively excluding the diversity branches and reducing the order of diversity (e.g. for maintenance purposes) is generally provided.

A diversity system always has some redundancy so that, if one of the branches fails, the communications are not interrupted but can proceed, perhaps with some degradation, through the other branches. However, some parts of the system may be single, as in the case mentioned above of the single transmitter in a dual space diversity terminal. Often, and particularly at unattended stations, all terminals are duplicated by the addition of redundant elements (a second transmitter in the above case). The unavoidable common parts are made of "passive" elements or of elements with a particularly low failure rate. The circuitry carrying the main information signal is generally duplicated, but this is not necessarily the case for circuits carrying secondary signals, such as the service channel.

The service and supervisory channels are extracted from the composite signal after combination. In more economical digital systems these channels are sometimes not added to the data stream but frequency modulate the r.f. carrier which is PSK modulated by the data.

An example of a completely solid state troposcatter terminal for use on oil platforms, including two transmitters, two solid state power amplifiers, four receivers, a quadruple diversity combiner, r.f. filters, a service channel and ancillary equipment, is shown in Fig. 7.3.

More sophisticated modulation–demodulation techniques can be applied, as we saw in Chapter 5, Section 5.8, to combat path distortion in the transmission of high data rates.

7.2 Transmitting system

The transmitting system is generally composed of two low power transmitters (also called "exciters") with output r.f. power of the order of 1 W, each followed by a high power r.f. amplifier which enhances the output power to the desired level. The normal power output is approximately standardized to values of 50 W, 100 W, 500 W, 1 kW and 10 kW. Different (lower or higher) values have also been used in some cases.

In analog systems FM with emphasis is used almost exclusively, primarily to maximize the output capability of the power amplifiers which generally operate in saturation. FM also enables the modulation index of the system to be optimized as we shall see later (Secion 7.4.4).

FSK and especially PSK modulations with two or four levels are used in digital systems. These systems require linearity of the amplifier stages to avoid distortion and spectrum smearing. Class C amplifiers are not very suitable for digital transmission.

The information signal (analog or digital) is normally added to the service channel and the supervisory signals used for the maintenance of the radio link itself. In analog systems the service signals are normally allocated to the lower part of the baseband, with the first channel in the physical band 0–4 kHz. However, if the receive system includes a baseband combiner, the physical band 0–4 kHz is reserved for the control signal of the combiner and the physical service channel is translated to higher frequencies.

7.2.1 Transmitters (exciters)

The transmitters (analog or digital) are almost the same as those used in conventional LOS radio links except for the more sophisticated cases described in Section 5.8. They are of the following types.

Fig. 7.3 Example of a 1.5 GHz troposcatter terminal (r.f. power, 50 W; quadruple diversity; 120 analog channel capacity; solid state amplifiers). The low power equipment, including two exciters, four receivers, a diversity combiner, r.f. filters, power supplies and ancillaries is in the left-hand cabinet. In the right-hand cabinet are two solid state power amplifiers with their power supply (50 W r.f. power) and ancillaries. (ARE, Castellanza, Italy.)

(a) *Multiplication type*: the information signal modulates a low frequency carrier which is then multiplied up to the operating frequency. This is an old technique and has the disadvantage of generating many strong spurious frequencies (as the multiplier stages require high signal levels) which cannot always be easily attenuated.

(b) *Conversion type*: the information modulates an intermediate frequency which is then up-converted to the operating frequency by a local oscillator. This is the most common system used and permits looping of modulator and demodulator at the intermediate frequency for testing, alignment etc.

(c) *Direct modulation type*: the main oscillator is a voltage-controlled oscillator directly modulated by the information signal and working at the operating frequency. This uses the least power and is the most economical. In digital transmitters with PSK modulation the main oscillator provides a fixed frequency which undergoes phase variations in a subsequent modulator (phase shifter). The greater simplicity is paid for by a wider r.f. spectrum, which arises because of the difficulty of shaping the spectrum with r.f. filters.

In analog transmitters the baseband signal goes through an optional emphasis circuit (which may or may not have CCIR characteristics) which improves the uniformity of the signal-to-noise ratio along the baseband. One or more service and supervisory channels are added at the bottom of the baseband together with a pilot tone (with or without CCIR characteristics) which monitors the continuity of the connection. The composite signal is then sent to the modulator, and the process continues as described above.

A difference in the mode of operation compared with corresponding LOS transmitters could be the use of a lower adjustable frequency deviation in the modulation process. This produces a narrower r.f. spectrum and optimizes the carrier-to-noise ratio at reception (see Secion 7.4.4).

In digital transmitters the input bit stream enters a multiplexing section in which the information signal, as well as the framing information, is added to the service channel(s) and the supervisory channel(s) streams. The composite signal, whose gross bit rate is slightly higher than that of the original bit stream (say by about 5%), is then scrambled to prevent the appearance of single lines in the radio spectrum and finally sent to the coder and the modulator. More sophisticated techniques involving special modems (Chapter 5, Section 5.8) may be used in digital systems of higher capacity.

7.2.2 Power amplifiers

A power amplifier is normally a separate apparatus. A class C amplifier, which is the most efficient, can be used in the transmission of a carrier modulated by an analog signal. However, if the modulation signal is digital, and particularly if PSK modulation is used, the power amplifier must be linear in order not to distort the signal, broaden the r.f. spectrum and create intersymbol interference. In this case the ideal amplifier is class A, but in practice class AB is also acceptable.

In the conversion of some old troposcatter links from analog to digital modulation the existing class C amplifiers have been retained by modifying their working points in order to linearize them as far as possible while accepting some degradation in the spectrum and performance. Class C amplifiers, although the most efficient, have power consumption and heat dissipation problems, and these problems are even greater with class A or class AB amplifiers in which all power not irradiated is dissipated.

At lower powers and lower frequency bands the amplifier can be completely solid state. By suitable series and parallel combinations of amplifying modules it is possible to construct amplifiers with output powers of 50–1000 W operating in frequency bands up to 1 GHz or more. Combination using hybrid circuits, circulators and isolators ensures that the amplifier itself remains operational if one or more modules fail. The output power simply decreases and there is no other damage. However, most power amplifiers still use tubes and are classified as follows.

(a) *Triodes or tetrodes*: these amplifiers are the simplest, the most rugged and the least expensive. They are also easy and cheap to maintain. They can be used in frequency bands up to 2000 MHz and are not very sensitive to power supply variations. However, the limited gain of these tubes (of the order of 10 dB) means that more than one stage is required to achieve the output power desired. This in turn allows the gain to be varied by the insertion or exclusion of stages. Also the bandwidth is limited and the efficiency is lower than for the tubes mentioned below, one of which may replace the complete multistage amplifier. This type is more suitable for low capacity and low cost systems.

(b) *Traveling-wave tubes* (TWTs): the power consumption is high and heat dissipation produces cooling problems because the tube works in class A. It is delicate and raises complex mechanical problems for mobile and transportable systems. However, it is satisfactory for digital transmission. It is wide band and is suitable for troposcatter military terminals in which frequent changes of operating frequencies may be required. Since its noise is also wide band, care must be taken with filtering to ensure that this noise is not picked up by the receivers at the terminal, thus degrading their performance.

(c) *Klystrons*: these are used most frequently although they are very expensive and have cooling problems. They generally work in class C. Klystrons with graduated knobs for tuning the cavities to different frequencies in the field are available. Output power is reduced by decreasing the input signal and increasing the dissipation. Reduction of the anode voltage would not be very efficient.

All power amplifiers have cooling problems. Air cooling (fans, blowers) is used for power up to 1 kW and water cooling is used for higher powers. Protection against the high voltages used (e.g. 7000 V in 1 kW klystrons) is also important. Interlocks are incorporated to cut off these voltages if the amplifier bay is inadvertently opened.

An example of a power amplifier is shown in Fig. 7.4. The power supply, which provides an anode voltage of up to 7 kV, is located at the bottom. The klystron (1 kW, 1.5 GHz, four cavities), which is focused by a permanent magnet, is located in the center. The pumped air cooling system can easily be

(a) (b)

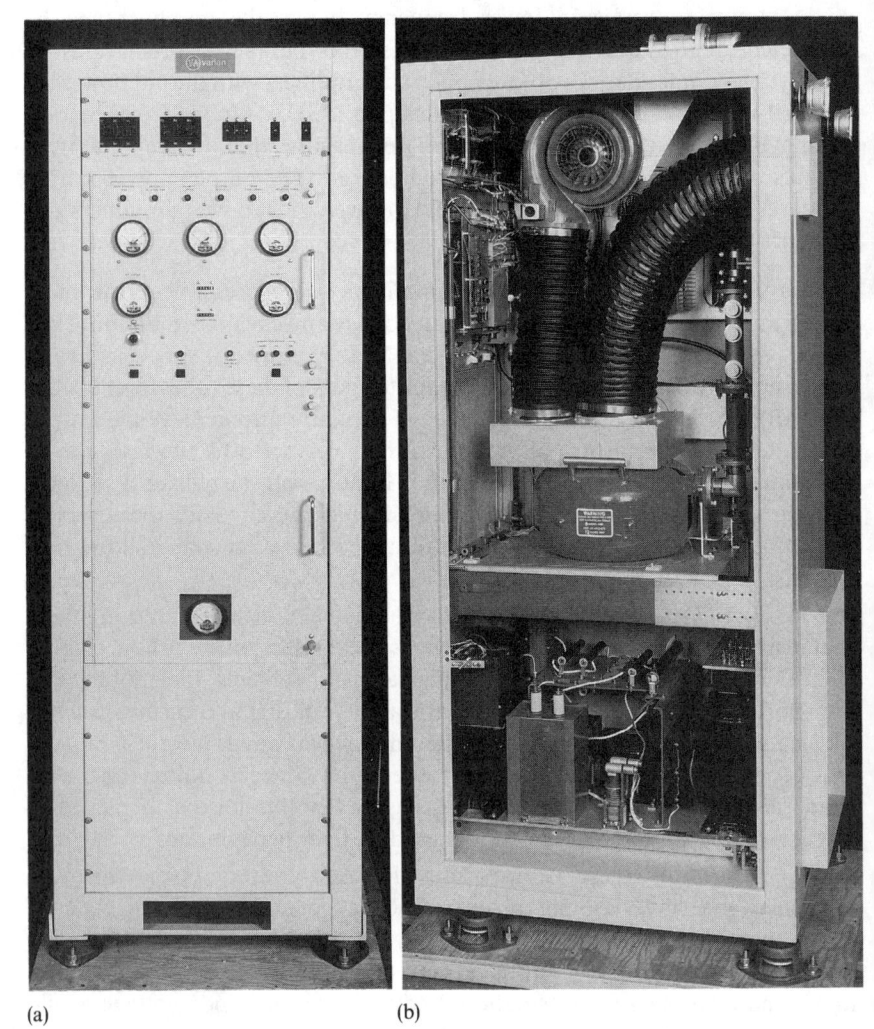

Fig. 7.4 Power amplifier (1.5 GHz, 1 kW): (a) external view; (b) internal view.

seen. The coaxial output is also visible: after an elbow there is a coaxial low pass harmonic filter followed by a stub section for impedance matching (the klystron output is strongly mismatched). The air inlets and outlets can be seen as well as the shock absorbers on the sides and bottom (the amplifier is for a mobile terminal).

7.3 Receiving system

The receiving system is generally composed of two (dual diversity) or four (quadruple diversity) radio receivers followed by a diversity combiner. The receivers may terminate at the i.f. level if a predetection combiner is adopted or at the baseband level if a postdetection combiner is used. These combiners were described in Chapter 5.

The diversity branches are almost independent up to the output of the combiner so that the equipment can be considered as fully duplicated (dual diversity) or quadruplicated (quadruple diversity); the only unavoidable common elements in the combiner are generally "passive" or have a very low failure rate.

If a baseband combiner is used there is duplication up to the output of the radio terminal, with a few minor circuit additions. However, after an i.f. combiner only a single limiter–discriminator–baseband chain, or the corresponding digital circuitry, is necessary. Therefore if duplication of the equipment is required for reliability reasons, it is necessary to add a second chain connected in some way to the first (e.g. with automatic switchover) so that the link is not interrupted if failure occurs.

After combination, the analog composed signal is subdivided into its components, i.e. baseband, service channel(s) and optional supervisory channels. The digital composed signal is first descrambled and then demultiplexed into its components.

We shall now first consider the general characteristics of the troposcatter receivers and then study in detail the input thermal noise of the receiving system which is a major problem affecting both analog and digital receivers.

7.3.1 Receivers

A conventional FM analog superheterodyne receiver is composed of the following parts.

(a) In the *front-end section* the received r.f. signal is first amplified by a low noise amplifier and then sent to a mixer which receives from another input an r.f. signal generated by a local oscillator. The output of the mixer is the intermediate frequency, which is the difference between the two radio frequencies.

(b) *An i.f. amplifier chain* amplifies the signal and includes an AGC which follows the amplitude variations of the signal and reduces their dynamics to a convenient level. An i.f. filter, which determines the receivers' bandwidth, is also included. Double-conversion superheterodynes include a second mixer and a second local oscillator, which generates a second lower intermediate frequency more suitable for operations such as filtering.

(c) *A limiter plus discriminator* cuts the signal to a fixed amplitude before it enters the discriminator (demodulator) where the modulating signal is recovered.

(d) In the *baseband section* the baseband, the service channel(s) and the pilot tone are separated, amplified, de-emphasized, filtered etc.

The corresponding conventional digital receiver is composed as follows:

(a) front-end as in point (a) above;

(b) an i.f. amplifier chain with i.f. filter and AGC;

(c) a digital demodulator, including clock recovery;

(d) a decoder and descrambler;

(e) a demultiplexer, which separates the data stream from the service channel and auxiliary streams and recovers the framing information;

(f) a line coder for the data stream (e.g. for HDB3 conventional coding, addition of AIS etc.).

Troposcatter receivers differ from the corresponding conventional LOS receivers in that they make use of some or all of the following:

(a) low noise input circuitry (the noise figure of the receiver (see Section 7.3.2) must be the minimum permitted by technology (today 0.5–3 dB generally); otherwise low noise preamplifiers (masers, parametric amplifiers) should be added before the input);

(b) a narrower i.f. bandwidth for minimizing the noise and thus improving the carrier-to-noise ratio;

(c) bandwidth compression and threshold extension techniques (see Section 7.4.1) for minimizing the noise bandwidth and lowering the threshold of the receiver;

(d) AGC curcuits with small time constants to ensure that they follow rapid fades;

(e) an emphasis–de-emphasis range that is sometimes larger than that recommended by the CCIR;

(f) an intermediate frequency that is sometimes different from the values recommended by the CCIR (35 and 70 MHz);

(g) a double-conversion superheterodyne, particularly for lower capacity receivers or for receivers in which the capacity can be changed by substituting i.f. filters (these are more easily realized at lower intermediate frequencies) (it is also convenient when using bandwidth compression receivers or some types of i.f. combiner or phase-lock detectors, all of which operate better at lower frequencies (see also *CCIR Report 939*));

(h) a greater dynamic range for the r.f.–i.f. signal;

(i) demodulator clock-recovery circuitry (in coherent digital receivers) provided with a flywheel operation for maintaining synchronicity after long fades;

(j) other features related to special adaptive modem techniques (see Chapter 5, Section 5.8) (these are particularly useful for higher capacity digital receivers).

We shall not describe all the above techniques, but aspects of particular interest to the system designer will be studied below.

7.3.2 Input thermal noise

The performance of a receiving system is measured by the noise level at its output. The output noise is generated by thermal or equivalent sources, which are conventionally considered to be present at the receiver's input, and intermodulation noise, which is present only when there is traffic in the channels and depends on the nonlinearities of the circuits. Only the first component is considered in this section. Note that the receiver's input is normally taken as the reference point for the received r.f. signal level, the noise level and the carrier-to-noise ratio.

It is known that the thermal noise power present in a frequency band B can always be written in the form KTB, where K is Boltzmann's constant and T is the "noise temperature" in kelvins. The noise power present at the input to the receiver is the sum of the following three terms.

(a) the equivalent noise power generated internally in the receiver.

It can be shown that an ideal receiver which generates no internal noise has an input noise power KTB where T is the ambient temperature in kelvins and B is the i.f. bandwidth of the receiver, or more exactly its noise bandwidth. A real receiver increases this minimum theoretical noise by a factor F_r (F_r is the noise figure of the receiver) when the ambient temperature is conventionally taken as $T_0 = 290$ K (17 °C). The real noise power is therefore $F_r K T_0 B$. The receiver noise temperature is then defined as $T_r = T_0 (F_r - 1)$ and the equivalent noise power generated internally can be written as $KT_r B$. The noise figure can also be written as $F_r = 1 + T_r/T_0$.

(b) the noise power within the band B of the receiver which is generated in the receiving feeder line.

This attenuates the signal by a factor L_f and has a noise figure L_f and a noise temperature $T_f = T_0(L_f - 1)$ at its input. This noise power at the output can be written $KT_f B/L_f$.

(c) the noise power received by the antenna within the band B of the receiver which is due to cosmic noise, the absorption of oxygen and water vapor, and the effective ground temperature.

When it reaches the receiver it is decreased by a factor L_f by the feeder line losses and can therefore be written $KT_a B/L_f$, where T_a is the antenna noise temperature, i.e. the average temperature of the sky and the ground covered by the antenna beam (note that the antenna sidelobes directed toward the ground contribute to an increase in T_a).

The overall input thermal noise N of the receiving system can therefore be written in the form

$$N = KT_e B = \frac{KT_a B}{L_f} + \frac{KT_f B}{L_f} + KT_r B$$

where T_e is the effective input thermal noise temperature of the receiving system. Dividing by KB gives

$$T_e = \frac{T_a}{L_f} + \frac{T_f}{L_f} + T_r \qquad (7.1)$$

The effective noise temperature T_e corresponds to an effective noise figure

$$F_e = \frac{T_e}{T_0}$$

$$= \frac{T_a/L_f + T_0(L_f - 1)/L_f + T_0(F_r - 1)}{T_0}$$

$$= \frac{T_a}{L_f T_0} + \frac{L_f - 1}{L_f} + F_r - 1$$

$$= F_r + \frac{T_a/T_0 - 1}{L_f} \qquad (7.2)$$

and the input thermal noise can be written

$$N = F_e KT_0 B \qquad (7.3a)$$

The above is a correct and complete calculation of the input thermal noise to the receiver. However, the simplified formula $N = F_r KT_0 B$, which actually represents only the noise of the receiver and disregards the contributions of the antenna and the feeder, is generally used. When (7.3a) is written in logarithmic units it becomes

$$N(\text{dBm}) = F_e(\text{dB}) + KT_0(\text{dBm}) + 10\log B(\text{Hz})$$
$$= F_e(\text{dB}) - 174 + 10\log B(\text{Hz}) \quad (7.3\text{b})$$

Consider for example a receiving system with the following characteristics: channel capacity, 60 telephone channels; noise bandwidth B, 2.4 MHz or 2.4×10^6 Hz; noise figure F_r, 3.2 (5 dB); antenna noise temperature T_a, 100 K; feeder line loss L_f, 2 (3 dB). The effective noise figure is

$$F_e = 3.2 + \frac{100/290 - 1}{2} = 2.87 \, (4.6 \text{ dB})$$

The equivalent overall input thermal noise of the receiving system is

$$N(\text{dBm}) = 4.6 - 174 + 10\log(2.4 \times 10^6) = -105.6 \, \text{dBm}$$

Note that F_e is less than F_r because in (7.2) T_a/T_0 is less than unity, and the last term is negative. The antenna, represented by its equivalent circuit (noise generator with a resistor in series), loads the receiver, represented by an ideal receiver fed by another noise generator with a series resistor. The impedance seen by the equivalent circuits is decreased, and this decreases the noise.

If the simplified formula had been used F_r, which is 0.4 dB higher than F_e, would have been used instead of F_e, giving an input thermal noise higher by the same amount. F_r is generally used in the calculations (as we shall do in subsequent chapters) because it is simpler and ignores the small gain obtained in the system. However, considerations of the cost per decibel (see Section 7.10), which is always high in a troposcatter system, encourage the designer not to ignore this small contribution.

The noise figure F_r of modern troposcatter receivers ranges from say 0.5 dB up to 6 dB or more, depending on the frequency band.

7.4 Design parameters for analog radio equipment

The parameters of the analog radio equipment which must be known for system design are as follows:

(a) the r.f. power available for transmission;

(b) the r.f. threshold level for the reception of telephony, and sometimes also that for the telegraphy carried over some telephone channels;

(c) the signal-to-noise ratio in the telephone channel band for various input r.f. signals;

(d) the noise performance of the overall system from baseband to baseband.

These and other associated parameters may of course be provided by the equipment manufacturer or can be directly measured, but some of them can also be calculated to a very good approximation. The transmitted power is a fixed parameter, although it can be made variable in steps as we shall see in Section 7.11 and does not need special study. The other parameters deserve attention and are discussed in detail below.

7.4.1 Analog receiver threshold for telephony

An FM or PM receiver has a threshold level. When the level of the received r.f. signal is above threshold, the output baseband noise decreases by 1 dB for each decibel increase in the received r.f. signal. In contrast, below threshold the output noise generally increases by 2–5 dB for each decibel decrease in the input r.f. signal.

The r.f. level corresponding to the threshold is normally taken as 10 times the input thermal noise,

$$t_d = 10N = 10F_e K T_0 B \qquad (7.4a)$$

or, expressing the parameters in decibels,

$$T_d(\text{dBm}) = 10 + F_e(\text{dB}) - 174 + 10 \log B(\text{Hz}) \qquad (7.4b)$$

The threshold level for the receiver considered at the end of Section 7.3.2 would be -95.6 dBm with the receiver inserted in the system. If the receiver is considered alone, as is usually the case, the threshold would be 0.4 dB higher, i.e. -95.2 dBm.

In troposcatter applications it is important to have a very low threshold. This can be obtained in the following manner:

(a) by using low noise devices, such as special mixers, tunnel diode amplifiers, very low noise preamplifiers etc., at the receiver input.

These devices can lower the threshold by several decibels.

(b) by using a parametric amplifier before the receiver.

A parametric amplifier is a delicate and complex apparatus which has an extremely low noise figure.

(c) by using a receiver provided with bandwidth compression which causes a threshold extension.

If the local oscillator follows the modulating signal so as to reduce the frequency deviation of the i.f. carrier, the i.f. bandwidth may be substantially reduced by a factor z, and therefore the equivalent input thermal noise and the threshold are reduced by a factor $10 \log z$ (in decibels). However, such a receiver is sensitive and complex, and there are physical limits to the possibility of compression.

Attention must be paid to the fact that most of these devices introduce intermodulation noise when the r.f. signal is high. In some cases they are switched on and off according to the variations in the r.f. signal, but this introduces further complications in the equipment.

7.4.2 Telegraph failure point (telegraphic threshold)

Telegraphic transmission on telephone multiplex channels can still take place when the received r.f. carrier is below the receiver's threshold defined above. This means that for an FM receiver the telegraphic threshold is lower than the telephone threshold. Normally, for FSK telegraphy, the telegraph failure point is reached when the voice frequency carrier-to-noise ratio is about 8 dB. The telegraphy threshold is different for the various telephone channels of the multiplex as will be seen below. Lower telephone channels in the baseband are less noisy but have a higher threshold, and vice versa for higher telephone channels. Therefore a voice frequency telegraph multiplex is generally allocated in an intermediate telephone channel. In practice the telegraphic threshold can be considered to be about 5 dB lower than the telephone threshold.

7.4.3 Thermal signal-to-noise ratio in a telephone channel

In an FM radio system the weighted thermal signal-to-noise ratio at the output of the top telephone channel can be calculated using the well-known formula

$$\frac{s}{n_v} = \frac{c}{n}\frac{B}{b}\left(\frac{\Delta f}{f_b}\right)^2 gp \tag{7.5a}$$

or, with the parameters expressed in decibels,

$$\frac{S}{N_v} = \frac{C}{N} + 10\log\left(\frac{B}{b}\right) + 20\log\left(\frac{\Delta f}{f_b}\right) + G + P \tag{7.5b}$$

where c/n (or C/N in decibels) is the r.f. carrier to equivalent thermal noise ratio at the receiver input, B is the i.f. bandwidth of the receiver, b is the bandwidth of a telephone channel (3.100 Hz), Δf is the r.m.s. deviation of the channel test tone, f_b is the top frequency of the baseband, g (or G in decibels) is the emphasis–de-emphasis gain and p (or P in decibels) is the psophometric weighting factor. Since at threshold $C/N = 10$ dB (see Section 7.4.1), for a 60 channel receiver we can put for example

$C/N = 10$ dB $\qquad B = 1.4 \times 10^6$ Hz $\qquad b = 3.1 \times 10^3$ Hz

$\Delta f = 50 \times 10^3$ Hz $\qquad f_b = 252 \times 10^3$ Hz $\qquad G = 4$ dB $\qquad P = 2.5$ dB and we obtain, at threshold for the top channel,

$$\left(\frac{S}{N_v}\right)_t = 10 + 10\log\left(\frac{1.4 \times 10^6}{3.1 \times 10^3}\right) + 20\log\left(\frac{50}{252}\right) + 4 + 2.5$$

$$= 10 + 26.6 - 14.1 + 4 + 2.5$$

$$= 29.0 \text{ dBp}$$

corresponding to a noise level of -29.0 dBm0p. A thermal signal-to-noise ratio curve for an FM receiver is shown in Fig. 7.5. It represents the signal-to-noise ratio in the worst channel, which is normally the top channel. At the r.f. level $T_d = 10 \log t_d$ corresponding to the threshold (see Section 7.4.1), the thermal signal-to-noise ratio in the top channel calculated using the formula given above has a value $(S/N_v)_t$. For an r.f. level m dB above threshold the ratio becomes (in decibels)

$$\frac{S}{N_v} = \left(\frac{S}{N_v}\right)_t + m$$

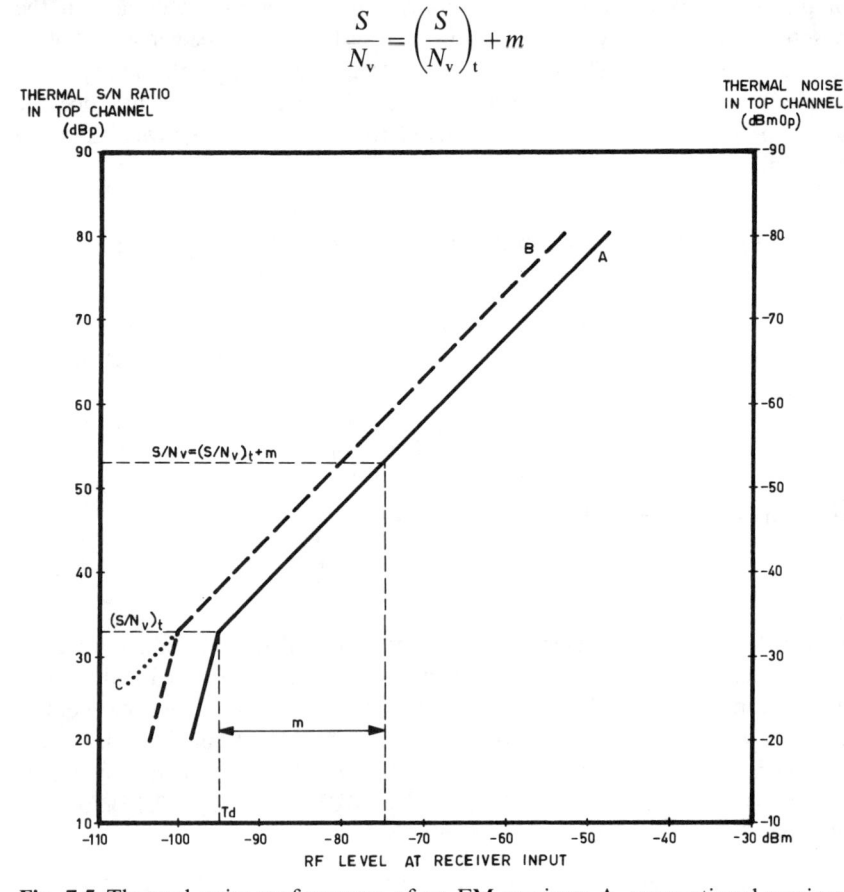

Fig. 7.5 Thermal noise performance of an FM receiver: A, conventional receiver; B, receiver with a low noise mixer or preamplifier; C, threshold extension.

TROPOSPHERIC SCATTER EQUIPMENT   171

The addition of a parametric amplifier or a low noise device to the receiver displaces the whole curve horizontally to the left because the threshold is lower, but $(S/N_v)_t$ remains the same. The result is an improvement in the signal-to-noise ratio at any r.f. level by an amount in decibels equal to the decrease in the threshold level. The addition of a threshold extension (bandwidth compression) device lowers the threshold and $(S/N_v)_t$ by the same amount, extending the straight line downwards.

The signal-to-noise ratio for the other telephone channels of the multiplex can be calculated from eqns (7.5) by substituting the corresponding value of f_b (the center frequency of the channel in the baseband) and G, which varies along the baseband (see *CCIR Recommendation 275-3*).

According to *CCIR Recommendations 398-3* and *399-3* it is not necessary to consider the signal-to-noise ratio of all channels. In practice, measurements are made in just a few "slots" in the baseband.

The signal level S in eqn (7.5) is constant as it depends on a constant frequency deviation Δf, and the r.f. noise power N in eqn (7.3) is also constant. The variables are the input r.f. carrier level and the output noise level at a zero reference point. This justifies the bottom and right-hand scales in Fig. 7.5. Note that above threshold the two variables have the same variations (in decibels) but with opposite signs.

7.4.4 Overall performance characteristics of analog radio equipment

The most important characteristics of analog radio equipment are summarized in Table 7.1. The frequency deviations for the channel test tone, the white signal load and the top modulating frequencies are those recommended by CCIR. The peak frequency deviation due to the multichannel signal is obtained by multiplying the channel frequency deviation by the traffic factor (CCITT). This is useful for calculating the approximately 40 dB width of the r.f. spectrum by application of Carson's formula, as is done in Table 7.1, column 8, where this width is assumed to be equal to the i.f. bandwidth of the receiver. The r.f. threshold and the signal-to-noise ratio at threshold in the top telephone channel are calculated using eqns (7.4) and (7.5) and are representative of the values obtainable.

Some compromises must be made in the design of analog troposcatter equipment. In order to ensure low input thermal noise, and therefore a low threshold, the i.f. bandwidth should be kept to a minimum. This means that the frequency deviation should be minimized so that the r.f. spectrum is narrow. However, a decrease in the frequency deviation results in an increase in the thermal noise in the telephone channels. It is good practice to make provision for varying the frequency deviation when the system is installed in order to find the value that minimizes the total noise in the telephone channels. Table 7.1

Table 7.1 Parameters of analog troposcatter equipment

Traffic capacity (channels)	CCIR r.m.s. deviation Δf (kHz)	CCIR load white signal[a] (dBm0 (ratio))	Peak factor white signal (dB (ratio))	Traffic factor T (dB (ratio))	Maximum deviation $T\Delta f$ (kHz)	Upper modulating frequency f_m (kHz)	I.f. bandwidth[b] (kHz)	R.f. threshold $10FKTB$ ($F = 5$ dB) (dBm)	Signal-to-noise ratio at threshold without emphasis (dBp)
12	35	3.3 (1.47)	13 (4.5)	16.3 (6.5)	230	60	580	−101.4	30.0
24	35	4.5 (1.68)	13 (4.5)	17.5 (7.5)	265	108	746	−100.3	26.0
36	35	5.2 (1.82)	13 (4.5)	18.2 (8.1)	285	156	882	−99.5	23.5
	50	5.2 (1.82)	13 (4.5)	18.2 (8.1)	405	156	1122	−98.5	27.7
48	35	5.7 (1.93)	13 (4.5)	18.7 (8.6)	300	204	1008	−99.0	21.8
	50	5.7 (1.93)	13 (4.5)	18.7 (8.6)	430	204	1269	−98.0	25.9
60	50	6.1 (2.02)	13 (4.5)	19.1 (9.0)	450	252	1404	−97.5	24.5
	100	6.1 (2.02)	13 (4.5)	19.1 (9.0)	900	252	2304	−95.4	32.7
	200	6.1 (2.02)	13 (4.5)	19.1 (9.0)	1800	252	4104	−92.9	41.2
	50	6.1 (2.02)	13 (4.5)	19.1 (9.0)	450	300	1500	−97.2	23.3
	100	6.1 (2.02)	13 (4.5)	19.1 (9.0)	900	300	2400	−95.2	31.3
	200	6.1 (2.02)	13 (4.5)	19.1 (9.0)	1800	300	4200	−92.8	39.8
120	50	7.3 (2.32)	13 (4.5)	20.3 (10.4)	520	552	2144	−95.7	19.5
	100	7.3 (2.32)	13 (4.5)	20.3 (10.4)	1040	552	3184	−94.0	27.2
	200	7.3 (2.32)	13 (4.5)	20.3 (10.4)	2080	552	5264	−91.8	35.5

[a] *CCIR Recommendation 393-4*.
[b] $B = 2(T\Delta f + f_m)$.

gives an idea of the variation in some parameters with the deviation and may help in the choice of suitable values at the design stage.

The performance characteristics are mainly based on the total noise generated in the equipment which is the sum of two main components: thermal noise, which is discussed above, and intermodulation noise, which is present only during traffic and is due to nonlinearities in the equipment. Although the thermal noise, the signal-to-noise ratio etc. can be calculated as shown above, the intermodulation noise is a parameter of the transmit–receive equipment that can be obtained only by direct measurement.

In performance evaluation it is important to consider the total noise curves of the radio equipment which are of the type shown in Fig. 7.6. They are measured from baseband to baseband on the complete radio terminals connecting the transmitters to the receivers through variable attenuators. The threshold in the top channel and the signal-to-noise ratio in three slots (corresponding to a low, a medium and a high telephone channel) in the linear range can be calculated as discussed above and normally agree with the measured values to within a fraction of a decibel. Note that the threshold line is not exactly a vertical segment yielding the same threshold value for each

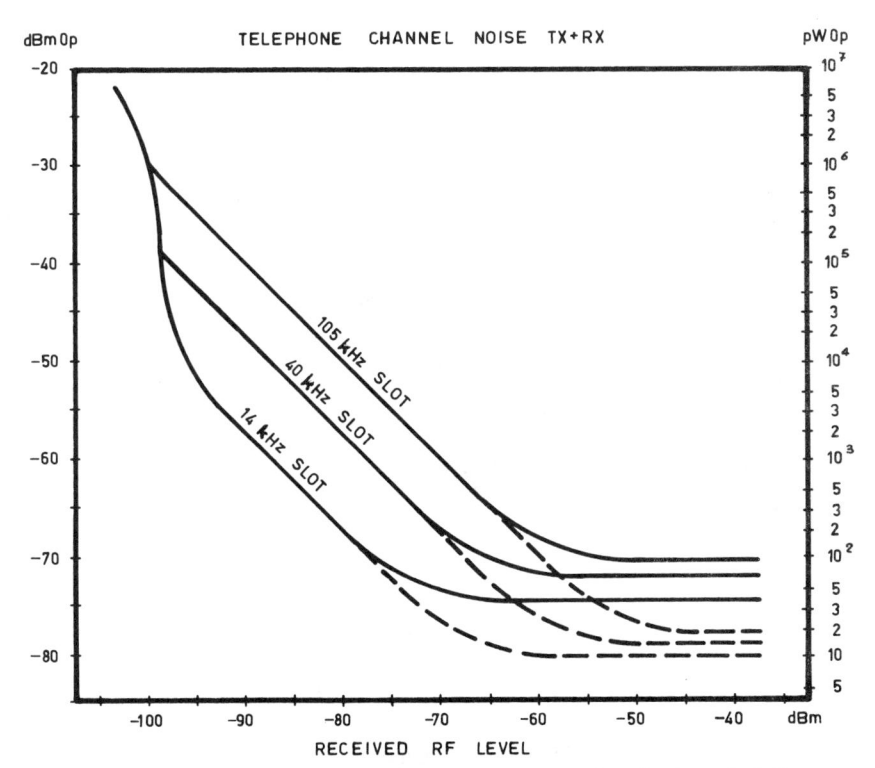

Fig. 7.6 Noise curves of a 24 channel troposcatter system (emphasis excluded): ———, total noise (CCIR load); - - -, thermal noise.

channel or slot but is slightly S-shaped, so that lower channels have a slightly higher threshold.

In practice this line is obtained by measurement. At the highest values of the received signal the noise becomes independent of the r.f. level and the curves become horizontal. This is the zone in which the equivalent thermal noise is a minimum (equal to the background noise), and it is easy to derive the equipment intermodulation noise by measuring the total noise and the background noise and taking the difference between the two (normally the intermodulation noise does not depend on the r.f. level).

For given r.f. signals, the variation in the noise level versus the variation in the frequency deviation is as shown in Fig. 7.7. The thermal noise alone is

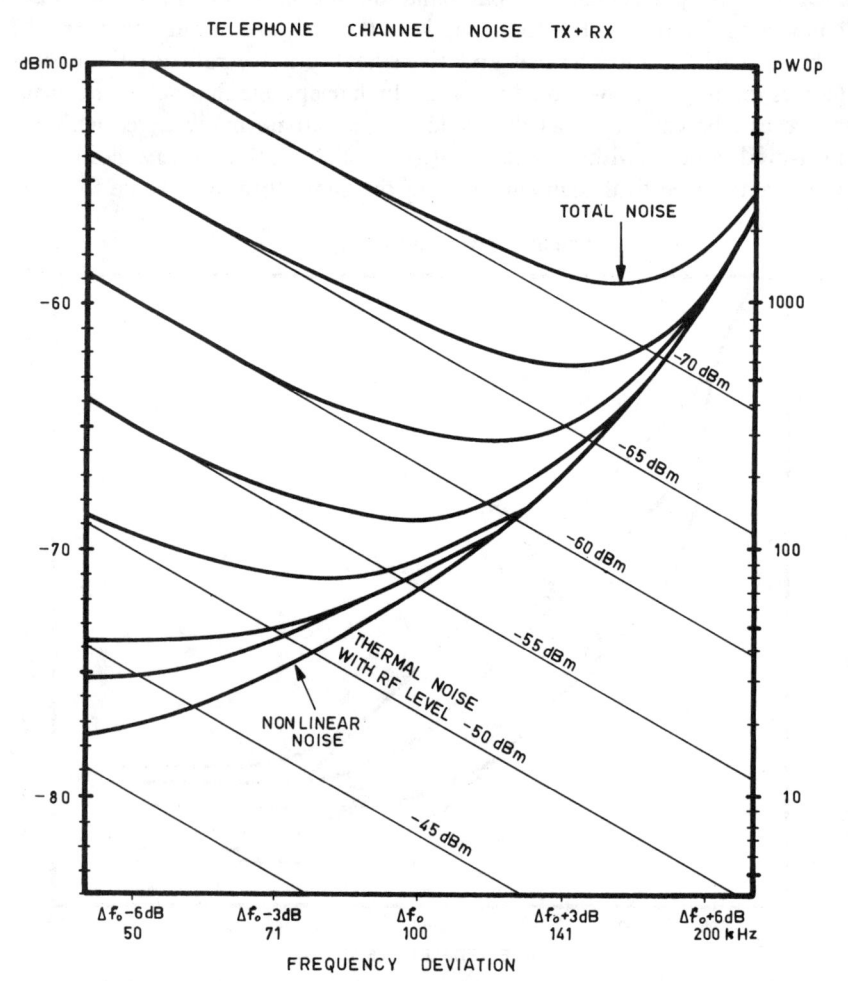

Fig. 7.7 Total noise curves in the upper telephone channel (600 kHz with emphasis) of a 120 channel troposcatter system with CCIR load.

represented by a set of straight lines, each corresponding to a particular r.f. level, while the nonlinear noise alone is roughly parabolic. The combination of the two yields a set of total noise curves with a minimum value which depends on the frequency deviation and the r.f. level. A corresponding set of curves in a different representation is shown in Fig. 7.8.

The effect of emphasis is to bring all the curves of Fig. 7.6 closer to the central curve, thus improving the noise characteristics of the top channels at the expense of that of the lower channels. Improvements are generally obtained in the intermodulation noise also. Recommended emphasis characteristics are specified in *CCIR Recommendation 275-3*. However, a much stronger emphasis is sometimes used in troposcatter radio equipment.

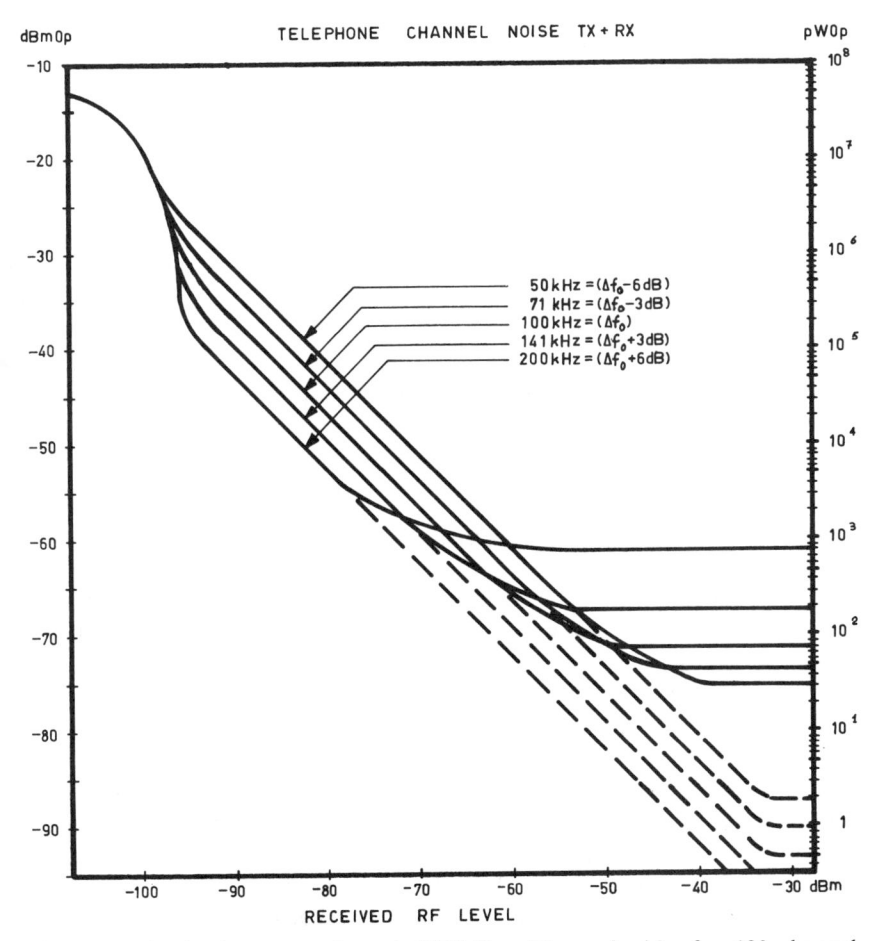

Fig. 7.8 Noise in the upper channel (600 kHz with emphasis) of a 120 channel troposcatter system: ———, total noise (CCIR load); - - -, thermal noise.

7.5 Design parameters of digital radio equipment

The main parameters needed for the design of digital radio equipment are the r.f. power available for transmission, the width of the i.f. receive filter, the duration of both the bit and the symbol, and the digital threshold which is derived from the BER curves. It is possible to calculate some of these parameters theoretically, but the actual r.f. spectrum, the bandwidth of the i.f. filter and the BER curve should normally be measured or obtained from the manufacturer.

In the next section we shall discuss the BER curves for the types of modulation used in troposcatter links, which are as follows.

(a) 2FSK frequency modulation or frequency-shift keying with two levels (actually two frequencies).

> The only practical system available uses limiter–discriminator detection in which the signal, continuously modulated by the two tones, passes through the i.f. filter and is demodulated incoherently. Many old troposcatter links of the analog type (with frequency modulation) which have been transformed into the digital mode use this technique, and it has also been used in new low capacity equipment. The width of the modulated carrier spectrum can be minimized by choosing the optimum frequency deviation. It may also be possible to alter the duty cycle for performance optimization. The drawback is the threshold, which even in the best case is several decibels higher than that of 2PSK modulation.

(b) 2PSK and 4PSK phase modulation or phase-shift keying with two or four levels (actually two or four carrier phases).

> The r.f. spectrum is sufficiently compact, and the threshold reaches the minimum values obtainable with digital systems. In order to reconstruct the data without the risk of bit inversions after demodulation they are coded differently (only the polarity variation of successive bits is transmitted), but this solution doubles the errors and the BER.

Other types of modulation are not generally used for various reasons. For example, 2DPSK requires almost as much hardware as 2PSK but has a degraded threshold. Multilevel modulations with more than four levels (as used in LOS digital links with 8, 16 or more levels) are unsuitable for troposcatter because of their low immunity to thermal noise, impulse noise, phase jitter and nonlinear distortion, together with their more complex hardware.

7.5.1 Bit error rate curve of a digital system

The performance of a digital system is given by the curve of the BER at the output of the receiver versus the level of the r.f. modulated carrier at the input. The theoretical BER curves provided for the types of modulation of interest are given as functions of the input carrier-to-noise ratio, and are as follows. For 2FSK frequency modulation

$$P = \tfrac{1}{2}\exp(-\gamma) \qquad (7.6a)$$

of which the inverse is

$$\gamma = -\ln(2P) \qquad (7.6b)$$

with direct coding–decoding, incoherent modulation–demodulation (limiter-discriminator), two modulation levels, a single i.f. filter containing the two tones separated by $2\Delta f$ (Δf is also the peak deviation) and an optimized peak frequency deviation Δf (equal to the r.m.s. frequency of the noise spectrum). The signal spectrum is optimized with a width B_0 when $2\Delta f = 0.65 B_0$ and B_0 is equal to the bit rate. For 2PSK and 4PSK phase modulation

$$P = \operatorname{erfc} \gamma^{1/2} \qquad (7.7)$$

with differential coding–decoding (which doubles the BER), incoherent modulation (transitions on any phase), two or four modulation levels and coherent demodulation. The symbols in eqns (7.6) and (7.7) are defined as follows: P is the BER and γ is the input (modulated) carrier-to-noise power ratio C/N. From (7.3a) $N = FKTB$ and therefore

$$\gamma = \frac{C}{F_r K T B} \qquad (7.8)$$

where C is the average power of the modulated carrier, F_r is the noise figure at the receive input, $KT = -174\,\mathrm{dBm}$ at conventional ambient temperature and B is the 3 dB noise bandwidth of the receiver (in practice the i.f. bandwidth). The erfc function is defined in Chapter 2, Section 2.2.1. It can be shown theoretically that eqn (7.6) is also valid for 2DPSK modulation.

Equations (7.6) and (7.7) are shown in Fig. 7.9 and are valid in ideal or nearly ideal conditions of filtering and sampling for a single bit or for a bit stream in which the intersymbol interference is ignored. Curve 1 of Fig. 7.9 is almost the same as curve 2 shifted 0.5 dB to the left. As the equation of curve 2 is simpler, we can use it for our calculations and transfer the results to the 2PSK and 4PSK case simply by decreasing γ by 0.5 dB.

The BER is minimum in matched-filter conditions, i.e. when the ideal filter equivalent to the set of transmit and receive filters (baseband, i.f. and r.f.) shaping the signal and the noise has a response curve maximizing the signal-to-noise ratio. In order to compare the various modulation systems it is usual

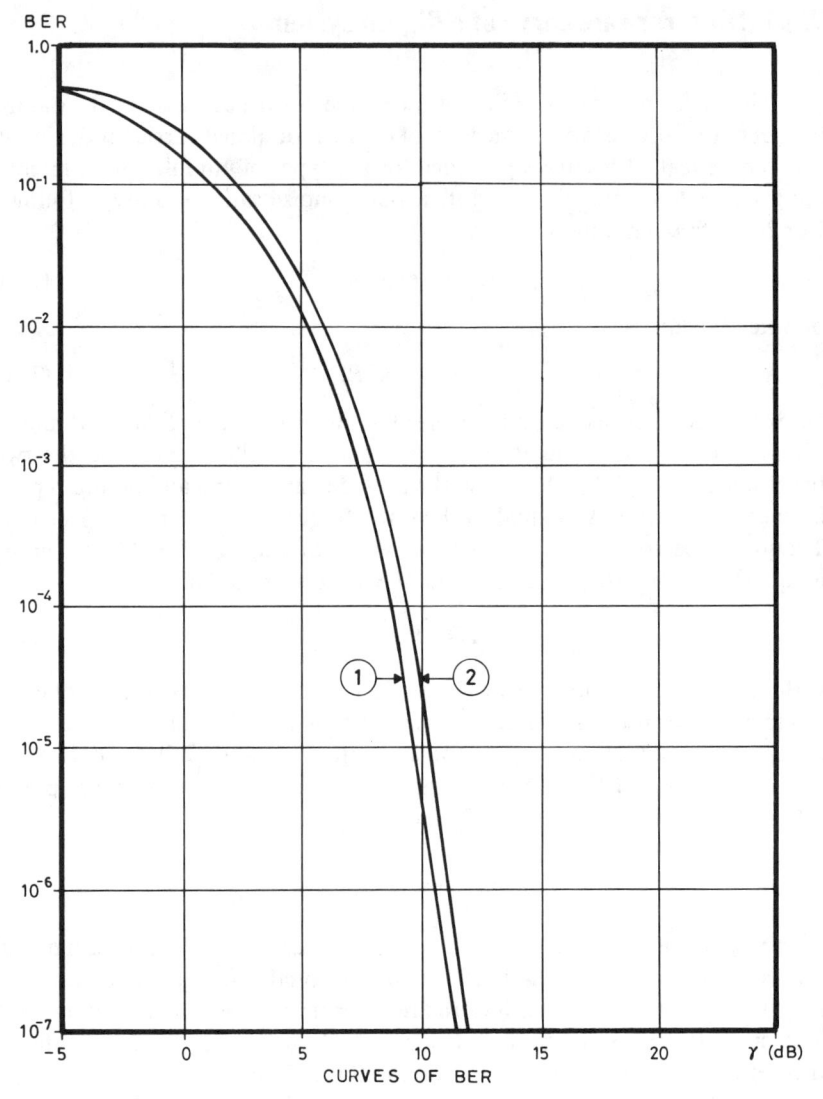

Fig. 7.9 BER curves for digital systems with different types of modulation: curve 1, 2PSK or 4PSK with differential coding–decoding (BER = erfc $\gamma^{1/2}$); curve 2, 2FSK with a single filter and optimized frequency deviation (BER = $\frac{1}{2}\exp(-\gamma)$) or 2DPSK.

to refer to the normalized carrier-to-noise ratio, which is measured in a band B_0 and is given by

$$\gamma_0 = \frac{C}{F_r KTB_0} = \frac{E}{\eta} \tag{7.9}$$

where E is the energy of a single bit, i.e. the mean power of the signal C divided by the bit rate B_0 ($E = C/B_0$), and η is the factor $F_r KT$ which, when multiplied by a 3 dB bandwidth B_0 (the numerical value of the bit rate), yields the noise power $F_r KTB_0$ in the band B_0. The relation between the two carrier-to-noise ratios is

$$\gamma = \gamma_0 \frac{B_0}{B} \tag{7.10}$$

In the real conditions of filtering, sampling, presence of intersymbol interference etc. there is a degradation in the optimum signal-to-noise ratio and in the BER. To account for this degradation we should substitute for γ in (7.6) and (7.7) a term $h\gamma$ with $h < 1$. The result for the BER is the same as having a lower carrier-to-noise ratio.

The curves $P = P(C)$ used in practice (C in dBm) are the same as those of Fig. 7.9 with the horizontal scale γ(dB) substituted by a scale C(dBm) derived from (7.8) and including the term h:

$$C(\text{dBm}) = \gamma(\text{dB}) + F_r(\text{dB}) + 10\log B + h(\text{dB}) - 174 \tag{7.11}$$

There is uncertainty in the value of the term h. The curves of the BER for the equipment being used should be directly measured from baseband to baseband on the complete radio terminals connecting the transmitters to the receivers through variable attenuators.

The above theoretical considerations are only meant to ensure that the chosen equipment is reasonably well designed and manufactured. The differences between the measured values and the theoretical minima are normally of the order of a few decibels (say up to 4–5 dB).

We conclude by observing that the BER curve of a digital system is very steep below some maximum value taken as a threshold which is not to be exceeded (e.g. BER = 10^{-4}), so that the BER vanishes for an increase of a few decibels in the carrier-to-noise ratio (see Fig. 7.9). This introduces the next section.

7.5.2 Threshold of a digital receiver

When the r.f. signal received by an individual receiver is above a certain threshold almost no bit errors occur. When the r.f. signal is below this threshold the BER drops to 0.5 which means that the bits are completely random.

With the receivers normally used in troposcatter equipment the passage from one condition to the other occurs in a range of a few decibels. Typically in a 2 dB range the BER can drop from 3×10^{-6} to 3×10^{-4}, which means from one error every 5 s to 20 errors s^{-1} in a 64 kbit s^{-1} transmission. Furthermore, the rate of variation in the level of the r.f. received signal in the presence of deep fades is normally very high, reaching tens of decibels per second, so that the threshold range is crossed in negligible time.

As the actual threshold depends on the r.f. level as well as on its rate of variation, it is natural to define a mean threshold value of the input r.f. level above which the transmission can be considered to be error free and below which it is unacceptable. This corresponds to substituting the actual BER curve of the individual receiver by an ideal step curve as shown in Fig. 7.10. The outage period is therefore equal to the duration of the r.f. signal below this threshold.

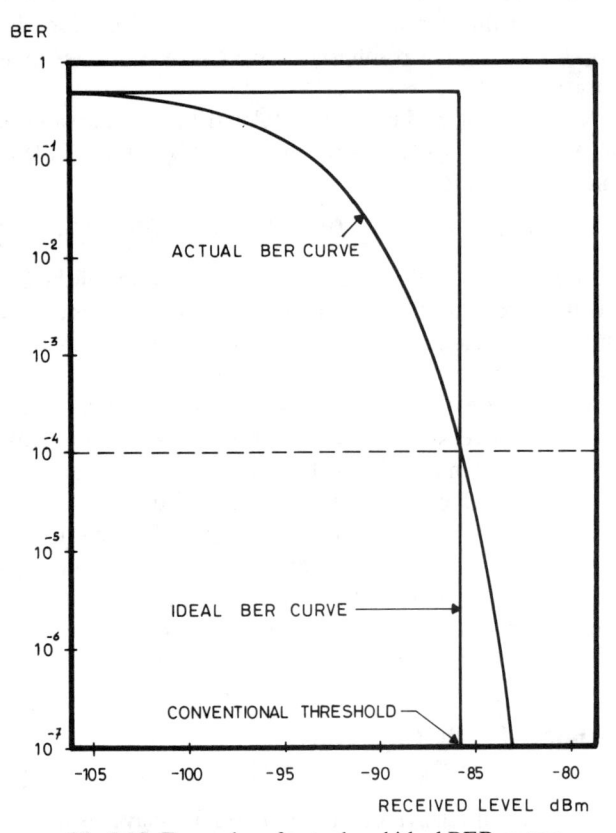

Fig.7.10 Examples of actual and ideal BER curves.

After accurate study DCS has proposed the input steady level (no fading present but only Gaussian noise) corresponding to a BER of 1×10^{-4} as the threshold. This is rather conservative for troposcatter, and values corresponding to 10^{-3} or less may be acceptable. Table 7.2 shows the values of the threshold level for some digital systems. These values have been calculated from (7.6), (7.7) and (7.11) for $P = 10^{-4}$, $F_r = 2.5$ dB, $h = 0$ and $B = B_0$. Thresholds measured on actual radio systems will be higher than those shown in Table 7.2 by some 2–5 dB.

TROPOSPHERIC SCATTER EQUIPMENT 181

Table 7.2 Radio frequency normalized threshold of some digital radio systems

Nominal bit rate B_0 (Mbit s^{-1})	Bit duration $1/B_0$ (µs)	Threshold (dBm) 2FSK or 2DPSK	2PSK or 4PSK
0.256	3.91	−108.1	−108.6
0.512	1.95	−105.1	−105.6
0.704	1.42	−103.7	−104.2
1.024	0.98	−102.1	−102.6
1.544	0.65	−100.3	−100.8
2.048	0.49	−99.1	−99.6
3.138	0.32	−97.2	−97,7
4.096	0.24	−96.1	−96.6
6.276	0.16	−94.2	−94.7
8.448	0.12	−92.9	−93.4
9.414	0.11	−92.5	−93.0
12.552	0.08	−91.2	−91.7

The thresholds are calculated from eqns (7.6), (7.7) and (7.11) for $P = 10^{-4}$, $F_r = 2.5$ dB, $h = 0$ and $B = B_0$. Thresholds measured on actual receivers may be 2–5 dB higher than the above theoretical normalized values.

7.5.3 Overall performance characteristics of digital radio equipment

The number of transmissible telephone channels is related to the nominal bit rate and to the standard types of digital multiplex available on the market. Table 7.3 shows the traffic parameters of some of this equipment. PCM and delta multiplex yield different qualities of transmission: the CCIR recommends PCM multiplex with analog–digital conversion at 64 kbit s^{-1}, but it does not recommend any type of delta multiplex which is much more suitable for military communications where it is widely adopted.

As a result of the choice we have a nominal bit rate, from which we can calculate the bit duration. In order to evaluate path distortion it is necessary to know the duration of the transmitted symbol. The data stream in the transmitter is added to other digital information (service channels, framing), and at the r.f. output the input nominal bit rate becomes a gross bit rate which is generally higher by a few per cent (say 5%, and never more than 10%). In a two-level modulation (e.g. 2FSK) the symbol duration and the bit duration are the same. In a four-level modulation (e.g. 4PSK) the symbol duration is double the bit duration. Table 7.2 gives the nominal bit durations, which are to be decreased by a percentage related to the gross bit rate.

It should also be possible to determine the width of the r.f. spectrum, the i.f. bandwidth of the receiver and therefore the threshold depending on the type of modulation and the gross bit rate. These parameters are very variable and

Table 7.3 Traffic parameters of some digital multiplex equipment

Nominal bit rate (Mbit s^{-1})	Number of voice channels	Analog–digital conversion Type	Rate (kbit s^{-1} channel^{-1})
0.256	14	Delta	16
0.512	14	Delta	32
0.512	30	Delta	16
0.704	10	PCM	64
1.024	30	Delta	32
1.024	60	Delta	16
1.544	23	PCM	64
1.544	46	Delta	32
1.544	90	Delta	16
2.048	30	PCM	64
2.048	60	Delta	32
2.048	120	Delta	16
3.138	46	PCM	64
3.138	90	Delta	32
4.096	60	PCM	64
6.276	92	PCM	64
8.448	120	PCM	64
9.414	138	PCM	64
12.552	184	PCM	64

therefore cannot be calculated easily as in the case of analog equipment. However, if B_0 is the numerical value of the gross bit rate in kilobits per second, the r.f. bandwidth (in kilohertz) can be estimated as follows: $1.3B_0$ kHz for 2FSK modulation; $1.4B_0$ kHz for 2PSK modulation; $1.0B_0$ kHz for 4PSK modulation. The value of the threshold can be obtained from these values. However, for exact calculations during system design, it is necessary to rely on measured values.

7.6 Typical troposcatter radio equipment

The radio equipment used for troposcatter links is a specialized and improved version of that used in conventional LOS links. We have already discussed the characteristics of the equipment from the point of view of system design. We shall now examine the specific features of troposcatter radio equipment and complete the information with a description of the physical features and other typical characteristics.

Apart from internal differences aimed at minimizing the bandwidth, the overall noise, the threshold, the BER etc., the main visible external differences with respect to conventional LOS equipment are as follows:

(a) the addition of high power amplifiers in transmission;

(b) the addition of low noise amplifiers in reception (however, today's technology enables low noise figures to be achieved by using suitable semiconductor devices);

(c) diversity reception (utilizing two or more receivers and a diversity combiner, generally of the maximal-ratio type);

(d) special modems for high digital rate transmission;

(e) much higher power consumption (e.g. 13 kV A for the terminal of Fig. 7.2 for 1 kW r.f. output power);

(f) the addition of cooling systems with ambient air inlets and hot air outlets for high power amplifiers.

The radio equipment is normally contained in racks or bays similar to those used in LOS links. Normally one bay contains all the low power equipment, including transmitters (exciters) up to an r.f. power of say 1 W, receivers, the diversity combiner, service channel(s), supervisory equipment and all related power supplies. The r.f. power amplifiers and their power supplies are normally contained in individual bays complete with cooling systems and various types of protection (against excessive r.f. reflected power from the antenna, overvoltages, overcurrents, overheating, risk of contact with high voltages etc.).

Different packing may be required for special applications (military, mobile). In mobile systems the equipment should be capable of withstanding vibrations, shocks and bumps, and it is normally mounted on suitable shock absorbers.

The branching system for connecting the equipment to the antennas is composed of filters and duplexers which can be included in the bays or installed externally.

We now examine separately some specific features of analog and digital equipment.

7.6.1 Analog radio equipment

Analog radio equipment is now classic and standardized, and little more need be added to what has been said so far. Frequency modulation is normally used, as in conventional radio links. Channel capacities, CCIR frequency deviations, upper modulating frequencies, i.f. bandwidths, r.f. thresholds, signal-to-noise ratios at threshold and other parameters for use in system design are given in Table 7.1.

The equipment normally includes one or more service channels in the sub-baseband (below the baseband). It is not convenient to use a physical service channel (0.3–4 kHz) in a troposcatter terminal with a baseband diversity combiner as its band is occupied by the control signal of the combiner, which follows the fades and has its spectrum just in that band. Limiting this spectrum to below 0.3 kHz means that the response of the combiner to fast fades becomes too slow. The lowest service channel is therefore translated to, for example, the band 4–8 kHz.

7.6.2 Digital radio equipment

Digital troposcatter has undergone further developments, starting with conventional digital techniques and continuing over the last decade with detailed studies of the special adaptive techniques mentioned in Section 5.8 for overcoming path distortions. The distortions, which are characterized by the time delay spread 2σ of the multipath transmission, are related to the duration T of the transmitted symbols through the dimensionless *path dispersion parameter* $2\sigma/T$. The performance of the equipment in the presence of path distortion is given in terms of BER versus $2\sigma/T$ for any given received r.f. level (Chapter 9, Section 9.1.2). A good performance accepts high values of the parameter without introducing intolerable degradation of the BER.

The techniques of conventional digital LOS systems can be exploited in low capacity equipment when the path distortion is negligible and the dispersion parameter is very low. When path distortions appear which are not serious enough to require adaptive modems or similar devices, specially designed forward error-correcting equipment can be added at the two ends, yielding a kind of time diversity which improves the transmission quality in a relatively simple way (see Chapter 5, Section 5.7). More sophisticated adaptive technology (Chapter 5, Section 5.8) is reserved for high capacity digital systems used for high quality transmission in the medium to long range, where the dispersion parameter reaches high values (even greater than unity). However, for reasons of economy several old analog troposcatter links are being re-used and converted to digital systems. Therefore we can subdivide the types of equipment used in digital troposcatter links as follows.

7.6.2.1 Conventional digital equipment

The transmitters and receivers are basically of the same types as those used in LOS radio links, with the following additional characteristics.

(a) The receivers have a low noise figure (say 0.5–5 dB).

(b) The diversity combiner is generally of the maximal-ratio type and predetection combination is used.

(c) FSK or PSK modulation with no more than four levels is used.

(d) In coherent receivers a flywheel action is often introduced in the demodulator clock-recovery circuitry in order to maintain synchronism in the presence of long fades. Otherwise one or more bits could be skipped on clock recovery, originating loss of synchronism and misframing.

(e) Power amplifiers need more linearity than for analog transmission in order not to spread the r.f. spectrum.

(f) The service channels for voice and data can be digital or analog as for LOS equipment.

Conventional equipment can generally be used for bit rates of up to 2 or 3 Mbit s^{-1} for which it is easier to design paths with negligible or little multipath dispersion.

7.6.2.2 Adaptive digital equipment

Adaptive equipment includes sophisticated modems specially designed to overcome the effects of high dispersion. It is used in medium or long troposcatter paths with high bit rates (3, 6, 12 Mbit s^{-1}) where the dispersion parameter $2\sigma/T$ ranges from less than 0.5 up to 2 or more. The modems take advantage of the frequency or time diversity effect present when the r.f. bandwidth is broader than the medium correlation bandwidth. Therefore they add to the advantages of the explicit diversity system used the further advantage of an in-band implicit frequency or time diversity. They are expensive but they permit high quality transmission on paths where conventional equipment would fail.

7.6.2.3 Conventional FM–FDM analog radio equipment

Conventional equipment is equipment that has been converted to transmit baseband digital signals. A terminal with 120 FDM channels has a baseband frequency band of at most 6–552 kHz. Therefore it can transmit a bit stream of up to 0.5 Mbit s^{-1} corresponding for example to 15 channels delta modulated at 32 Mbit s^{-1}. The r.f. modulation becomes a two-level frequency modulation of the 2FSK type. The modulation index, defined as the difference $2\Delta f$ between the two frequencies divided by the bit rate, has an optimum value for the system error rate when it is equal to 0.65. In these conditions if the radio modem remains linear it is possible to transmit a bit stream of 1544 kbit s^{-1} corresponding to 24 PCM channels at 64 kbit s^{-1} with an r.f. spectrum occupancy of about 2 MHz.

The capacity of these terminals is small and their bandwidths are generally smaller than the medium correlation bandwidth of a path of average characteristics, so that particular problems of intersymbol interference do not normally arise. Since the detection is incoherent, there are no problems of synchronization loss.

7.6.2.4 Special additional equipment

Additional equipment consists of special forward error-correction devices designed for use in troposcatter transmission. Several forward error-correction methods have been proposed for combating the error bursts due to fading. Redundant error-correction bits are added to the bit stream. This increases the gross bit rate, the r.f. bandwidth and the BER but the degradation is completely overcome by the error-correcting capability of the codes used, so that the overall performance improves significantly. As the errors occur in bursts, a suitable interleaving technique is introduced to displace adjacent bits by a suitable time interval. This implies a delay in the transmission time as well as the necessity of bit storage, but it has been shown in practice that a time delay of 250 ms is sufficient for obtaining the full dual time diversity improvement [7.1].

This solution, which is valid for bit rates of up to 2 Mbit s^{-1}, is particularly convenient when intersymbol interference is not high enough to justify the employment of adaptive modems. It can also be used to improve the quality of poor links.

It should be noted here that the use of transmultiplexers permits other solutions. A transmultiplexer is a device that transforms a CCITT digital baseband into the corresponding CCITT analog baseband, and vice versa for the same telephone channels. For example, the baseband of two groups of 30 digital channels, i.e. 2×2048 kbit s^{-1}, is transformed into the baseband of 60 FDM channels, i.e. 312–552 kHz, and vice versa. This is done by integrated circuits which interconnect the channels at the voice level (with digital multiplexing on one side and analog multiplexing on the other side) with digital filtering.

A digital baseband can thus be processed through a transmultiplexer and transmitted via an analog troposcatter radio link with a narrower bandwidth and a lower threshold. The inverse transformation is made on the other side. Despite the presence of this additional equipment, this solution could still prove economical in some cases.

7.6.3 Radio frequency filters and duplexers—isolators

R.f. filters are normally used in troposcatter terminals. In transmission they are necessary for attenuating the spurious and the harmonic frequencies generated in the transmitters and in the power amplifiers to a sufficiently low level before they reach the transmit antenna. The harmonics of a power amplifier are often suppressed by a coaxial filter placed near the output as part of the coaxial output line. When wideband tubes (e.g. TWTs) are used in the power amplifier it is necessary to suppress the noise transmitted in the reception band before it reaches the receiver's filter, which cannot eliminate it.

On the reception side r.f. filters are used to suppress the transmitted frequency so that its residuals reach the receiver at a sufficiently low level (e.g. below -30 dBm; remember that the transmit power may be more than 1 kW). They also suppress the other internal or external unwanted frequencies.

The duplexer connects a transmitter and a receiver to the same antenna through a single line. In the presence of a transmitted power of say 1 kW

(+60 dBm) the receiver must be able to detect a level of say −100 dBm at its frequency. The response curves of the filters and the duplexer determine the minimum frequency separation between transmitter and receiver. As there is generally a bad impedance match at the output of the power amplifier, an isolator is sometimes introduced to decrease the reflection coefficient and the intermodulation noise generated by reflections in the feeders (see Chapter 8, Section 8.2.2).

7.6.4 Coaxial and waveguide feeders

The feeders are the transmission lines carrying the r.f. signals from the radio equipment to the antenna and vice versa. For frequencies below 2 or 3 GHz the feeders are normally coaxial cables, whereas waveguides are preferred for higher frequencies.

Neither the coaxial cables nor the waveguide feeders should be of the rigid type, which require multiple joints and elbows or flexible sections between the equipment and the antennas, but the type with an outer conductor of corrugated copper protected by a polyethylene jacket should be chosen. The waveguides should be of the elliptical type. Thus long, continuous and flexible feeders in a single length, terminated by appropriate flange connectors, are obtained with obvious advantages.

In general the dielectric is air so that the r.f. attenuation is minimized. Thus there is continuity with the air enclosed in the dielectric protection of the antenna illuminator, so that the feeder plus the illuminator become a kind of air pipe closed at both ends. However, the sealing is never perfect, particularly at the illuminator side, so that moisture often condenses inside the feeder.

Depending on the way in which it is installed, the feeder could be heated by the sun during the day and cooled by the air during the night. This has the effect of pumping humid air from outside into the feeder. After a few months the increase in the attenuation and the standing wave ratio (SWR) and the loss of isolation may cause problems or faults which are due to the feeder's being partially full of water. The illuminator is often provided with an outlet closed by a valve for draining the condensed water. This problem is avoided by "pressurizing" the air-dielectric feeders and the antenna illuminators, i.e. pumping dry air into them at a pressure of some tens of centimeters of water above the external atmospheric pressure. Thus air can only flow from the inside to the outside.

As the air losses are normally very low for one or more sets of feeders, only one small air drier–pressurizer is required per station. This drier has several outlets each connected by a thin plastic pipe to an inlet at the output connector of the radio equipment. After some time depending on the air losses (normally weeks or months), the dessicant changes color because of water absorption and must be regenerated by heating.

Coaxial cables with foam dielectric are also available commercially. They present a higher attenuation and a lower power-handling capability but do not require pressurization. Of course in this case the antenna illuminator container should be sealed if possible, or drained from time to time. This approach can be used for a mobile troposcatter station as solid dielectric feeders present a greater mechanical resistance to frequent handling and the illuminators may be less subject to absorption of moisture.

In general the feeders chosen and their connectors should present the minimum attenuation, should be laid down for a minimum length and should be capable of carrying the maximum r.f. power at the maximum temperature reached (which could be say 70 °C if the feeder is in the sun). The designer can find all the necessary data in the manufacturers' catalogs.

The most commonly used coaxial cable is the $1\frac{5}{8}$ in type. At 2 GHz and a temperature of 30–40 °C it can handle 2–4 kW, depending on the type of climate, with an attenuation of 3.5 dB per 100 m and a velocity factor of 92% if the dielectric is air. With a foam dielectric the power-handling capacity becomes about 2.4 kW with an attenuation of 4.1 dB per 100 m and a velocity factor of 88%. For a $\frac{7}{8}$ in coaxial cable the values are 1.3–2 kW with an attenuation of 6 dB per 100 m and a velocity factor of 92.5% for air dielectric, and 1 kW with an attenuation of 6.5 dB per 100 m and a velocity factor of 89% for a foam dielectric.

At 2 GHz an elliptical waveguide at a temperature of 40 °C can handle 45 kW with an attenuation of 1.2 dB per 100 m. The group propagation velocity is 73% of the velocity of light in a vacuum. At 5 GHz the above values become 6.8 kW, 3.7 dB per 100 m and 71% respectively.

The manufacturers' catalogs give all data necessary for an appropriate choice of feeders, connectors and ancillaries, and instructions for installation.

7.7 Antennas

7.7.1 General characteristics of fixed antennas

Troposcatter antennas are usually of the parabolic type and consist of an illuminator (or feedhorn), which is actually the real antenna, and a parabolic reflector with a diameter ranging from a few meters up to 30 or even 40 m. Parabolic antennas with diameters up to about 18 m are normally mounted on a self-supporting steel tower and can be oriented within at least ±20° in the horizontal plane (Fig. 7.11).

Parabolic antennas with diameters greater than about 20 m are normally of the billboard type, i.e. the reflecting surface is directly mounted on the supporting structure and the antenna is completely fixed on the terrain

Fig. 7.11 Parabolic antenna (450 MHz, 10 m diameter) mounted on a self-supporting tower.

(Fig. 7.12). In a space diversity system the two parabolic reflectors can be designed in such a way that their surfaces belong to the same paraboloid. It is then possible to locate both illuminators at the common focus, thus minimizing the length of the feeder lines and the corresponding attenuations.

Billboard antennas raise problems of correct orientation. In troposcatter antennas the orientation of the beam, and particularly the vertical orientation, is sometimes obtained by moving the illuminator. However, this practice should be avoided as it produces variations in the sidelobe level, thus increasing the possibility of interference. The illuminators can be designed for single or dual polarization. Special "shaped-beam" illuminators have been developed for illuminating the reflector more efficiently, thus increasing the antenna gain and at the same time decreasing the illumination at the edges and

Fig. 7.12 Billboard type center-fed antenna (900 MHz, 90 ft (27 m)). (GTE Comelit, Milan, Italy.)

keeping the sidelobes low. It is important that the sidelobes associated with troposcatter antennas are as low as possible for the following reasons: because of the high power radiated, interference could take place through them; sidelobes directed towards the Earth pick up thermal noise and introduce some performance degradation.

For diameters of up to say 10 or 12 m the manufacturer measures the complete radiation pattern on a sample antenna. The radiation pattern for larger antennas is calculated because of the difficulty or impossibility of measuring it. In any case the manufacturer provides masks of radiation levels that are not exceeded in the various directions. In the spirit of *CCIR Report 614-2* they are used for coordination with other radio relay services in an attempt to avoid the possibility of interference.

When the antenna is intended for an angle diversity path, more than one illuminator is provided for the same parabolic reflector and all illuminators are slightly out of focus. However, the design of such an antenna is critically related to the geometrical characteristics of the path, which in general is long and requires large antennas with beamwidths between 0.5° and 1° [7.2]. The illuminators should withstand the maximum r.f. power used in the system, with margins for temperature variation and for future requirements. Together with the feeders they are often pressurized by dry air to prevent the ingress of moisture.

Each input to the antenna has a certain bandwidth, which must be taken into account when allocating the working frequencies and studying the diversity configurations. For example, at 450 MHz the bandwidth is of the order of 40 MHz with an SWR below 1.3, while at 900 MHz the above values may become 50 MHz and 1.15 respectively. The decoupling between the two polarizations may be 35 dB or higher.

The antennas and their supporting tower or structure are normally designed for a high wind speed, which may reach 150 or 200 km h^{-1} with or without ice loading. For reducing the wind pressure in countries where snow and ice are absent, the reflecting surface may be made of mesh panels instead of the continuous steel or aluminum sheets generally used.

7.7.2 Mobile antennas

So far we have dealt with fixed antennas. However, in mobile troposcatter applications the antennas are also mobile. In this case they cannot have diameters exceeding 6–8 m. The antenna is composed, for example, of a foldable parabolic reflector mounted on a four-wheel trailer. The reflector has a main center section and two or four folding side-mounted sections (Fig. 7.13).

The reflector may be of either the mesh type (for climates without snow and for lower resistance to wind) or the solid surface type (e.g. fiberglass or aluminum panels). The folded antenna with the illuminator and the supporting trailer must remain within the maximum dimensions allowed for transportation by road, rail or air. To ensure a sufficiently high gain with a limited antenna diameter, the operating frequency is generally high (e.g. in the 4.5 GHz band).

The mobile antenna can be adjusted on site in azimuth and elevation and compensated for nonhorizontal terrain. In operation it should withstand a wind of about 120 km h^{-1}. The antenna is erected using a hand-driven or motor-driven hydraulic system and should require a minimum of time and manpower.

7.7.3 Gain and gain degradation: path gain

A good approximation of the free-space gain of a parabolic antenna over the isotropic radiator is given by

$$G(\text{dB}) = 20 \log f + 20 \log D - 42.2 \qquad (7.12)$$

where f is the frequency in megahertz and D is the diameter in meters of the aperture of the paraboloid. The guaranteed gain value given by the

Fig. 7.13 (a) Mobile antenna (2 GHz, 6.4 m diameter); the same antenna from the side. (GTE Comelit, Milan, Italy.)

manufacturer may differ slightly from that obtained using eqn (7.12). It is very difficult to measure the gain of large antennas, and therefore this is generally calculated, to a very good approximation, from the radiation pattern of the illuminator which can easily be measured.

The 3 dB beamwidth is given by eqn (3.15), and the gain degradation for the two antennas in the path is given by (4.6) (this is also adopted by the CCIR). The overall gain of the transmitting and receiving antennas after degradation is called the *path antenna gain* or the *total effective antenna gain* and is given by

$$G_p = G_t + G_r - \Delta g \quad (7.13)$$

The results of calculations of beamwidths and path gains using the above formulae for various antennas and frequency bands are given in Table 7.4.

7.8 Multiplex equipment

The multiplex equipment, either analog or digital, used in troposcatter systems is the same as that used in conventional civil or military LOS systems. The reader is assumed to have at least a basic knowledge of this subject. We shall therefore limit ourselves to identifying specific topics which are of particular importance in troposcatter applications.

7.8.1 Analog multiplex

Analog multiplex equipment has been in use for many years and has reached a high degree of standardization. The characteristics of the equipment normally used are as specified in the CCITT Recommendations, at least for capacities of 24 channels or more. Multiplex equipment for lower capacities often uses simpler techniques and may have nonstandard parameters, except for the channel frequency response which is normally required to comply with the CCITT masks.

Telephone channels have a nominal bandwidth of 4 kHz, and the voice band actually transmitted is 300–3400 Hz. The 4 kHz channels are packed together using the frequency-division multiplex (FDM) technique to form the baseband in which they are allocated side by side. The following basebands are frequently used in troposcatter applications: 6 channel capacity, 0–24 kHz or 4–28 kHz; 12 channel capacity, 6–54 kHz or 12–60 kHz; 24 channel capacity, 6–108 kHz or 12–108 kHz; 60 channel capacity, 12–252 kHz or 60–300 kHz; 120 channel capacity, 12–552 kHz or 60–552 kHz.

The CCITT multiplex is often provided with a pilot frequency for monitoring the continuity of the link. When such a multiplex is adopted in some troposcatter links it may be advisable to delay the pilot failure alarm to

Table 7.4(a) Beamwidth and path gain of parabolic troposcatter antennas

450 MHz band

Antenna diameter (m)	3 dB beamwidth (deg (mrad))	Path gain $G_t + G_r - \Delta g = G_p$ (dB)
4	11.7 (204)	22.9 + 22.9 − 0.9 = 44.9
6	7.8 (136)	26.4 + 26.4 − 1.3 = 51.5
8	5.8 (102)	28.9 + 28.9 − 1.7 = 56.1
10	4.7 (81)	30.9 + 30.9 − 2.1 = 59.7
12	3.9 (68)	32.4 + 32.4 − 2.5 = 62.3
15	3.1 (54)	34.4 + 34.4 − 3.1 = 65.7
18	2.6 (45)	36.0 + 36.0 − 3.7 = 68.3
24	1.9 (34)	38.5 + 38.5 − 4.8 = 72.2
27	1.7 (30)	39.5 + 39.5 − 5.4 = 73.4

900 MHz band

3 dB beamwidth (deg (mrad))	Path gain $G_t + G_r - \Delta g = G_p$ (dB)
5.8 (102)	28.9 + 28.9 − 1.7 = 56.1
3.9 (68)	32.4 + 32.4 − 2.5 = 62.3
2.9 (51)	34.9 + 34.9 − 3.3 = 66.5
2.3 (41)	36.9 + 36.9 − 4.0 = 69.8
1.9 (34)	38.5 + 38.5 − 4.8 = 72.2
1.6 (27)	40.4 + 40.4 − 6.0 = 74.8
1.3 (23)	42.0 + 42.0 − 7.1 = 76.9
1.0 (17)	44.5 + 44.5 − 9.3 = 79.7
0.9 (15)	45.5 + 45.5 − 10.5 = 80.5

Table 7.4(b) Beamwidth and path gain of parabolic troposcatter antennas

2700 MHz band

Antenna diameter (m)	3 dB beamwidth (deg (mrad))	Path gain $G_t + G_r - \Delta g = G_p$ (dB)
4	1.9 (34)	38.5 + 38.5 − 4.8 = 72.2
6	1.3 (23)	42.0 + 42.0 − 7.1 = 76.9
8	1.0 (17)	44.5 + 44.5 − 9.3 = 79.7
10	0.8 (14)	46.4 + 46.4 − 11.6 = 81.2
12	0.6 (11)	48.0 + 48.0 − 13.8 = 82.2
15	0.5 (9)	50.0 + 50.0 − 17 = 82.9
18	0.4 (8)	51.5 + 51.5 − 20.3 = 82.7
24	0.3 (6)	54.0 + 54.0 − 26.7 = 81.3

4700 MHz band

3 dB beamwidth (deg (mrad))	Path gain $G_t + G_r - \Delta g = G_p$ (dB)
1.1 (20)	43.3 + 43.3 − 8.2 = 78.4
0.7 (13)	46.8 + 46.8 − 12.1 = 81.5
0.6 (10)	49.3 + 49.3 − 15.9 = 82.7
0.4 (8)	51.2 + 51.2 − 19.6 = 82.8
0.4 (7)	52.8 + 52.8 − 23.4 = 82.2
0.3 (5)	54.8 + 54.8 − 28.9 = 80.7
—	
—	

avoid useless intervention in the case of very short interruptions due to deep fades.

7.8.2 Digital multiplex

The types of digital multiplex utilized are as follows.

(a) In *PCM conventional multiplex*, for example that specified in *CCITT Recommendation G-703*, the analog-to-digital conversion is effected at 64 kbit s^{-1} to ensure an optimum signal-to-noise ratio. The first hierarchical level can be either 1544 kbit s^{-1} (24 channels) or 2048 kbit s^{-1} (30 channels), and the second level can be either 6312 kbit s^{-1} (96 channels) or 8448 kbit s^{-1} (120 channels). Higher levels are not considered for troposcatter transmission.

(b) *Delta-modulated multiplex*, with analog-to-digital conversion effected at 32 or 16 kbit s^{-1}, is more convenient for low bit rates (e.g. 256, 512, 1024 kbit s^{-1}) and links whose quality is not high. Continuously variable slope delta modulation (CVSD) multiplex equipment is available with a very good voice intelligibility for a BER as high as 1×10^{-2}, and a performance identical with that at error-free operation can be achieved at a BER of 1×10^{-3}. Synchronization can be achieved and maintained at a BER approaching 1×10^{-1} [7.3]. Delta multiplex is not recommended by the CCITT.

When using digital multiplex equipment in troposcatter radio links some attention should be paid to the possible loss of synchronization or *bit count integrity* (BCI) as a consequence of deep fades. The time for re-acquisition of synchronization and frame realignment (not longer than 100 ms) adds directly to the fade duration.

This is not generally a problem for the radio link as long as it is provided with a flywheel action in its demodulator if it is coherent or if incoherent demodulation is used (see Section 7.6.2). The problem arises for the multiplex where errors may cause misframing or incorrect justification. The mean time between loss of BCI or between framing or justification errors can be evaluated. In general if the radio link is designed according to DCS criteria this mean time is sufficiently long and the overall disturbance can be neglected. Otherwise a greater margin should be taken for the system gain or a more sophisticated multiplex should be used with a longer frame (or justification) word, an appropriate number of checks before the misframing alarm etc. [7.1, 7.4].

7.9 Auxiliary equipment

A troposcatter station is often provided with some auxiliary equipment, as is the case for conventional LOS stations. For example, the electronic equipment is usually provided with local alarm indications, at least for the most important faults.

When a station is unattended it is usual to provide it with a remote-monitoring system which transmits alarms and other indications to a maintenance center. The remote-monitoring signal can be transmitted through a telegraph channel allocated above the speech band of the analog service channel, or through a digital service channel provided for the purpose in a digital terminal. Many types of remote-monitoring equipment are available commercially.

When the feeders use air dielectric they have to be pressurized with dry air or dry nitrogen. As the pressurization generally involves the antenna illuminator also, the air pressure should be maintained relatively low (a few tens of centimeters of water above the normal atmospheric pressure). Special dehydrator and pressurizing devices with the necessary installation kits for providing a complete system are available commercially, and are sometimes provided by the feeder manufacturers.

7.10 Ability of combined equipment to overcome path loss

When the various types of equipment described in this chapter are assembled in the appropriate manner they form the two end stations of the troposcatter path. The main parameter characterizing this assembly and used in the design of the link is the capability of the combined equipment to overcome the maximum path loss. By summing the power levels, gains and losses of the components in the selected combination it it is possible to calculate the value of the overall maximum path loss that the combination can overcome for a given minimum acceptable performance (characterized by the signal-to-noise ratio for analog equipment or the BER for digital equipment).

The capability $E_q(\text{dB})$ of the combined equipment to overcome the maximum path loss is given by the following formula, in which all variables are expressed in decibels (or dBm):

$$E_q(\text{dB}) = P_t + G_t + G_r - \Delta g - l_i - T_d + g_d \qquad (7.14)$$

where P_t is the level (in dBm) of the r.f. power at the transmitter (high power amplifier) output, G_t and G_r are the gains of the transmitting and receiving antennas, Δg is the gain degradation of the antennas (Section 7.7.3), l_i is the loss of feeders and filters between the transmitter output and any distant receiver input (also known as the installation loss), T_d is the threshold level (in dBm) of the r.f. carrier at the input of the receiver of any diversity branch and g_d is the diversity gain. The threshold level at the input of the receiver of any diversity branch is the minimum r.f. level for which the transmission performance is acceptable. This means that for an analog path the signal-to-noise ratio in the worst (highest) telephone channel is not less than a given minimum (e.g. 32 dBp), which cannot be lower than that calculated using eqn (7.5) or given in Table 7.1, and that for a digital path the maximum BER does not exceed a given value (e.g. 10^{-4}).

The capability of the equipment for transmission of telegraphy over telephone channels is higher than that for telephony because the telegraphy threshold is some 5 dB lower than the telephony threshold (see Section 7.4.2).

Equation (7.14) is derived assuming that there is one transmitter on one side and one receiver on the other side. The effects of diversity are accounted for in the term g_d. Depending on the approach adopted, we shall see in Chapters 8 and 9 that the term g_d can be interpreted as follows:

$g_d = 0$ if the effects of diversity are included in the path attenuation or in an equivalent r.f. received signal;

g_d is the median diversity gain (the value exceeded for 50% of the time) if E_q is to be compared with median path losses;

g_d is the $q\%$ diversity gain (e.g. the value exceeded for 99.9% of the time) if E_q is to be compared with the maximum instantaneous path loss.

As the same value of E_q can be obtained with different combinations of equipment, it is necessary in practice to determine the most economical solution. For example an increase of 3 dB in the gain can be obtained by doubling the r.f. transmitted power (higher costs, greater power consumption, more cooling problems etc.) or by increasing the diameters of all antennas by about 20% (which is a further investment cost but saves several operation and maintenance costs) or by using a low noise mixer or preamplifier in the first stage of the receivers, which, when applicable, is the least expensive solution. Therefore the "cost of the decibel" for the same gain variation may be very different for each solution, and a careful study of this problem is required in each practical case.

The combined equipment capability may reach values of say 240 dB. Table 7.5 shows an example of a calculation for two families of analog equipment with various channel capacities. The median value of the diversity gain is considered here, as this capability is intended to be compared with CCIR median path losses.

7.11 Systems with adaptive capability

The equipment capability E_q is normally a fixed quantity, but in some cases it can also be made to vary in a stepwise manner. For example, during good propagation conditions high power amplifiers could be partially or

Table 7.5 Calculation of the maximum loss that can be overcome by various configurations of analog equipment

450 MHz band

Traffic capacity	6+1 channels				12+1 channels				24+1 channels			
Antenna diameter (m)	6	9	12	18	6	9	12	18	6	9	12	18
R.f. transmitted power (300 W) (dBm)	+54.8	+54.8	+54.8	+54.8	+54.8	+54.8	+54.8	+54.8	+54.8	+54.8	+54.8	+54.8
Gain of TX antenna (dB)	+27.0	+30.5	+33.0	+36.5	+27.0	+30.5	+33.0	+36.5	+27.0	+30.5	+33.0	+36.5
Gain of RX antenna (dB)	+27.0	+30.5	+33.0	+36.5	+27.0	+30.5	+33.0	+36.5	+27.0	+30.5	+33.0	+36.5
Antenna gain degradation (dB)	−1.4	−2.0	−2.6	−3.9	−1.4	−2.0	−2.6	−3.9	−1.4	−2.0	−2.6	−3.9
Loss of cables, filters etc. (dB)	−3.5	−3.5	−3.5	−3.5	−3.5	−3.5	−3.5	−3.5	−3.5	−3.5	−3.5	−3.5
Threshold of receiver (dBm)	+107.0	+107.0	+107.0	+107.0	+103.0	+103.0	+103.0	+103.0	+101.0	+101.0	+101.0	+101.0
Maximum path attenuation overcome without diversity (dB)	210.9	217.3	221.7	227.4	206.9	213.3	217.7	223.4	204.9	211.3	215.7	221.4
Dual diversity												
Median gain (dB)	+3.0	+3.0	+3.0	+3.0	+3.0	+3.0	+3.0	+3.0	+3.0	+3.0	+3.0	+3.0
Maximum attenuation overcome with dual diversity (dB)	213.9	220.3	224.7	230.4	209.9	216.3	220.7	226.4	207.9	214.3	218.7	224.4
Quadruple diversity												
Median gain (dB)	+5.0	+5.0	+5.0	+5.0	+5.0	+5.0	+5.0	+5.0	+5.0	+5.0	+5.0	+5.0
Maximum attenuation overcome with quadruple diversity (dB)	215.9	222.3	226.7	232.4	211.9	218.3	222.7	228.4	209.9	216.3	220.7	226.4

900 MHz band

Traffic capacity	24+1 channels				60+1 channels				120+1 channels			
Antenna diameter (m)	6	9	12	18	6	9	12	18	6	9	12	18
R.f. transmitted power (500 W) (dBm)	+57.0	+57.0	+57.0	+57.0	+57.0	+57.0	+57.0	+57.0	+57.0	+57.0	+57.0	+57.0
Gain of TX antenna (dB)	+33.0	+36.5	+39.0	+42.5	+33.0	+36.5	+39.0	+42.5	+33.0	+36.5	+39.0	+42.5
Gain of RX antenna (dB)	+33.0	+36.5	+39.0	+42.5	+33.0	+36.5	+39.0	+42.5	+33.0	+36.5	+39.0	+42.5
Antenna gain degradation (dB)	−2.6	−3.9	−5.1	−7.5	−2.6	−3.9	−5.1	−7.5	−2.6	−3.9	−5.1	−7.5
Loss of cables, filters etc. (dB)	−3.0	−3.0	−3.0	−3.0	−3.0	−3.0	−3.0	−3.0	−3.0	−3.0	−3.0	−3.0
Threshold of receiver (dBm)	+100.0	+100.0	+100.0	+100.0	+95.4	+95.4	+95.4	+95.4	+94.0	+94.0	+94.0	+94.0
Maximum path attenuation overcome without diversity (dB)	217.4	223.1	226.9	231.5	212.8	218.5	222.3	226.9	211.4	217.1	220.9	225.5
Dual diversity												
Median gain (dB)	+3.0	+3.0	+3.0	+3.0	+3.0	+3.0	+3.0	+3.0	+3.0	+3.0	+3.0	+3.0
Maximum attenuation overcome with dual diversity (dB)	220.4	226.1	229.9	234.5	215.8	221.5	225.3	229.9	214.4	220.1	223.9	228.5
Quadruple diversity												
Median gain (dB)	+5.0	+5.0	+5.0	+5.0	+5.0	+5.0	+5.0	+5.0	+5.0	+5.0	+5.0	+5.0
Maximum attenuation overcome with quadruple diversity (dB)	222.4	228.1	231.9	236.5	217.8	223.5	227.3	231.9	216.4	222.1	225.9	230.5

totally removed from the circuit, thus decreasing the transmitted power, the possibility of interference and the risk of receiver saturation. This is sometimes done manually on some installations during the best months.

However, a troposcatter system may be required in an environment in which there is a high risk of interference, e.g. in marine oil fields where ducting is prevalent. In this case the transmit power should be kept at the minimum necessary from the design stage onwards. Furthermore, as the path attenuation varies with time, the r.f. power should also be made variable, possibly by an automatic system, so that it can at least follow slow variations in path loss.

Fast fading is expected to disappear during ducting or anomalous propagation and the variation in path loss should be slow, but the real situation is more complex. For example, during periods of ducting over the North Sea fades of up to 36 dB where observed at 2 GHz with a steepest slope of 14 dB min^{-1}, and an overall change of 25 dB took up to 5 min to complete. In the Middle East variations of 16 dB min^{-1} were observed at 1.9 GHz during the onset of ducting, while 10 dB variations took place in only 7 s [7.5].

Adaptive systems have been described in the literature [7.5, 7.6] and are particularly suitable in zones where anomalous propagation conditions occur frequently. They are based on a motor-driven high power attenuator which controls the amplifier output power (in this case limited to 20 W for example) in 5 dB steps for a total variation of 45 dB. However, they cannot follow fast Rayleigh fading. Each step has a duration of a few seconds. The control signal is taken from the average received r.f. signal and the first step is initiated when the r.f. received signal is more than 40 dB above threshold. The system is therefore protected against deep fast fades, and the highest r.f. levels, which cause the greatest interference, are limited.

Adaptive systems have also been used in receivers in order to avoid very strong r.f. signals that may saturate stages in the receivers and create unacceptable distortion. In this case a variable attenuator is put in the receiver's front-end, and its loss is controlled by the level of the received signal. The main disadvantage of this solution is that the minumum loss of this attenuator is not zero but about 2 dB, which increases the threshold of the receiver.

7.12 Comments and suggestions for further reading

The characteristics of the various types of equipment discussed in this chapter are given in manufacturers' catalogs. The larger manufacturers sometimes describe their equipment and methods in their own technical reviews [7.3, 7.7]. Articles describing new equipment sometimes appear in the literature. The theory is given in books dealing with radio links.

References

7.1 Cairns, J. B. S. A review of digital troposcatter links, *Conf. Publ. Telecom 79, 3rd World Telecommunications Forum, September 1979*, Part 2, pp. 23.10.1–23.10.10, ITU, Geneva, 1979.

7.2 Gough, M. W. Angle diversity applied to tropospheric scatter systems, *AGARD Conf. Proc.*, **70**, 32-1–32-15, 1970.

7.3 Brand, T. E., Connor, W. J., Sherwood, A. J., Unkauf, M. G., Tagliaferri, O. A., Liskov, N., Curtis, R., Boak, S., Bagnell, R., Abele, R. J., Smith, G. E. and Zawislan, F. *Technical Publications on Digital Troposcatter*, Raytheon, Sudbury, MA, June 1980.

7.4 Osterholz, J. L. Design considerations for digital troposcatter communications systems, *AGARD Conf. Proc.* **244**, 22-1–22-15, 1977.

7.5 Johnson, J. K. Automatic power control of transmitters, *Electron. Eng.*, 69–77, December 1978.

7.6 Skingley, B. S. Level control in tropospheric scatter systems, *AGARD Conf. Proc.*, **244**, 23-1–23-8, October 1977.

7.7 Special issue on tropospheric scatter, *Rev. Tech. Thomson Houston Electron.*, **34**, June 1961.

Chapter 8

Path Design and Prediction of Performance for Analog Troposcatter Links

In this chapter we show how to calculate an analog troposcatter radio path with the desired transmission characteristics. The performance objectives normally required for these links are analyzed and simple criteria for performance prediction are derived. On the basis of these elements the path calculations can be optimized, taking the various parameters involved into consideration.

8.1 Performance objectives

The performance objectives for an analog radio link refer in principle to the achievement of a specified minimum signal-to-noise ratio in the worst telephone channel which must be exceeded for a given high percentage of time. Occasionally, when telegraph channels are also transmitted on a telephone channel of the multiplex, a maximum character error rate which must not be exceeded for a high percentage of time is also given.

When the path under study is to be integrated with a public telecommunications network its performance is usually specified in terms of CCIR Recommendations. Other specifications may be involved for other types of network (e.g. military specifications), with requirements of greater or lesser stringency ranging from the very severe DCS conditions based on an 11 000 km reference circuit to the requirement that it is just possible to talk for an unspecified percentage of time with reasonable quality.

The more severe specifications are based on the performance of the *hypothetical reference circuit*. This is an ideal radio link constructed with hops equal to the one under study and with a length related to the maximum distance over which it is expected that a communication will be established. At the end of this long circuit the signal-to-noise ratio should exceed given minimum values for given high percentages of time. The path under study should be designed in such a way that, when it is introduced in the hypothetical reference circuit, the performance of this circuit meets the specified requirements. If the troposcatter path is d km long, the CCIR hypothetical reference circuit for transhorizon radio relay systems is composed of a cascade of $2500/d$ such paths, and the value $2500/d$ is taken to the nearest whole number (*CCIR Recommendation 396-1*).

A more complex reference circuit 11 000 km long has been specified by the American DCS for global communications. We shall examine in detail later the CCIR Recommendations which specify the maximum allowable noise power at a point of zero reference level for given percentages of time.

Less severe specifications (implying lower costs) are often applied to radio links which are not to be connected to public networks but service limited areas. Examples of such specifications are as follows.

(a) The hourly median signal-to-noise ratio in the worst telephone channel should exceed 30 dBp for 99.9% of the year (or of the worst month). This means that for 0.1% of the year (about 9 h) the signal-to-noise ratio could be lower than 30 dBp for 50% of the time.

(b) The instantaneous signal-to-noise ratio in the worst telephone channel should exceed 30 dBp for 99.9% of the year (or of the worst month).

(c) The telegraph character error rate should not exceed 10^{-4} for 99.9% of the worst month.

It is sometimes specifically required that the performance objectives should be met without considering the improvement due to any compandors present in the multiplex channels.

The performance objectives are intended as figures that the designer should meet in calculating a troposcatter system in the most accurate way and should not be used for acceptance tests of the implemented system. Owing to the intrinsic uncertainty of some parameters, the actual results obtained on the real links may differ from those expected. It is certainly not advisable to "guarantee" a certain performance unless sufficient margins are built into the design.

8.2 Noise

The performance specifications deal with the signal-to-noise ratio or, equivalently, with the noise level at a point of zero relative level. The noise considered is of course the total radio noise from all sources which appears at the baseband output of the radio link in slots corresponding to the telephone channels of the multiplex. This noise consists of the following components.

(i) *Thermal noise* can be calculated exactly using the formulae given in Chapter 7. It has the same statistical distribution, with opposite variations, as the received r.f. signal, if this is above the threshold.

(ii) *Intermodulation noise* is present during traffic in the other telephone channels and reaches its maximum during the *busy hour* (about 11 a.m. on working days). This noise and its statistics are difficult to predict.

Intermodulation noise consists of the following:

(a) noise generated by nonlinearities in the equipment (see Chapter 7) which should be measured directly;

(b) noise generated by reflections in the r.f. feeders, which can be calculated to an acceptable approximation as we shall see in Section 8.2.2.2;

(c) noise due to multipath propagation, which can be calculated to a rough approximation as we shall see in Section 8.2.2.3 (its statistics depend on the traffic and on the variability of the propagation conditions).

It is relatively easy to predict the behavior of thermal noise as it follows log-normal or Rayleigh statistics, but it is much more difficult to anticipate the behavior of intermodulation noise. It has become a general convention for the designer to assign half the allowed noise to the thermal component and half to the intermodulation component. More exactly, this subdivision is made for the 1 min noise which must not be exceeded for 80% of the time. The intermodulation component is assumed to be that of the busy hour.

According to *CCIR Recommendation 393-4* the multiplex signal during the busy hour can be represented by a steady white noise signal (see also Table 7.1, column 3) with a level

$$Q(\text{dBm0}) = -1 + 4\log C \qquad (8.1)$$

where C is the number of telephone channels for which the radio system is designed. When this signal is applied to an actual troposcatter path under normal operating conditions and total noise tests are made, the results can be represented in a form similar to that of Fig. 7.7. The total noise is a minimum for a particular frequency deviation which, in the case of a well-designed path, should be as close as possible to the nominal deviation. However, this minimum is not fixed in practice but varies with the multipath distortion.

We shall now consider thermal and intermodulation noise separately.

8.2.1 Thermal noise

We saw in Chapter 7, Section 7.4.3, that the output thermal noise and the input r.f. signal have identical but opposite variations (in decibels), and consequently (Chapter 6, Section 6.1) the output thermal noise and the path loss have identical variations. The study of the behavior of thermal noise can be simplified by considering that most analog troposcatter links use a maximal-ratio diversity combiner at either baseband or i.f. level. In view of Chapter 5, Section 5.6, and the above considerations, we can write a simple formula relating the variations in path loss to the distribution of the mean noise with diversity.

For a single path the distribution of the 1 min mean noise level $N_1(q\%)$ in the band of a telephone channel (normally the upper or worst channel) during a period of time (normally the worst month) can be written as

$$N_1(q\%)(\text{dBm0p}) = N_1(50\%) - Y(q\%) - h_d \qquad (8.2)$$

where $N_1(50\%)$ is the median noise level (dBm0p) without diversity in the period considered (worst month) corresponding to $L(50\%)$ of eqns (6.6) and (6.8), $q\%$ is the percentage of time under consideration, $Y(q\%)$ is the slow fading term (in decibels) of (6.6) and (6.12), which may or may not include the prediction error, and h_d is 1.6 dB for dual diversity with maximal-ratio combination and 6.4 dB for quadruple diversity with maximal-ratio combination. This is the noise distribution for a single path, but in general it is necessary to know the noise distribution at the end of a hypothetical reference circuit composed of z sections in cascade along which the noises add in power. Each section is identical with the path under study but has uncorrelated fading. In this case the term $Y(q\%)$ is substituted by the term $G(\sigma; z; q)$ which represents the sum (convolution) of the z uncorrelated terms $Y(q\%)$ normally distributed with the same standard deviation σ (see Chapter 2, Section 2.5.3). Therefore at the end of the hypothetical reference circuit the noise distribution (8.2) becomes

$$N_z(q\%)(\text{dBm0p}) = N_1(50\%) + G(\sigma; z; q) - h_d \qquad (8.3)$$

where the function $G(\sigma; z; q)$, calculated using approximate methods, is shown in Fig. 8.1.

DESIGN OF ANALOG TROPOSCATTER LINKS 207

(a)

PERCENT OF TIME ORDINATE IS NOT EXCEEDED

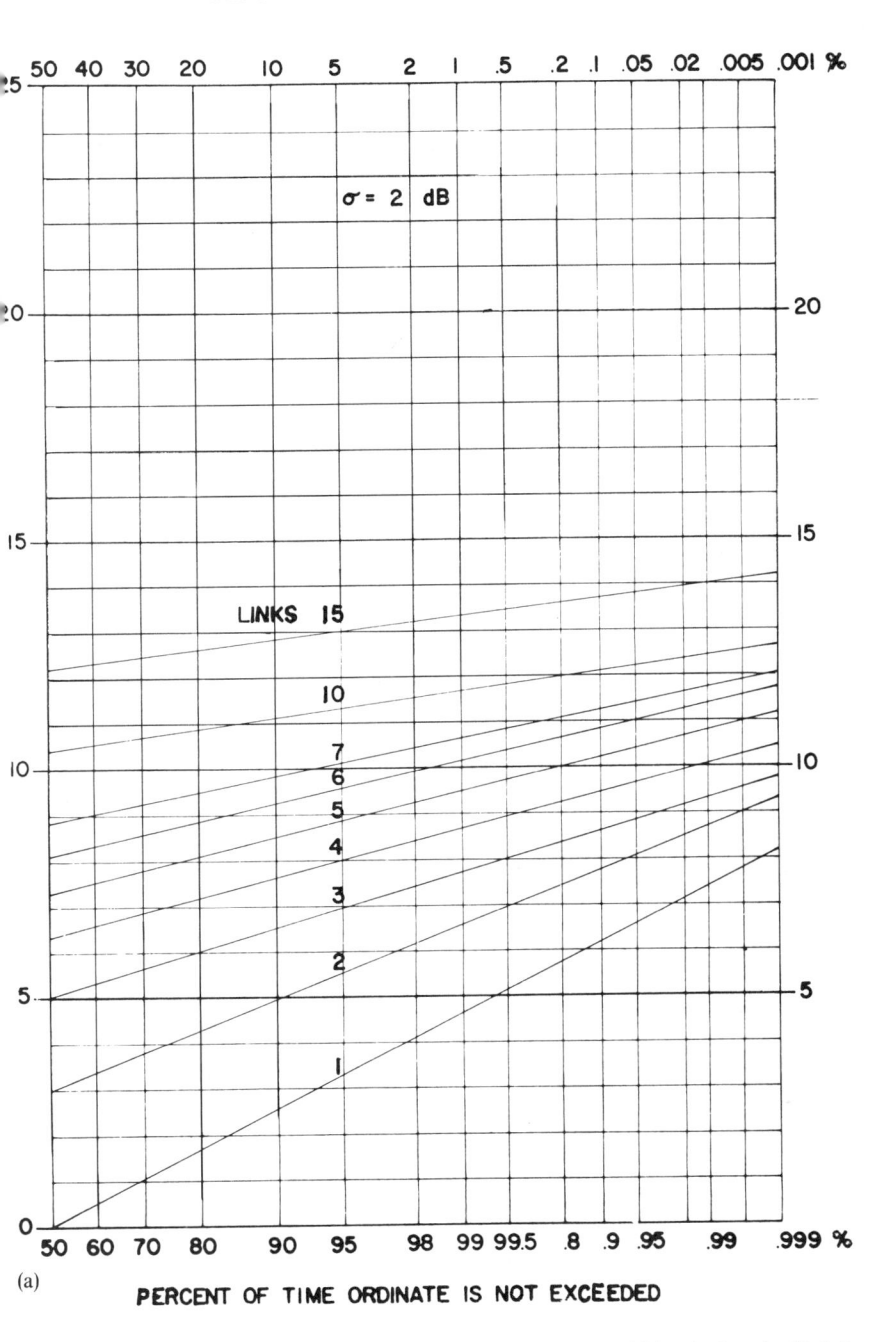

Fig. 8.1 Function $G(\sigma;z;q)$—distribution of noise over several identical paths (links) with independent log-normal fades: (a) $\sigma = 2\,\text{dB}$; (b) $\sigma = 4\,\text{dB}$; (c) $\sigma = 5\,\text{dB}$; (d) $\sigma = 6\,\text{dB}$; (e) $\sigma = 7\,\text{dB}$.

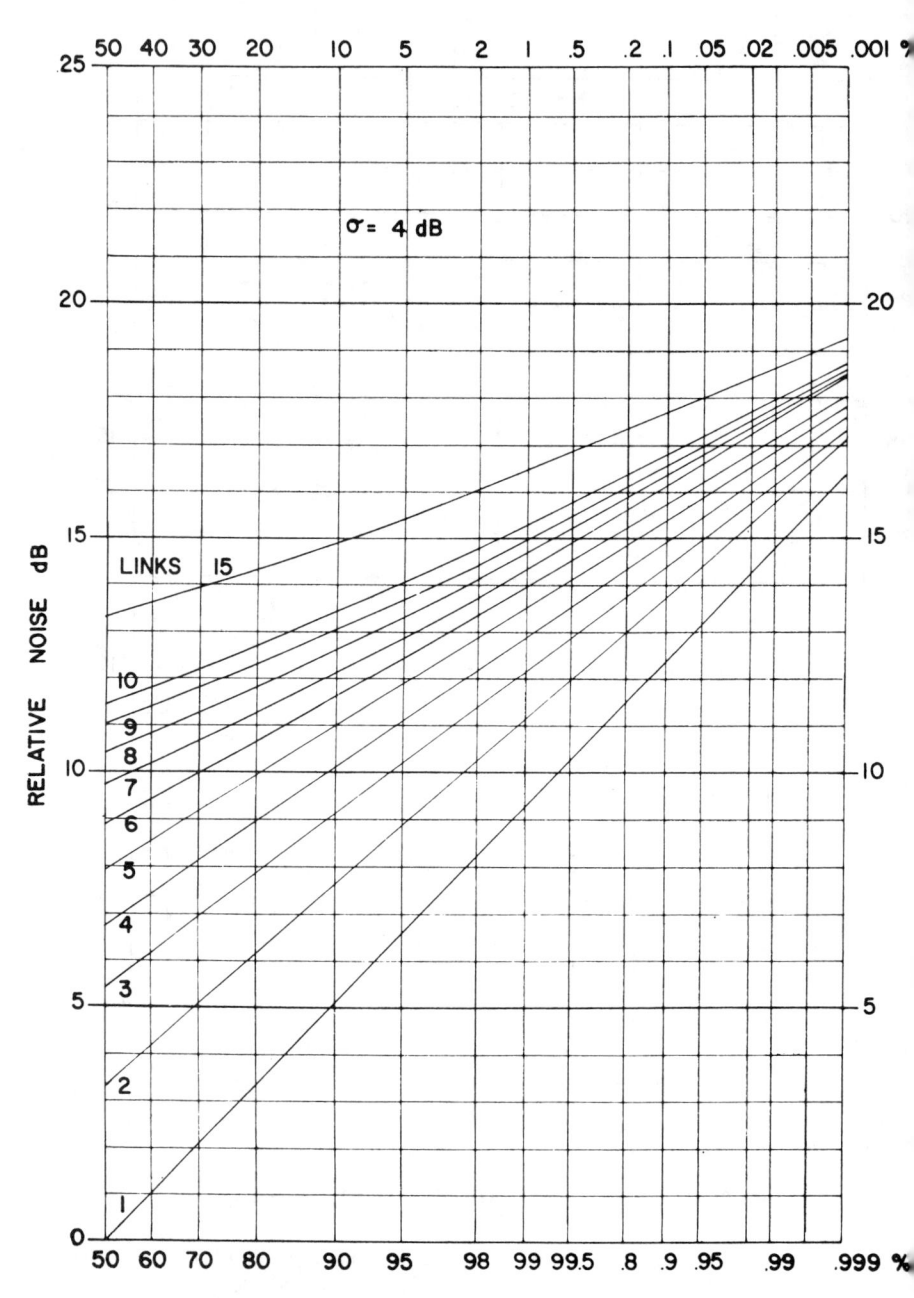

(b)

DESIGN OF ANALOG TROPOSCATTER LINKS 209

(c) PERCENT OF TIME ORDINATE IS NOT EXCEEDED

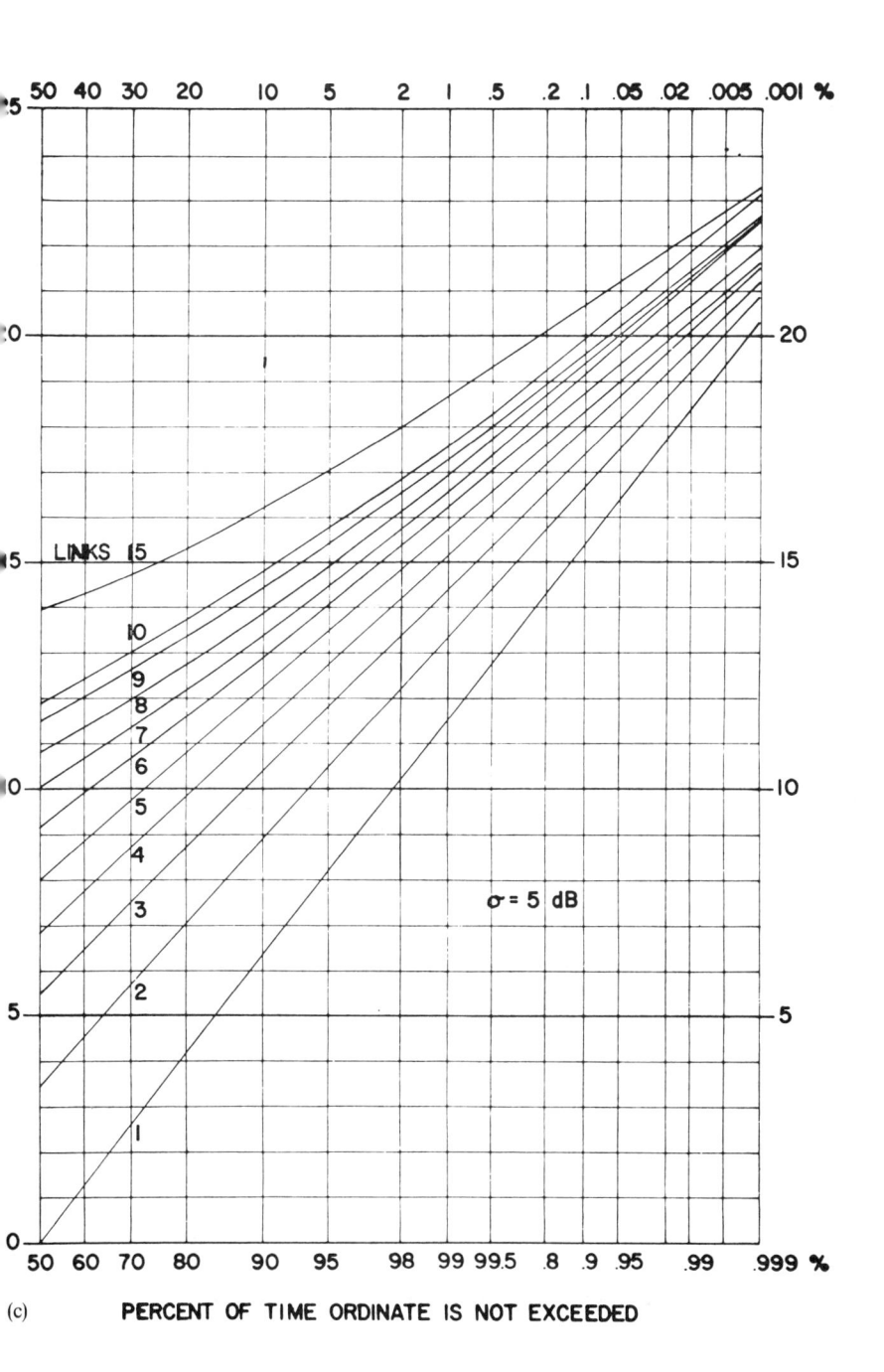

(d) PERCENT OF TIME ORDINATE IS NOT EXCEEDED

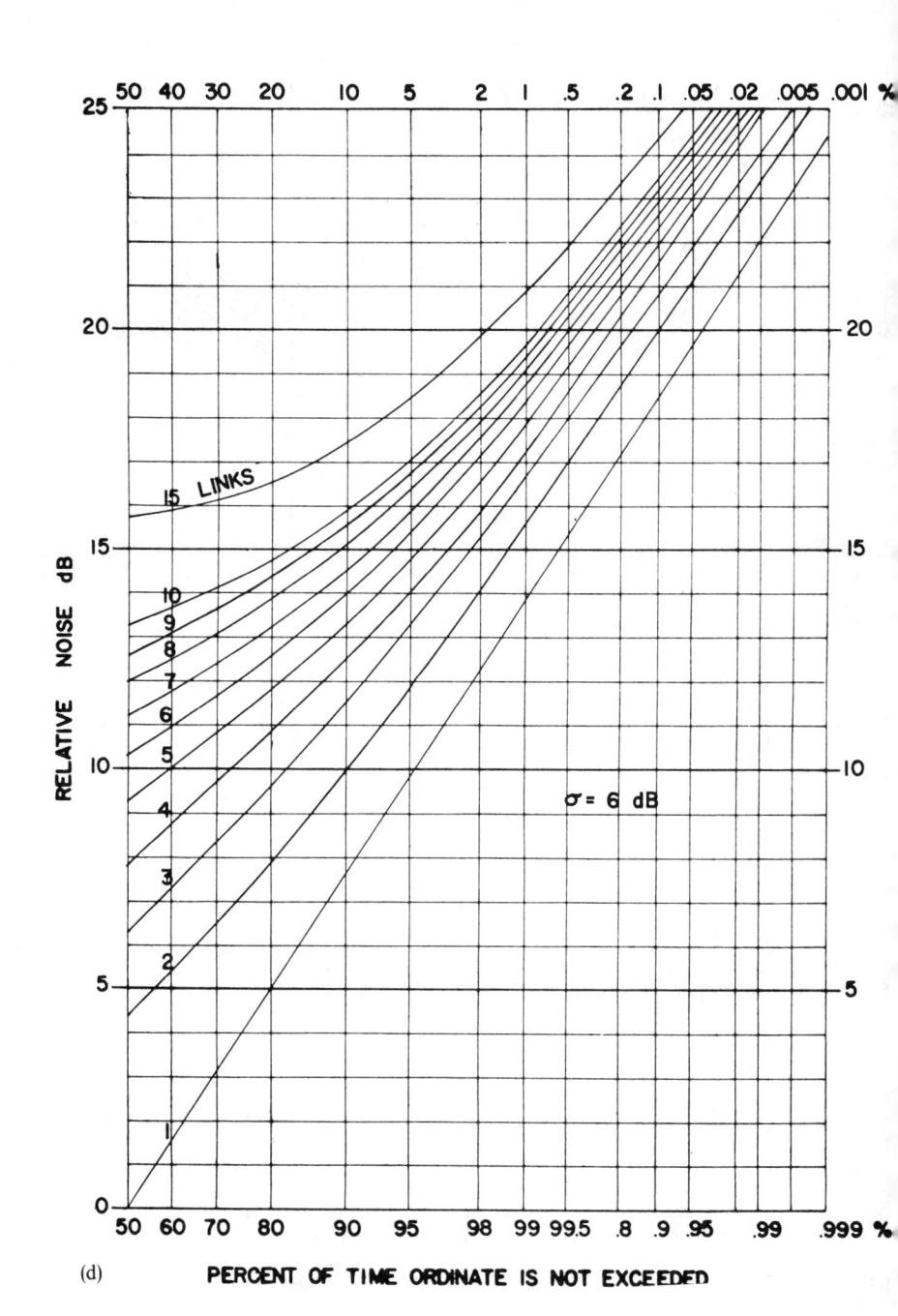

DESIGN OF ANALOG TROPOSCATTER LINKS

(e) **PERCENT OF TIME ORDINATE IS NOT EXCEEDED**

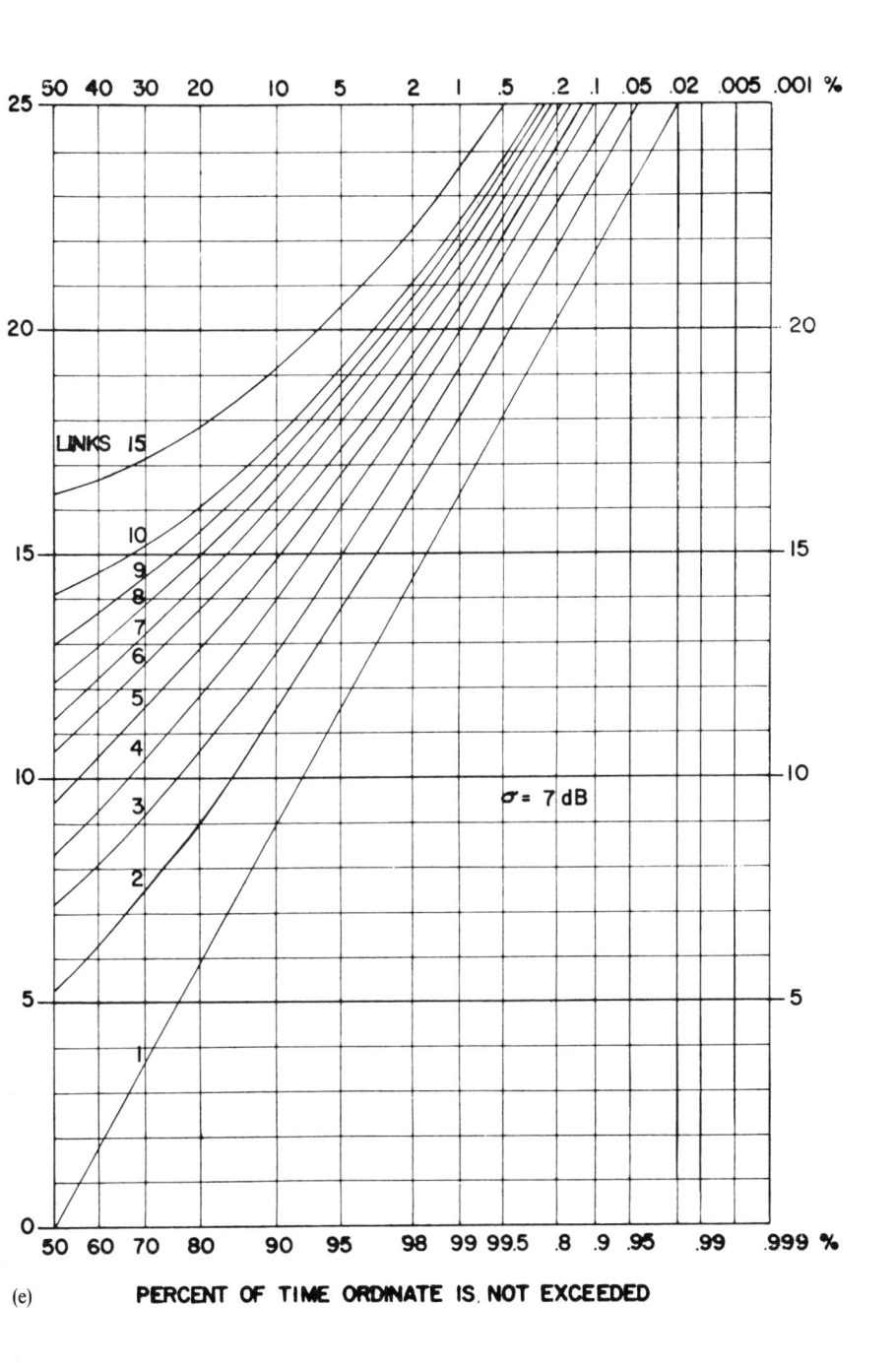

The above equations are valid if, in the presence of fast fading (Rayleigh fading), the 1 min median r.f. level at the receiver input is above the threshold level by at least 12 dB for dual diversity or at least 2 dB for quadruple diversity for the same $q\%$ of time. This condition can be written as a relation between noises:

$$N_1(50\%) + G(\sigma; 1; q) \leqslant N_t - k_t \tag{8.4}$$

where N_t is the noise at threshold (in dBm0p) for a receiver of any diversity branch and is obtained from eqn (7.5) or from Table 7.1, column 10, and k_t is 12 dB for dual diversity with maximal-ratio combination and 2 dB for quadruple diversity with maximal-ratio combination. Note that $G(\sigma; 1; q) = -Y(q\%)$.

So far we have considered the 1 min mean noise, but the "instantaneous" noise values are also of interest for performance calculations. We have seen in Chapter 6, Section 6.5 and Fig. 6.9, that the distribution of the instantaneous equivalent path loss for a single hop with diversity is given by

$$L(q\%) = L(50\%) - Z_c(q\%)$$

Instead of $Z_c(q\%)$ we can write more explicitly $Z(D; \sigma; q)$, where D is the order of diversity and σ is the standard deviation of the log-normal component. Some of these distributions are shown in Fig. 8.2 with different scales and represent the variations in the instantaneous output noise after a maximal-ratio diversity combination.

The distribution of the instantaneous noise levels in a telephone channel at the end of the path during the worst month can therefore be written as

$$N_1(q\%)(\text{dBm0p}) = N_1(50\%) - Z(D; \sigma; q) \tag{8.5}$$

It can be assumed that in practice $q\%$ is always very small for the hypothetical reference circuit with z sections equal to the path under study. Therefore it can be assumed to a good approximation that only one section at a time presents a deep fade. The corresponding high noise is almost the same at the end of the faded section and at the end of the reference circuit, but the percentage of time considered for the single path is a factor of z smaller. Therefore for the hypothetical reference circuit eqn (8.5) becomes

$$N_z(q\%)(\text{dBm0p}) = N_1(50\%) - Z(D; \sigma; q/z) \tag{8.6}$$

Equations (8.3), (8.4) and (8.6) will be used in the calculations of performance in Section 8.3.2.

8.2.2 Intermodulation noise

8.2.2.1 Intermodulation noise of the equipment

The value of the intermodulation noise of the equipment, as measured using a CCIR white noise signal, is normally provided by the equipment

DESIGN OF ANALOG TROPOSCATTER LINKS 213

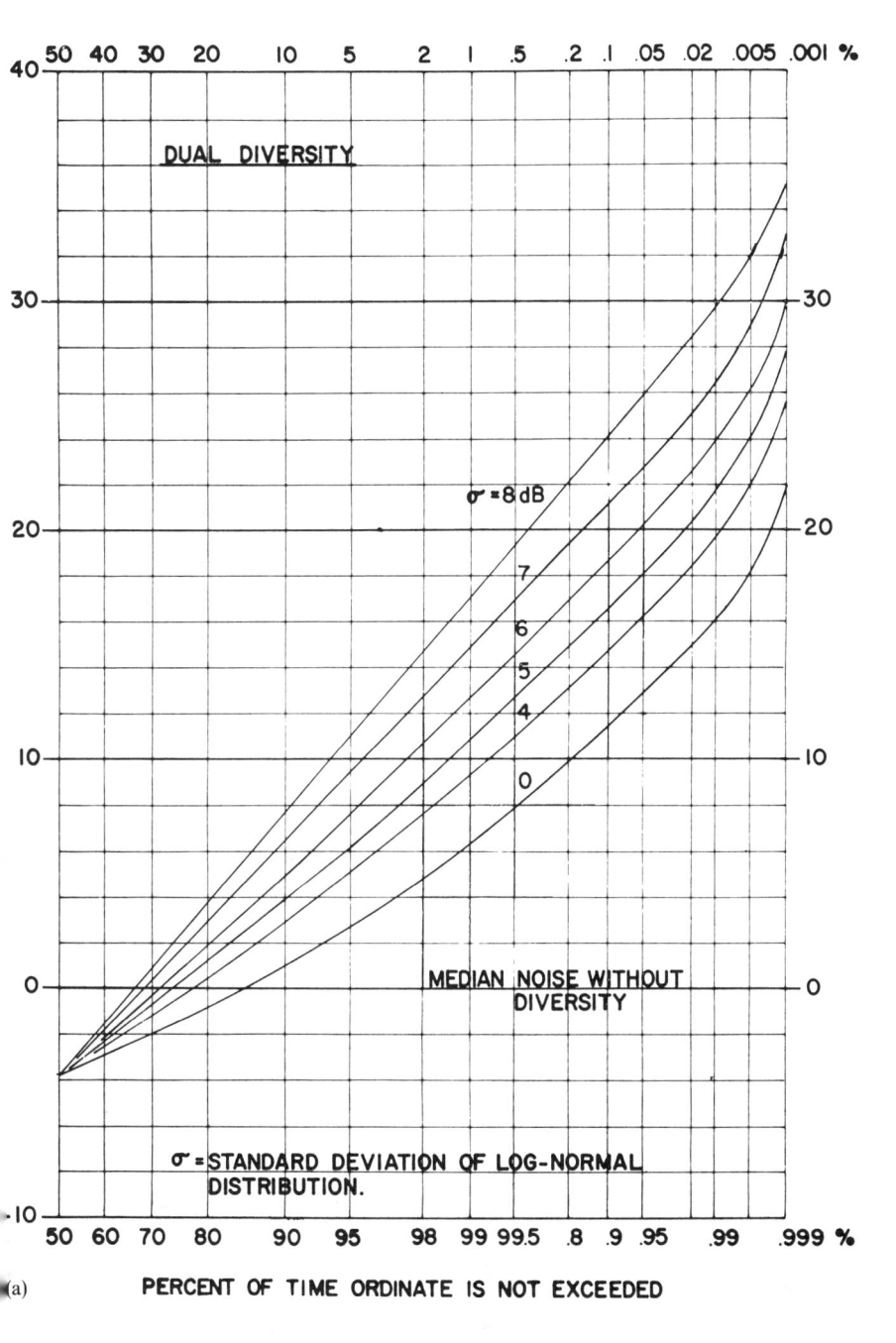

Fig. 8.2 Function $F = -Z(D;\sigma;q)$—distribution of noise for combined log-normal and Rayleigh fading: (a) dual diversity with maximal-ratio combination; (b) quadruple diversity with maximal-ratio combination.

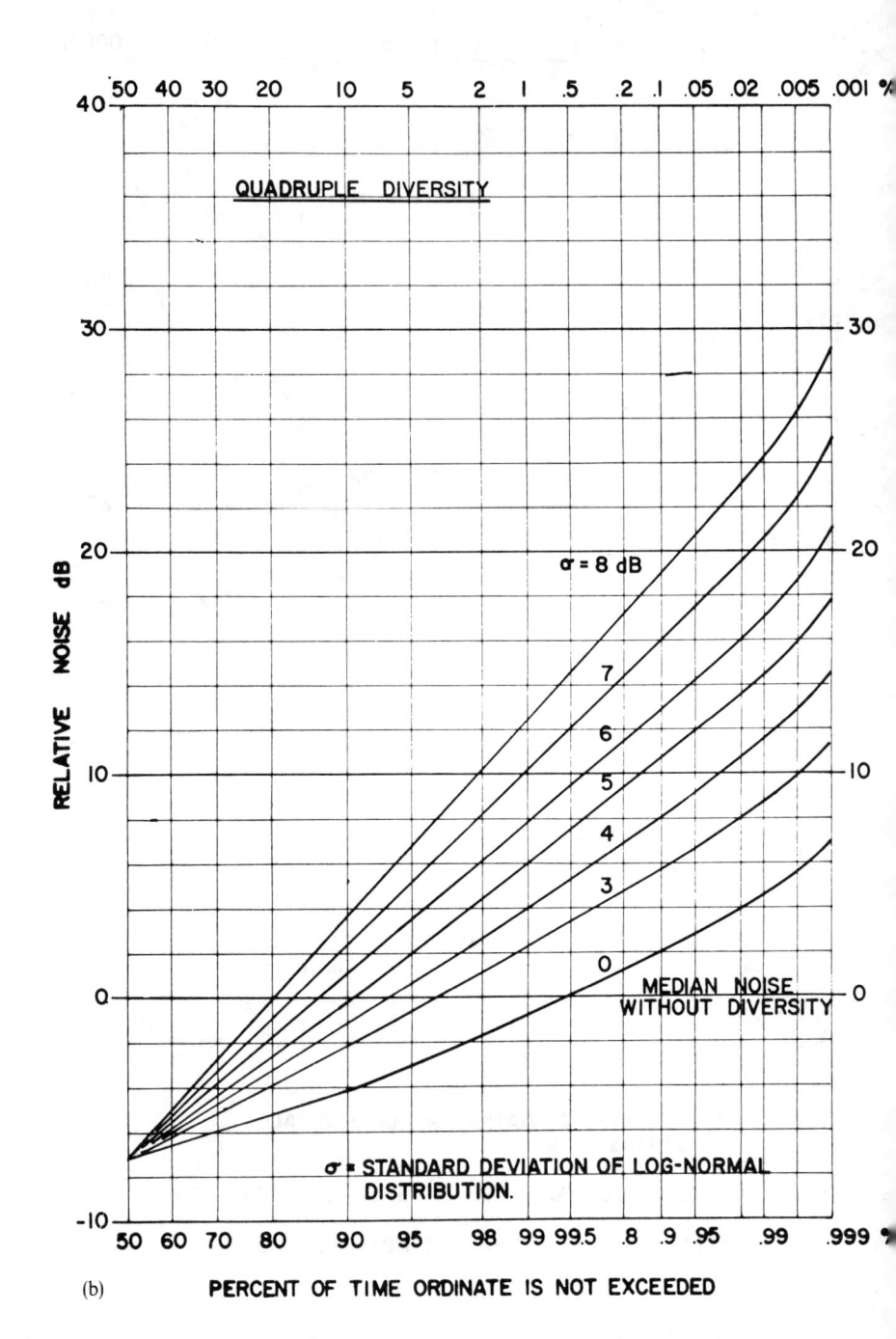

(b)

DESIGN OF ANALOG TROPOSCATTER LINKS 215

manufacturer. It is often defined using the concept of "slot ratio" or "noise power ratio" (NPR) as follows.

Consider the level diagram shown in Fig. 8.3. A CCIR white noise signal of appropriate level representing the baseband signal during the busy hour (remember eqn (8.1) and Table 7.1) is applied through the baseband of a loop consisting of a transmitter, a power amplifier, a receiver and a combiner. Its level is indicated by Q dBm0.

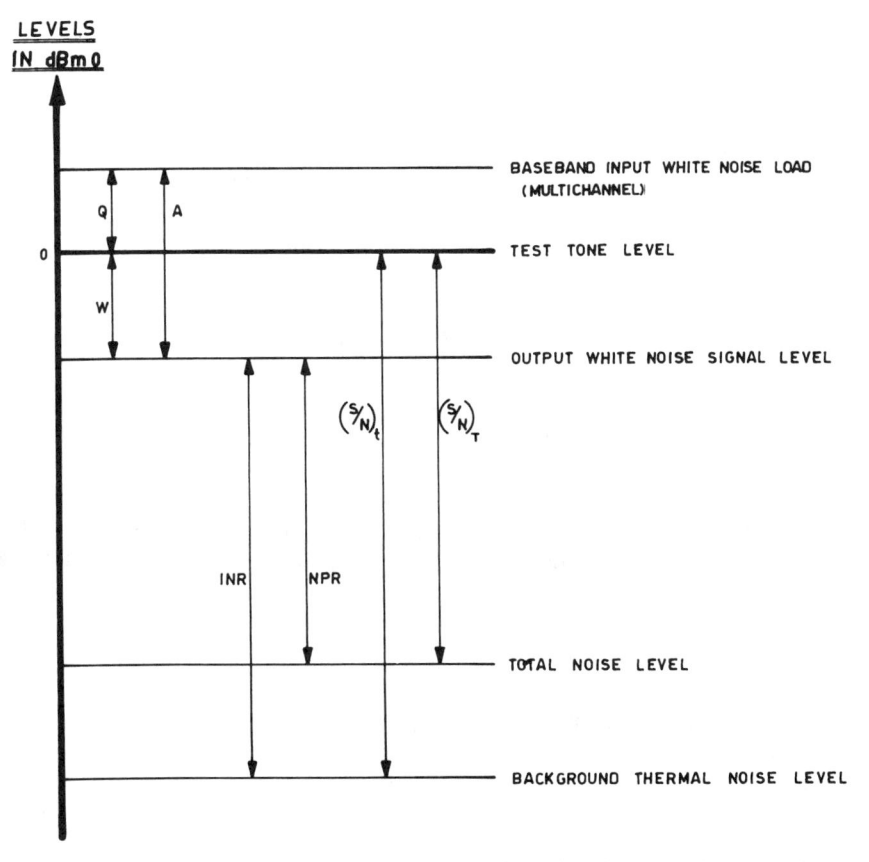

Fig. 8.3 Parameters and levels in the CCIR white signal test method: Q, CCIR baseband noise level (dBm0); A, 10 log(baseband/channel bandwidth) (dB); W, white signal level equivalent to voice (dBm0); INR, intrinsic noise ratio (dB); NPR, noise power ratio (dB); $(S/N)_t$, signal-to-thermal-noise ratio (dB); $(S/N)_T$, signal-to-total-noise ratio (dB).

In the 4 kHz slot of a telephone channel, this level becomes W dBm0 (approximately equal to $Q - 10 \log C$ where C is the number of channels). If the white signal generator is turned off, the noise decreases by INR dB, reaching the background noise level $-(\text{INR}+W)\,\text{dBm0}$. The same result should in principle be obtained if a band-stop filter in the white signal generator eliminates the noise in the slot. In practice, however, the noise is decreased by only NPR dB owing to the presence of intermodulation.

The total noise level is $-(\text{NPR}+W)\,\text{dBm0}$ and the difference INR $-$ NPR gives the intermodulation noise contribution. Normal values of NPR for troposcatter equipment may range from about 40 dB up to 60 dB.

The intermodulation noise power can also be given directly in pW0p. In any case it is measured with high r.f. levels at the receiver input, corresponding to the horizontal part of the noise curve (Fig. 7.6) where the thermal noise is a minimum.

8.2.2.2 Noise due to reflections in radio frequency feeders

Reflections in feeders cause the addition of electromagnetic waves frequency modulated by the same signal but with different modulation phases. The additional waves therefore have different instantaneous frequencies and this leads to intermodulation distortion. For example the transmit side of the radio equipment sends a wave to the antenna through the coaxial cable, which attenuates it by a factor a (or A dB). The antenna impedance is never perfectly matched to the cable, but it has a given SWR.

A fraction $r_a = (\text{SWR}-1)/(\text{SWR}+1)$ of the incident wave is reflected back as an echo of relative amplitude ar_a and, after another attenuation by the same factor a, again reaches the transmit output (in practice consisting of a duplexer or a filter). Again, there is no perfect impedance matching at this point but a given SWR. Thus another fraction r_e of the incident wave $a^2 r_a$ is reflected back. Therefore the transmit output consists of the sum of the original wave plus the echo of relative amplitude $R = a^2 r_a r_e$ or, in decibels, $R(\text{dB}) = 20\log R = 20\log(r_a r_e) + 2A$. This echo is not in phase with the original wave but is delayed by the round-trip propagation time and causes intermodulation distortion.

The example in Table 8.1 shows a practical calculation of the maximum expected intermodulation noise in the worst (top) telephone channel owing to reflections in the feeders. The input data are given in lines 1–10 and 13. Some of these data are taken from Table 7.1. The CCIR white noise signal load Q is that of eqn (8.1). The r.m.s. deviation of the composite signal is Q dB higher than the single-channel r.m.s. deviation (Q is 20 times the logarithm of the ratio of the two deviations).

The velocity factor of the coaxial cable multiplied by the velocity of light in a vacuum yields the propagation velocity inside the cable. The round-trip attenuation and propagation time, as well as the relative amplitude of the echo, are obtained immediately.

Figure 8.4 was calculated by Medhurst [8.1] and enables the slot ratio under the worst conditions (echo phased for maximum distortion) to be derived. If we add the nominal voice level in the channel (W in Fig. 8.3) and the

DESIGN OF ANALOG TROPOSCATTER LINKS 217

Table 8.1 Calculation of nonlinear feeder echo noise in the top telephone channel

1. Frequency band (MHz)	900	900
2. Number of channels	24	120
3. Highest baseband frequency f_b (kHz)	108	552
4. CCIR white noise signal load Q (dBm0)	+4.5	+7.3
5. R.m.s. deviation Δf of composite signal (kHz)	59	232
6. Modulation index $\Delta f / f_b$ of composite signal	0.55	0.42
7. SWR		
Antenna end	1.15	1.15
Equipment end	3.0	3.0
8. Reflection coefficient $r = (SWR-1)/(SWR+1)$		
Antenna end r_a	0.07	0.07
Equipment end r_e	0.50	0.50
9. Specific attenuation of coaxial cable (dB m^{-1})	0.02	0.02
10. Coaxial cable length (m)	40	40
11. Coaxial cable amplitude attenuation		
a	0.91	0.91
A (dB)	−0.8	−0.8
12. Round trip attenuation $2A$ (dB)	−1.6	−1.6
13. Velocity factor of coaxial cable	0.92	0.92
14. Round trip propagation time t (μs)	0.3	0.3
15. Echo amplitude $R = r_a r_e a^2$ or $20 \log R = 20 \log(r_a r_e) + 2A$ (dB)	−30.8	−30.8
16. Parameter $2\pi f_b t \times 10^{-3}$	0.20	1.0
17. Factor $\gamma + 20 \log R$ (Fig. 8.4) (dB)	41.0	14.0
18. Slot ratio γ in top channel (dB)	71.8	50.3
19. Correction for white signal level w (Fig. 8.3): $10 \log(f_b/3.1) - Q$ (dB)	10.9	15.2
20. CCITT weighting factor (dB)	2.5	2.5
21. Nonlinear noise contribution in one feeder (TX or RX):		
Level (dBm0p)	−85.2	−68.0
Power (pW0p)	7.2	160
22. Nonlinear noise contribution in two feeders (TX+RX) (pW0p)	14.4	320

psophometric factor to the slot ratio, we obtain the noise in dBm0p. The total intermodulation noise for the two feeders can then be calculated immediately.

In practice the impedance matching at the antennas and at the receive connector is relatively good (e.g. the SWR is less than 1.2), whereas at the transmit connector it is easy to find a bad SWR because of the intrinsic mismatch of the power tube in the final power amplifier. In this case, if the noise generated is excessive, an isolator is installed at the transmit output connector. It increases the feeder loss on the transmit side by about 0.5–1 dB, but the reflected wave from the antenna finds quasi-perfect impedance matching, thus preventing further reflection.

8.2.2.3 Noise due to multipath effects during propagation

The nonlinear noise is generated by the addition of electromagnetic waves with out-of-phase frequency modulation and therefore with different

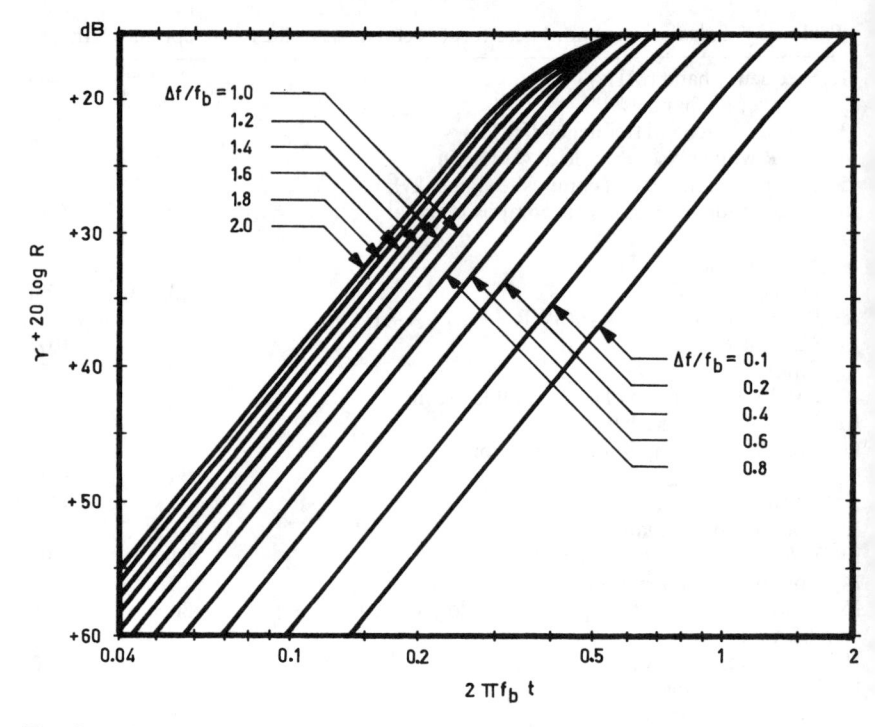

Fig. 8.4 Curves for the top telephone channel of an FM system showing the intermodulation due to the echo phased for maximum distortion in the antenna feeder lines. It is used in conjunction with Table 8.1.

instantaneous frequencies. The main difference with respect to the former case (reflections in feeders) is that multiple echoes are present instead of a single echo. The problem of calculating the noise generated in this case is complex and has not yet been solved. A number of methods have been proposed by various workers, and we shall follow that of Beach and Trecker [8.2] which is also accepted by the CCIR. This method is approximate and is based on the effect of a single echo. Multiple echoes are accounted for by an additional factor. The formulae are derived from a mathematical theory of troposcatter propagation.

The example in Table 8.2 shows a simplified practical calculation of the multipath intermodulation noise in the top telephone channel. The input data are given in lines 1–15, and the first six lines are the same as those of Table 8.1.

In the propagation model it is assumed that the axes of the antennas are tangential to the horizons and that they constitute the main signal path to which all other paths are referred for the determination of attenuation and delay. The *echo loss* defines the magnitude of an echo (signal scattered in some part of the common volume) with respect to the magnitude of the main signal. The *echo delay* is defined similarly.

DESIGN OF ANALOG TROPOSCATTER LINKS 219

Table 8.2 Calculation of multipath nonlinear noise in the top telephone channel

1. Frequency band (MHz)	900	900
2. Number of channels	24	120
3. Highest baseband frequency f_b (kHz)	108	552
4. CCIR white noise signal load Q (dBm0)	+4.5	+7.3
5. R.m.s. deviation Δf of composite signal (kHz)	59	232
6. Modulation index $\Delta f/f_b$ of composite signal	0.55	0.42
7. Path length d (km)	385	385
8. Path angular length $\theta_0 = d/8.5$ (mrad)	45.3	45.3
9. Elevation angle θ_1 at site a (mrad)	−6.2	−6.2
10. Elevation angle θ_2 at site b (mrad)	−5.4	−5.4
11. Difference Δh in height between sites a and b (m)	426	426
12. Angle $\alpha = \theta_0/2 + \Delta h/d + \theta_1$ (mrad)	17.6	17.6
13. Angle $\beta = \theta_0/2 - \Delta h/d + \theta_2$ (mrad)	16.1	16.1
14. Diameter of antennas (m)	12	18
15. Antenna 3 dB beamwidth Ω (mrad)	33.4	22.8
16. Parameter $\rho = 1 + \Omega/\theta_0$	1.73	1.50
17. Parameter $\rho^2 - 1$	1.99	1.25
18. Multipath delay $t = 1.66 d\alpha\beta(\rho^2 - 1) \times 10^{-6}$ (μs)	0.36	0.22
19. Parameter $f_b t \times 10^{-3}$	0.039	0.12
20. Echo loss $R = 70 \log \rho$ (dB)	+16.6	+12.3
21. Interference excess E (Fig. 8.5) (dB)	+46.0	+27.0
22. Off-axis loss z of antenna gain (dB)	+6.0	+6.0
23. Diversity factor D (diversity order 2) (dB)	+3.0	+3.0
24. Correction e for multiple echo (dB)	−9.0	−9.0
25. Slot ratio $\gamma = R + E + z + D + e$ (dB)	62.6	39.3
26. Correction w for white signal level (Fig. 8.3): $10 \log(f_b/3.1) - Q$ (dB)	+10.9	+15.2
27. CCITT weighting factor (dB)	+2.5	+2.5
28. Nonlinear noise contribution		
Level (dBm0p)	−76.0	−57.0
Power (pW0p)	25.0	2,000

The calculations start by considering a single echo corresponding to the 3 dB upper edges of the antenna patterns, so that its magnitude with respect to the main signal is $-3-3 = -6$ dB fixed. The theoretical echo loss is $R = 70 \log \rho$, where ρ is given in Table 8.2, line 16. The echo delay (Table 8.2, line 18) is calculated using a less general formula than that derived from our eqn (3.14). In the case of a single echo Fig. 8.5, which is derived from a rearrangement of Fig. 8.4, yields the interference excess factor E given in Table 8.2, line 21.

The slot ratio or NPR for the top telephone channel is the algebraic sum of the following terms:

(1) the echo loss of the off-axis rays with respect to the axial rays;
(2) the interference excess;
(3) the loss of gain for the chosen off-axis rays (6 dB fixed);

Fig. 8.5 Curves for the top telephone channel of a pre-emphasized FM system representing the excess intermodulation above the echo loss due to multipath propagation. It is used in conjunction with Table 8.2.

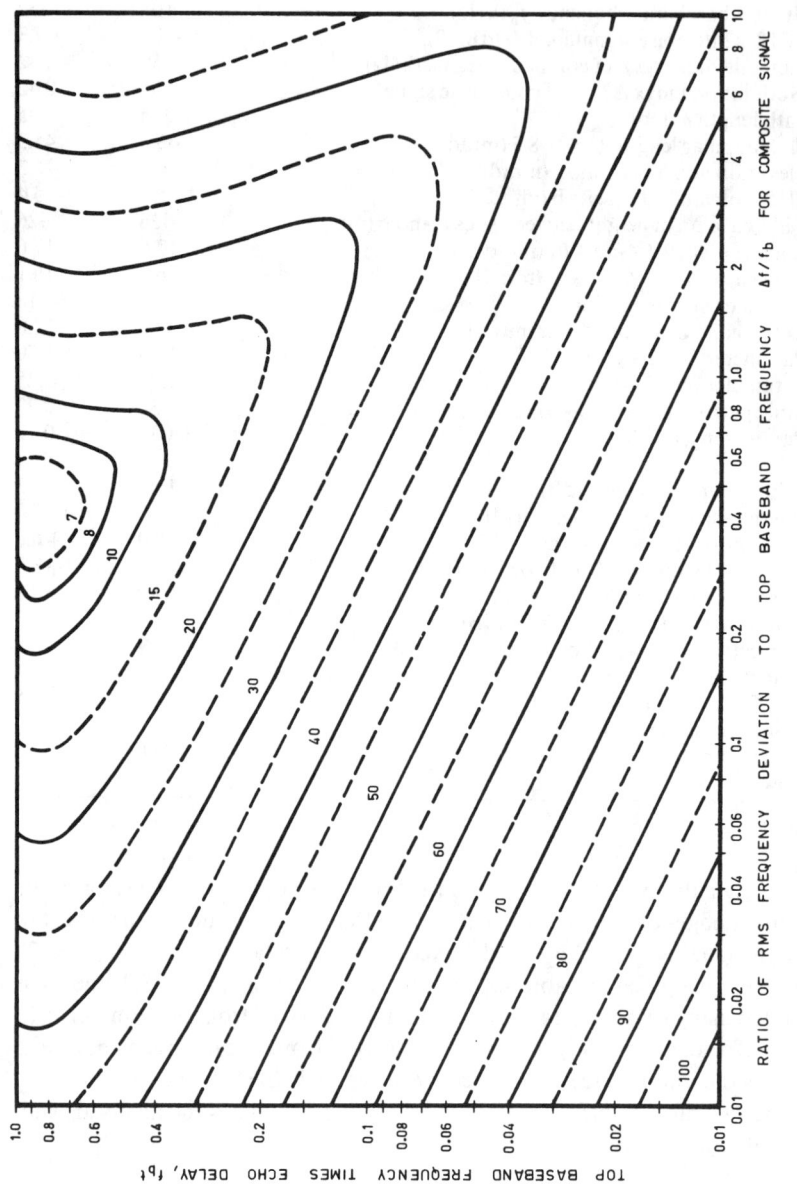

(4) the noise decrease due to diversity (3 dB for dual diversity and 6 dB for quadruple diversity);

(5) a correction factor for accounting for multiple echoes (9 dB fixed).

Once the slot ratio is known, the calculation proceeds as in Table 8.1. A further refinement could be to repeat the calculations for different off-axis rays and to look for the maximum noise.

Paths near the main path yield strong echoes but with a very small differential delay, whereas paths far from the main path yield weak echoes with a large differential delay. Between these two extreme conditions, in which the noise is almost negligible, there is an intermediate situation (found by calculation) in which the noise reaches a maximum. In the above calculation we have assumed that this maximum noise corresponds to the path of the two 3 dB upper edges of the antenna patterns.

The multipath noise calculated in this way is valid for the worst month. For other periods we should subtract 0.7 dB for each decibel decrease in path loss.

8.3 Path calculation

In principle the path calculation consists in comparing the equipment capability to overcome attenuation with the maximum expected path loss. This comparison should leave a sufficient margin for achieving the required performance.

In general the most appropriate technical and economical solution is found after several solutions with different choices for the various parameters have been investigated. The starting data are as follows.

(a) The equipment capability E_q in decibels (see Chapter 7, Section 7.10) is given by

$$E_q(\text{dB}) = P_t + G_p - l_i - T_d + g_d - i \qquad (8.7)$$

where the symbols are those used in eqn (7.14) and $G_p = G_t + G_r - \Delta g$ is the path antenna gain, g_d is the diversity gain, whose value is chosen as explained below but is generally set 1 or 2 dB lower than the theoretical value, and i is an implementation margin of a few decibels that we have added to account for various types of degradation in the plant (aging, misalignment, inaccuracies etc.).

(b) The path loss not exceeded for the percentages of time $q\%$ stated in the performance objectives (Section 8.1) is given by

$L(q\%) = L(50\%) - Y(q\%)\,\text{dB}$ for median values

$L(q\%) = L(50\%) - Z(q\%)\,\text{dB}$ for instantaneous values

The final terms Y and Z may or may not include the prediction error with the required degree of confidence (eqns (6.11), (6.13)).

(c) The signal-to-noise ratio $(S/N)_t$ at threshold in dBp or the noise level N_t at threshold in dBm0p in the top telephone channel without diversity is used (Section 7.4.3). Remember that $N(\text{dBm0p}) = -S/N$ dBp.

(d) The signal-to-noise ratio S/N in dBp or the noise level N in dBm0p required by the performance objectives for not less than $q\%$ of time in the top telephone channel (obviously with diversity) is needed. According to the objectives this noise level, better indicated as $N(q\%)$, can be median (or mean) or instantaneous.

The equipment capability must exceed the given path loss by at least

$$N_t - N(q\%) = S/N - (S/N)_t \text{ dB}$$

for the specified time percentage q. We can therefore write the following inequalities:

for median values

$$E_q(\text{dB}) \geqslant L(50\%) - Y(q\%) + N_t - N(q\%) \tag{8.8}$$

with the median value $g_d(50\%)$ in E_q and the median value of $N(q\%)$; for instantaneous values

$$E_q(\text{dB}) \geqslant L(50\%) - Z_c(q\%) + N_t - N(q\%) \tag{8.9}$$

with $g_d = 0$ in E_q (the diversity gain is already included in Z_c) and the instantaneous value of $N(q\%)$.

Several sets of system parameters (transmit power, antenna diameter, receiver threshold, filter losses etc.) can satisfy the path inequality and each set leads to a different cost. The system parameters can then be optimized by considering the performance requirements and the different "cost of the decibel" involved in the choice of the various parameters.

We shall now show how to calculate the path in two typical cases.

8.3.1 Conventional troposcatter links

When a troposcatter radio link composed of one or more paths is part of an independent communications network the high quality of public links is not generally required and the link can be implemented at a significantly lower cost. In this case the performance specifications could be (a) or (b) of Section 8.1 for example: the hourly median or the instantaneous signal-to-noise ratio in the worst telephone channel should exceed 30 dBp for 99.9% of the worst month. Since such values are near threshold, the total noise is almost equal to its thermal component (see Fig. 7.8). However, it is also convenient to calculate the thermal noise not exceeded for 80% of time and to verify that the intermodulation noise does not exceed this value.

We consider first the specifications for median values and calculate the single path using inequality (8.8). Figure 8.6 shows an analysis of such a path in diagrammatic form. The example shown is a dual diversity system with a capacity of 12 FDM telephone channels, an r.f. power of 300 W and antennas of diameter 12 m working at 400 MHz.

Starting from a zero reference level the gains and losses of the equipment are added algebraically until the equipment capability $E_q = 219$ dB is found. E_q is represented as a vertical vector starting from the telephony threshold level and reaching some high level. The median (178 dB) and the slow fading losses are shown as vectors pointing downwards from that level. The broken lines and the scales on the right-hand side represent the set of signal levels received for various percentages of time, as well as the corresponding performance in the worst telephone channel and the margins above threshold.

All scales in the figure are calculated for a single receiver without diversity. The effect of diversity is represented by the small vector $g_d = 3$ dB at the left-hand top corner of the figure which shifts the whole path loss diagram upward and at the same time increases the output signal-to-noise ratio. For example, the monthly median level of the received r.f. equivalent signal is -60 dBm, corresponding to 41 dB above threshold and to a thermal noise level of -71 dBm0p in the worst telephone channel. The median level actually received by each diversity receiver is lower by the median diversity gain and the thermal noise is higher by the same amount. In this case the true level is $-60 - 3 = -63$ dBm and the thermal noise is $-71 + 3 = -68$ dBm0p.

For 80% of the worst month the hourly median equivalent level is -69 dBm, which is 32 dB above threshold, and the thermal noise is -62 dBm0p. It should now be verified that the intermodulation noise due to the equipment, the feeders and the multipath is not higher than the thermal noise, so that the total noise does not exceed -59 dBm0p.

For 99.9% of the month (corresponding to a total outage of 43 min) the slow fading will not exceed 33 dB and the equivalent signal will be higher than -93 dBm, with an hourly median noise (almost all thermal) not exceeding -38 dBm0p. The binary and the character error rates for a telegraph channel carried by a telephone channel are also shown in Fig. 8.6. They were calculated using a procedure explained later. The fast fading loss shown at the bottom of the 99.9% hourly median loss gives an idea of the performance during the worst hour. The values of the fast fading losses (in decibels) take into account the upward shift due to the median diversity gain.

A more complete visualization of path performance is obtained by combining Fig. 8.6 with Fig. 6.5 (calculated for the path under study), which gives an idea of the uncertainties of prediction.

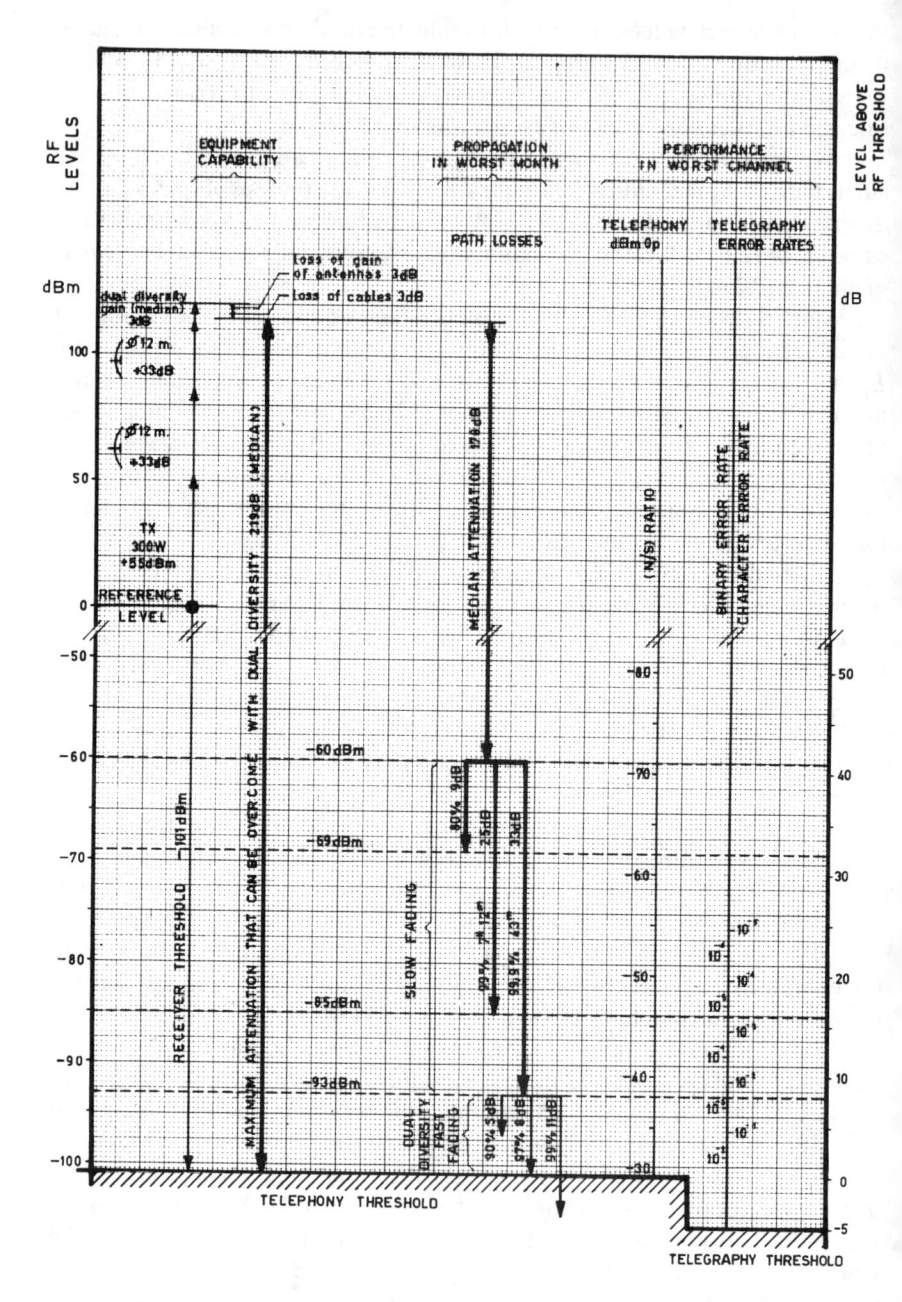

Fig. 8.6 Graphic design and performance evaluation of a troposcatter path.

The above results show that the path considered is overdimensioned with respect to the performance specifications. Therefore we could decrease the transmit power or the antenna diameters or some other parameter. Alternatively, in view of the prediction errors, we could keep the system as it is if the costs are acceptable. The margins may be useful for future expansion of the system.

When the troposcatter radio link is composed of z paths in cascade there is the additional problem of deriving the specifications for each path from those of the complete link. In the case of the 1 min median thermal noise the problem is solved with the aid of Fig. 8.1. It is assumed that the z paths all behave in a similar manner and that their loss distribution has the same standard deviation. If, for example, the requirement for a link of three hops with $\sigma = 6$ dB is a median noise of -35 dBm0p (almost all thermal), which is not exceeded for 99% of time, Fig. 8.1(d) shows that a single hop should have a median noise 2.7 dB lower for the same 99% of time. The (maximum) intermodulation noise is assumed to be fixed on each path and equally subdivided between paths. In practice this noise varies, but its maxima generally do not appear at the same time as the thermal noise maxima.

On other occasions the performance specifications may be based on "instantaneous" values. For example, it may be required that the noise in a single path should not exceed -42 dBm0p for 99.9% of the worst month. This noise is almost all thermal. The inequality to be applied is (8.9). In the former example with $E_q = 219$ dB and $L(50\%) = 178$ dB, Fig. 6.9(c) for 99.9% yields $Z_c = -17$ dB if the standard deviation σ of the slow fading is 5 dB. In this case, if the noise at threshold is -30 dBm0p, eqn (8.9) becomes

$$219 > 178 + 17 - 30 + 42 = 207 \text{ dB}$$

With respect to this requirement the system is overdimensioned by 12 dB, but part of this margin could be taken up by the prediction error or we can say that there is a greater degree of confidence than expected. In either case we could reduce some parameters.

If the above specifications are requested for the performance of a link composed of three approximately equal hops for example, the 0.1% outage time of the link becomes an outage time of $0.1/3 = 0.033\%$ for each individual hop for the same noise level, and from Fig. 6.9(c) we obtain $Z_c = -20$ dB.

8.3.2 High quality troposcatter links

When a troposcatter radio link composed of one or more paths (hops) is integrated in a public communication network or an extended military network, the quality of the transmission must satisfy stringent requirements. We shall consider in particular the quality required by the CCIR Recommendations, as this is normally specified in most international Requests for Bids.

The following two different classes of troposcatter radio relay system are considered in *CCIR Recommendation 397-3*.

(a) *First class*, including systems intended to operate between points for which other transmission systems could be used without great difficulty: the performance recommendations for this class are the same as those for LOS radio relay systems.

(b) *Second class*, including systems to be used between points for which other transmission systems cannot be used without great difficulty or for which the recommendations of the first class cannot be met: broader performance limits are accepted for this class.

Since the cost of a high quality troposcatter system is generally greater than that of an LOS system of the same length and capacity, most of the systems to be considered will fall into the second class.

The CCIR Recommendations are given in terms of allowable noise power in any telephone channel at the end of the hypothetical reference circuit during the worst month. The CCIR Noise Clauses and the subdivision of the noise components in a form suitable for further calculations are given in Table 8.3. The factor 1.8 (2.5 dB) in Clauses I(c) and II(c) is the CCITT weighting factor introduced to allow comparison with the other clauses.

If these CCIR Clauses are satisfied, the path under study is considered to be acceptable as part of a circuit carrying international traffic. For a troposcatter link composed of more than one path, each path can be individually analyzed and compared with CCIR Recommendations.

For design purposes the total noise of the CCIR Clauses is first split into thermal and nonlinear (intermodulation) components. The allowable nonlinear noise is conventionally taken as constant for all the time and equal to half the total noise of Clauses I(a) and II(a).

If it is assumed that the noise powers are additive, the following inequalities for the various components of the nonlinear noise can be applied to the hypothetical reference circuit:

$z(N_e + N_f + N_p) \leqslant 3750$ pW0p   first-class systems

$z(N_e + N_f + N_p) \leqslant 12\,500$ pW0p   second-class systems

where z is the number of hops in the hypothetical reference circuit, N_e is the noise due to the radio equipment (one hop), N_f is the noise due to the r.f. feeders (one hop) and N_p is the noise due to multipath propagation (one hop). These three types of nonlinear noise can be calculated or measured for the path under study, and their agreement with the above inequalities can be checked.

The allowable thermal noise is obtained from the total allowable noise after deduction of the nonlinear component (see the right-hand side of Table 8.3). The inequalities for the thermal noise are written immediately (Table 8.4)

DESIGN OF ANALOG TROPOSCATTER LINKS 227

Table 8.3 CCIR performance objectives for troposcatter radio relay links[a]

CCIR noise clauses[b]	Derivation of thermal noise from total noise minus intermodulation noise	
I First-class systems		
(a) 1 min mean power exceeded for not more than 20% of minutes in the worst month: 7500 pW0p	7500 − 3750 = 3750 pW0p	= −54.3 dBm0p
(b) 1 min mean power exceeded for not more than 0.1% of minutes in the worst month: 47 500 pW0p	47 500 − 3750 = 43 750 pW0p	= −43.6 dBm0p
(c) 5 ms mean power exceeded for not more than 0.01% of time in the worst month: 1 000 000 pW0	$10^6/1.8 - 3750$ = 551 806 pW0p	= −32.5 dBm0p
II Second-class systems		
(a) 1 min mean power exceeded for not more than 20% of minutes in the worst month: 25 000 pW0p	25 000 − 12 500 = 12 500 pW0p	= −49.0 dBm0p
(b) 1 min mean power exceeded for not more than 0.5% of minutes in the worst month: 63 000 pW0p	63 000 − 12 500 = 50 500 pW0p	= −43.0 dBm0p
(c) 5 ms mean power exceeded for not more than 0.05% of time in the worst month: 1 000 000 pW0	$10^6/1.8 - 12\,500$ = 543 056 pW0p	= −32.6 dBm0p

[a] Allowable noise power in an FDM telephone channel of the hypothetical reference circuit during the worst month.
[b] *Recommendation 397-3.*

if we suppose that a maximal-ratio combiner is used (which is true in most cases) so that we can apply eqns (8.3), (8.4) and (8.6).

Each inequality relates the allowable thermal noise $N_z(q)$ in the hypothetical reference circuit to the corresponding allowable monthly median

Table 8.4 Conditions for CCIR thermal noise

CCIR Clause	Corresponding inequality (all noise levels in dBm0p)	Diversity order 2	Diversity order 4
I First-class systems			
I(a)	$N_1(50) + G(\sigma; z; 20\%) \leqslant -54.3 + h_d$	-52.7	-47.9
I(b)	$N_1(50) + G(\sigma; z; 0.1\%) \leqslant -43.6 + h_d$	-42.0	-37.2
I(c)	$N_1(50) - Z(D; \sigma; 0.01/z) \leqslant -32.5$	-32.5	-32.5
II Second-class systems			
II(a)	$N_1(50) + G(\sigma; z; 20\%) \leqslant -49.0 + h_d$	-47.4	-42.6
II(b)	$N_1(50) + G(\sigma; z; 0.5\%) \leqslant -43.0 + h_d$	-41.4	-36.6
II(c)	$N_1(50) - Z(D; \sigma; 0.05/z) \leqslant -32.6$	-32.6	-32.6
Validity conditions			
I(d)	$N_1(50) + G(\sigma; 1; 0.1\%) \leqslant N_t - k_t$	$N_t - 12$	$N_t - 2$
II(d)	$N_1(50) + G(\sigma; 1; 0.5\%) \leqslant N_t - k_t$	$N_t - 12$	$N_t - 2$

For dual diversity $h_d = 1.6$ and $k_t = 12$ (dB) and for quadruple diversity $h_d = 6.4$ and $k_t = 2$ (dB) (assuming maximal-ratio combination).

noise $N_1(50)$ without diversity in a single path. In fact, the only unknown is $N_1(50)$. The value chosen for $N_1(50)$ is the minimum of the values obtained from the single inequalities and represents the maximum allowable median noise without diversity in the worst month.

Simple calculations show that, for first-class systems in the various CCIR climatic zones, Clause I(a) is generally the most stringent for both dual and quadruple diversity except for the desert climate when Clause I(b) sometimes prevails. For second-class systems either Clause II(a) or Clause II(b) may prevail [8.3]. The path inequality, instead of (8.8) and (8.9), becomes the simpler

$$E_q(\text{dB}) \geqslant L(50) + N_t - N_1(50) \quad (8.10)$$

with $g_d = 0$ in E_q (diversity excluded). Substituting the chosen minimum value of $N_1(50)$ in (8.10) we find the minimum necessary value of E_q without diversity, from which we can dimension the system.

An example of the calculation is given in Table 8.5. A system with $z = 9$, $\sigma = 6$ dB and a tentative value of 225.4 dB for E_q is checked for operation in dual and quadruple diversity with maximal-ratio combination. The example is self-explanatory. It is assumed that the intermodulation noise has been calculated elsewhere.

8.4 Telegraphic transmission performance

Sometimes one or more telegraph channels are transmitted through a

DESIGN OF ANALOG TROPOSCATTER LINKS

CCIR Clause	Order of diversity	Allowed median thermal noise $N_1(50)$ (dBm0p)	Hop $X-Y$ ($\sigma = 6\,\text{dB}; z = 9$)		Dual diversity	Quadruple diversity
Thermal noise conditions						
II(a)	2	$-47.4 - G(6;9;20)$	=	$-47.4 - 14.4$	-61.8	
	4	$-42.6 - G(6;9;20)$	=	$-42.6 - 14.4$		-57.0
II(b)	2	$-41.4 - G(6;9;0.5)$	=	$-41.4 - 20.6$	-62.0	
	4	$-36.6 - G(6;9;0.5)$	=	$-36.6 - 20.6$		-57.2
II(c)	2	$-32.6 + Z(2;6;0.05/9)$	=	$-32.6 - 25.5$	-58.1	
	4	$-32.6 + Z(4;6;0.05/9)$	=	$-32.6 - 18.1$		-50.7
II(d)	2	$-26.0 - 12.0 - G(6;1;0.5)$	=	$-38.0 - 15.4$	-53.4	
	4	$-26.0 - 2.0 - G(6;1;0.5)$	=	$-28.0 - 15.4$		-43.4
Maximum allowable median noise without diversity (dBm0p)		$N_1(50)$	=		-62.0	-57.2
Noise at threshold in worst channel without diversity (dBm0p)		N_t	=		-26.0	-26.0
Margin above threshold for meeting CCIR Recommendation 397-3 (dB)		$N_t - N_1(50)$	=		36.0	31.2
Maximum loss overcome without diversity (dB)		E_q	=		225.4	225.4
Worst-month median path loss (dB)		$L(50)$	=		192.5	192.5
Effective margin above threshold (dB)		$E_q - L(50)$	=		32.9	32.9
Residual (if $\geqslant 0$ CCIR Recommendation is met) (dB)		$E_q - L(50) - \{N_t - N_1(50)\}$	=		-3.1	$+1.7$
Intermodulation noise conditions						
Intermodulation noise contribution of the equipment (pW0p)		zN_e	=		4500	3000
Intermodulation noise contribution of feeders (pW0p)		zN_f	=		710	710
Intermodulation noise contribution of propagation (pW0p)		zN_p	=		8600	4300
Total intermodulation noise (pW0p)		$zN_i = z(N_e + N_f + N_p)$	=		13810	8010
Residual (if $\geqslant 0$ CCIR Recommendation is met) (pW0p)		$12\,500 - zN_i$	=		-1310	$+4490$

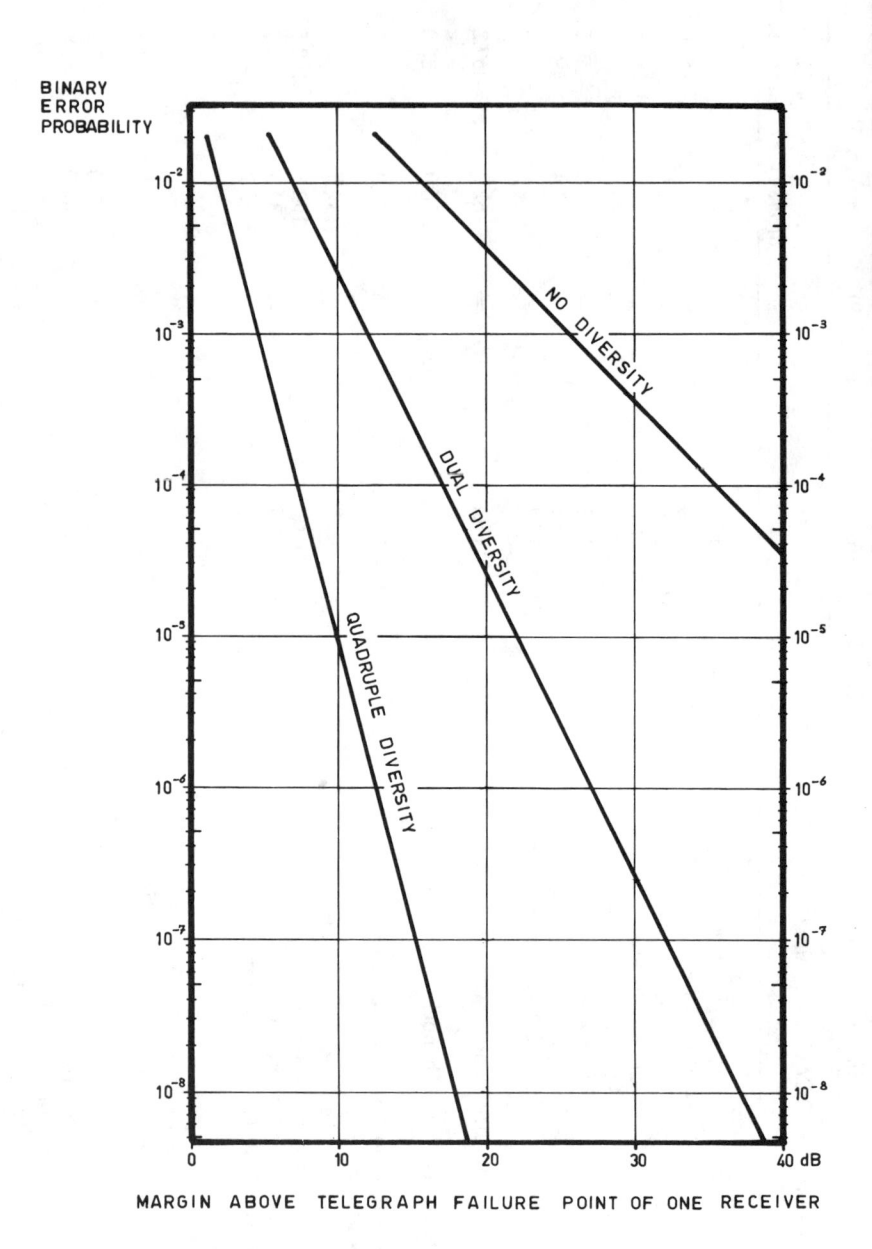

Fig. 8.7 Curves for the telegraphic error probability.

telephone channel of the multiplex. In general they are CCITT channels working at 50 or 100 bauds. The transmission quality needs to be evaluated in terms of the rate of received errors. These errors may refer to the transmitted bits, when a BER is defined, or to the characters of a teletype, when a character error rate is defined. There is a fixed ratio between the two error rates for any system.

Errors in telegraphic transmission are caused mainly by the disappearance of the telegraph signal into the noise due to fading. The thermal noise level in a telegraph channel with a nominal bandwidth of say 120 Hz (CCITT) is $10 \log(3100/120) = 14.1$ dB lower than that in a telephone channel of effective bandwidth 3100 Hz (CCITT). In addition, the telegraph signal is generally about 20 dB lower than the telephone test tone (CCITT).

The signal-to-noise ratio for telegraphy is therefore about 6 dB lower than that for telephony, but this is still higher than 20–25 dB at threshold. As in telegraphy the BER is about 10^{-4} for a signal-to-noise ratio of 10 dB, the result is that there are virtually no errors at the telephony threshold and the telegraphic threshold is lower (by say 5 dB) than the telephony threshold.

The only important cause of error is short-term fading, which makes the signal bounce rapidly upwards and downwards across the threshold. In the presence of Rayleigh fading, when the r.f. carrier is below threshold, BER $= 1/2$ because the two binary elements appear at random. When the carrier is above threshold there are no errors. The BER is then equal to half the probability that the r.f. signal is below threshold in all diversity branches at the same time.

The BER in the presence of Rayleigh fading is given by the following formula, derived from (2.20(a)):

$$\text{BER} = \tfrac{1}{2}\{1 - \exp(-0.7 \times 10^{0.1X})\}^D \qquad (8.11)$$

where D is the order of diversity and X (dB) is the difference in level between the telegraphic threshold and the median r.f. carrier. Figure 8.7 is a graphic representation of eqn (8.11). The character error rate is obtained by multiplying the BER by an appropriate factor depending on the telegraphic system used.

Experience with existing troposcatter links has shown that a satisfactory performance for telegraphy transmission which meets the usual requirement of a character error rate lower than 1×10^{-4} can be obtained only with quadruple diversity systems.

8.5 Comments and suggestions for further reading

The method of comparing the equipment capability with the path loss comes from ref. 8.4. Examples of path design can also be found in refs 8.5 and 8.6.

References

8.1 Medhurst, R. G. Echo distortion in frequency modulation, *Electron. Radio Eng.*, **36** (7), 253, 1959.

8.2 Beach, C. D. and Trecker, J. M. A method for predicting interchannel modulation due to multipath propagation in FM and PM tropospheric radio systems, *Bell Syst. Tech. J.*, **42** (1), 1–36, 1963.

8.3 Meek, T. J. and Roda, G. Tropospheric scatter radio systems and their integration with public communication networks, *14th Int. Conv. on Communications, Genoa, 12–15 October 1966*, Istituto Internazionale delle Communicazioni, Genoa, 1966.

8.4 du Castel, F. *Propagation Tropospherique et Faisceaux Hertziens Transhorizon*, Chiron, Paris, 1961.

8.5 Panter, P. F. *Communication Systems Design—Line-of-sight and Troposcatter Systems*, McGraw-Hill, New York, 1972.

8.6 Transmission systems—economic and technical aspects of the choice of transmission systems, *GAS 3 Manual*, Vols 1 and 2, ITU, Geneva, 1976.

Chapter 9

Path Design and Prediction of Performance for Digital Troposcatter Radio Links

In this chapter we show how to design a digital troposcatter radio path of given characteristics. The quality and the performance expected from digital transmission are first discussed and suitable performance objectives are derived. Simple criteria for calculating the path are given. The path calculations can then be optimized by a suitable choice of the various parameters involved.

9.1 Digital transmission on a troposcatter channel

We have seen in the preceding chapters that the main features of troposcatter digital transmission are as follows:

(a) an abrupt receive threshold above which the transmission is good and below which it is impossible (Chapter 7, Section 7.5.2);

(b) the presence of fast fading causing interruptions to the transmission which are of variable frequency and length;

(c) a multipath delay spread causing intersymbol interference.

Before establishing design criteria and performance objectives it is necessary to analyze in more detail how the above features influence the main transmission characteristics.

9.1.1 Signal-to-noise ratio in telephone channels

The quality of an analog telephone channel is characterized by the minimum signal-to-noise ratio which must be exceeded for a large percentage of time, or better by the time distribution of its noise level. In a digital system the signal-to-noise ratio is fixed, at least as long as the transmission can take place, and is determined by the analog-to-digital conversion process adopted in the multiplex equipment.

When a transmission is interrupted because a fade reduces the r.f. carrier level from above to below the receiver's threshold, the signal-to-noise ratio drops to complete failure. The transition zone in which the drop in quality takes place is a few decibels wide, as can be seen on the BER curve of the receiver (Fig. 7.9). During fades this zone is crossed very rapidly in almost zero time, and thus the BER curve can be assumed to be rectangular (Fig. 7.10) and the received data stream is either present with almost no errors or is absent with bits completely in error.

The quality of a digital telephone channel is therefore better characterized by the ratio of the presence to the absence of the data stream. In practice, if a digital and an analog troposcatter path work under similar conditions of fading margin, propagation, field strength etc., the user subjectively favors the digital link even if the designer considers this link to be the worst. In fact the user experiences very good reception with a very low noise, sometimes briefly interrupted by clicks (due to fading plus the realignment time). Sometimes he misses a word or a sentence and asks for repetition, but the transmission is always noiseless. In contrast the user of the analog link hears an annoying variable background noise with interruptions from time to time, which gives a bad impression.

9.1.2 Transmission degradation: digital errors

Degradation of digital transmission is due to the introduction of errors in the received bit stream. Two main causes of error must be considered when transmitting through a troposcatter channel:

(a) errors due to the disappearance of the r.f. signal below the receiver

DESIGN OF DIGITAL TROPOSCATTER LINKS 235

threshold during deep fades—this source of error affects all types of digital troposcatter links;

(b) errors due to excessive time delay spread which causes intersymbol interference—this error source is independent of the r.f. signal level and affects mainly wideband or high-bit-rate transmission.

In the first case the system is limited by the path attenuation, which cannot exceed a certain value. In the second case the system is limited by the path distortion even if the path attenuation is low. The designer of a digital troposcatter path always chooses the various parameters in order to avoid or combat intersymbol interference. In the next two sections we examine the effects of delay dispersion and how to combat it.

9.1.3 Multipath delay spread and pulse distortion

In Chapter 6 it was shown how a pulse is smeared by multipath transmission, and a time delay spread 2σ was defined in Chapter 4, Section 4.6. This spread may range from near zero to tenths of a microsecond according to the path geometry and the antenna beamwidth. At a transmission rate of 2 Mbit s^{-1} the bit duration is 0.5 µs; therefore it may easily reach the same order of magnitude as the spread.

When the delay spread is not negligible compared with the duration of the transmitted bits (or symbols in a transmission with more than two levels), the multipath distortion produces intersymbol interference which causes an irreducible error rate floor even in the presence of a strong r.f. signal (Fig. 9.1).

In the case of wideband systems, when the r.f. spectrum is wider than the correlation bandwidth, selective fading causes a further increase in the distortion. However, this problem only appears for bandwidths wider than about 3 MHz. In Chapter 7, Section 7.6.2, we defined the dimensionless path dispersion parameter $2\sigma/T$ which is used to evaluate the distortion in digital transmission. The time delay spread 2σ can be calculated (Chapter 6, Section 6.10) or measured. The duration T of the symbols depends on the transmitted bit rate and on the number of modulation levels (normally two or four in troposcatter).

The path dispersion parameter has the same poorly known statistical distribution as the delay spread. We can consider it provisionally as a normal distribution with a 99.9% value equal to twice its median value.

The performance of a typical digital modem versus the path dispersion parameter is shown in Fig. 9.2. It can be seen that the deterioration of the BER is negligible for values of the parameter up to 0.2–0.3, but as the distortion increases it rapidly becomes unacceptable. In a 4PSK digital system with a bit rate of 2 Mbit s^{-1} the symbol duration is 1 µs and $2\sigma/T$ may reach a value of 0.2–0.3 without producing significant intersymbol interference.

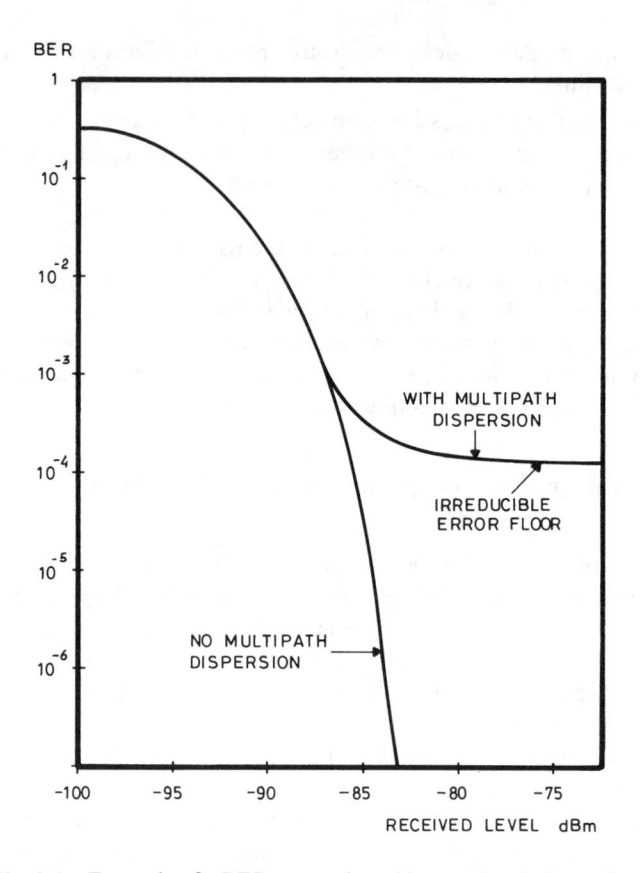

Fig. 9.1 Example of a BER curve altered by multipath dispersion.

9.1.4 Expedients against multipath delay dispersion

When the path is critical a careful choice of system parameters can reduce the effects of multipath dispersion. Consideration should be given to the following.

(a) *Scatter angle*: this should be kept as small as possible by siting the stations such that they minimize the delay spread (see also point (b)).

(b) *3 dB beamwidth of the antennas*: this should be kept as small as possible in order to decrease the size of the common volume, the height of which is low if the scatter angle is small. The time delay of the multipath signals is then reduced, thus minimizing the delay spread. This leads to the use of antennas with diameters greater than that strictly required by path loss considerations and therefore more expensive.

DESIGN OF DIGITAL TROPOSCATTER LINKS 237

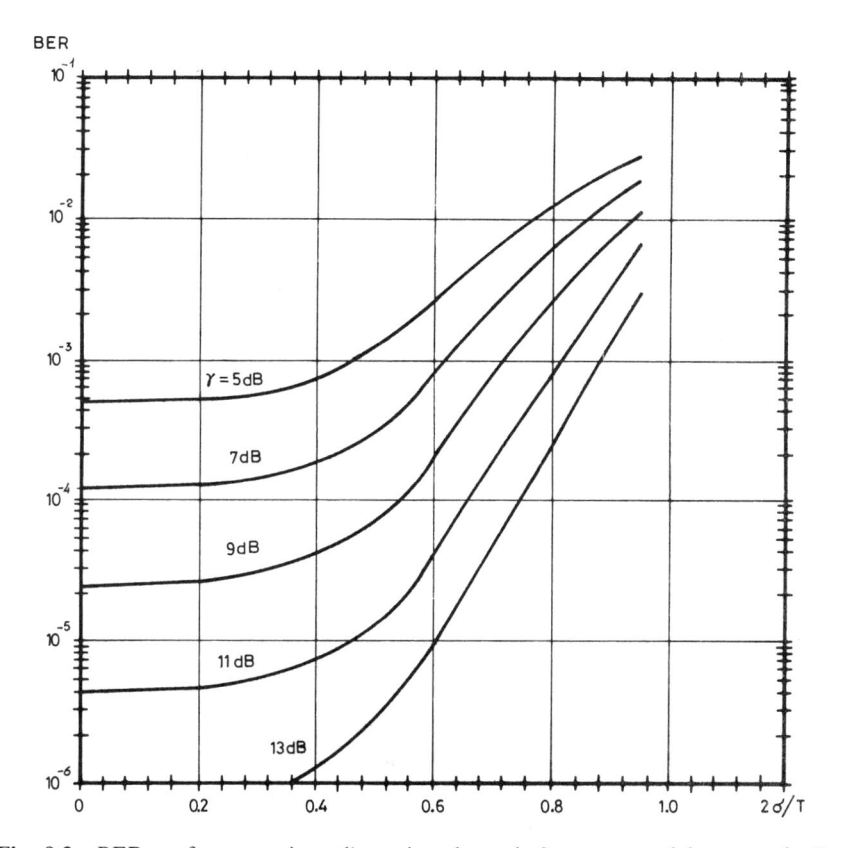

Fig. 9.2 BER performance in a dispersive channel: 2σ, average delay spread; T, symbol duration.

(c) *Frequency band*: the highest band available should be used, so that for a given antenna diameter the 3 dB beamwidth is reduced.

(d) *Diversity*: the use of diversity which is mandatory for troposcatter systems substantially improves the transmission quality. Quadruple diversity is normally recommended. The signal distortions on the various diversity branches are not simultaneous and not correlated. The combiner should drop the most dispersive diversity branch, thus obtaining a combined signal which is less distorted although it is affected by some additional thermal noise due to the temporary decrease in the order of diversity.

(e) *Type of modulation*: doubling the modulation levels (e.g. by choosing 4PSK instead of 2PSK) for the same bit rate would halve the symbol rate and double the r.f. pulse length for the same path delay spread, thus strongly reducing the intersymbol interference. However, more than four levels are unsuitable for troposcatter transmission owing to low immunity to various disturbances.

(f) *Bit rate versus traffic capacity*: the same number of telephone channels can be transmitted at a lower bit rate, thus strongly reducing the intersymbol interference. For example at 2 Mbit s^{-1} 32 PCM channels can be transmitted at 64 kbit s^{-1} per channel or 128 channels can be transmitted at 16 kbit s^{-1} per channel with delta modulation and at the expense of a lower signal-to-noise ratio.

(g) *Equipment with adaptive modems*: this is specifically designed for this purpose but is sophisticated and expensive (see Chapter 5, Section 5.8).

These expedients should be carefully considered during the design of the path. In practice the path and performance of a digital troposcatter system are calculated assuming that the problem of intersymbol interference has been solved previously and that the only source of error is fading. In other words the system is designed assuming that it is limited by path attenuation and not by intersymbol interference.

9.1.5 Behavior of diversity reception

In digital transmission we consider the order of total diversity, which is the product of the orders of explicit and implicit diversity (see Chapter 5, Section 5.7). For high bit rates (3, 6, 12 Mbit s^{-1}) and medium- to long-range systems for which the r.f. spectrum is wider than the correlation bandwidth, suitable adaptive modems capable of exploiting the effects of implicit diversity must be used. For bit rates up to about 2 or 3 Mbit s^{-1} there are generally no problems of bandwidth limitation, fading is flat in the r.f. band and there are no implicit diversity effects.

In any case diversity techniques combat fast fading, which is Rayleigh distributed over periods of up to a few minutes, equivalent to an average telephone call (5 min). In a single receiver the r.f. signal will be below threshold for a certain percentage of time and there will be a transmission outage in the corresponding branch.

In a set of N receivers of a diversity system a transmission outage due to fading begins when the highest r.f. signal level at the input of any receiver falls below its digital threshold and ends when the highest r.f. signal level at the input of any receiver again rises above the digital threshold.

The outage probability for selection diversity is given by

$$P_N(\gamma) = \{1-(2P_o)^{1/\gamma}\}^N \tag{9.1}$$

where γ is the carrier-to-noise power ratio at the receiver input (numerical value), N is the order of diversity and P_o is the BER at threshold (e.g. $P_o = 10^{-4}$). Remembering eqns (2.23) and (2.26), eqn (9.1) can easily be transformed into

$$U(\text{dB}) = 10\log\{-\ln(1-P_N^{1/N})\} \tag{9.2}$$

where $-U$(dB) is the average or mean r.f. carrier-to-threshold ratio (see eqn (2.16(b)) in which W is the threshold value). This formula yields directly the margin (in decibels) necessary for a given outage probability in a system with order of diversity N.

The outage probability for maximal-ratio combination is given by

$$P_N(U) = 1 - \exp(-s) \sum_{k=0}^{N-1} \frac{s^k}{k!} \qquad (9.3)$$

where $-U$(dB) is the average or mean r.f. carrier-to-threshold ratio, N is the order of diversity and $s = 10^{0.1U}$ (see eqn (2.18(a)). The curves in Fig. 9.3 represent eqns (9.2) and (9.3) and show the transmission outage probability for a diversity system using selection or maximal-ratio combination as a function of the average (or mean, which is 1.6 dB above the median) carrier-to-threshold ratio in the presence of Rayleigh fading. In the case of digital systems eqn (5.9) can also be taken into account.

The outage rate, the mean duration of outage and the distribution of durations have been calculated [9.1] and are shown in Figs 9.4, 9.5 and 9.6. Their mathematical expressions are as follows.

(a) The mean outage rate (in s^{-1}) is

$$J = 2.4 N c s^{1/2} \exp(-s)\left(1 - \frac{\exp(-s)}{1-\exp(-s)} \sum_{k=1}^{N-1} \frac{s^k}{k!}\right) \qquad (9.4)$$

where c is the mean fade rate in hertz, s is as defined above and N is the order of total diversity. The mean time between fade outages of any duration is therefore

$$v = 1/J \qquad (9.5)$$

(b) The mean outage duration is

$$t_o = P_N/J \qquad (9.6)$$

(c) The probability of an outage duration of up to t s is

$$P(t) = \frac{2}{u} I\left(\frac{2}{\pi u^2}\right) \exp\left(-\frac{2}{\pi u^2}\right) \qquad (9.7)$$

where $u = t/t_o$ and I is a modified Bessel function of the first order.

The above formulae allow us to calculate the following additional useful parameters which are utilized in subsequent sections.

(d) The mean time between fade outages lasting for a time $t_1 < t < t_2$ is

$$v(t_1, t_2) = \frac{v}{P(t_1) - P(t_2)} \qquad (9.8)$$

where for example $t_1 = 0.2$ s, $t_2 = 5$ s.

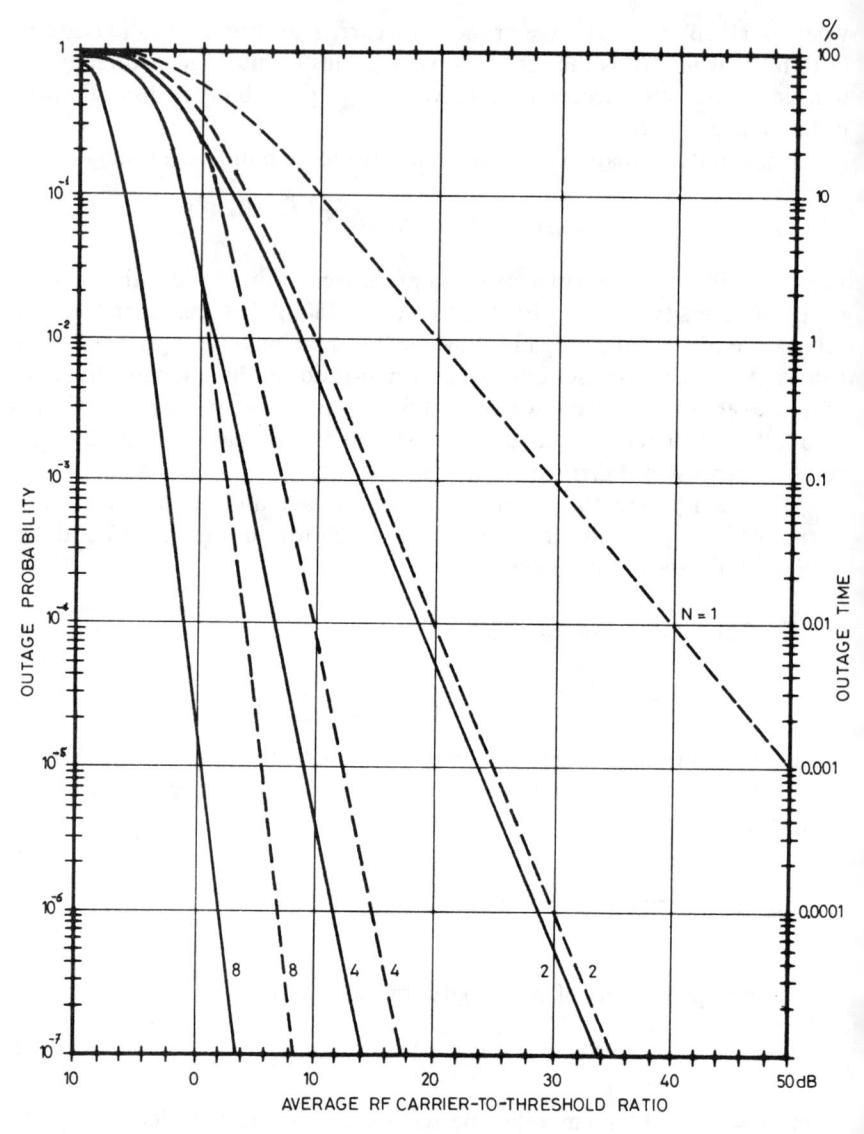

Fig. 9.3 Outage probability in digital diversity reception in the presence of Rayleigh fading: N, order of total diversity; - - -, selection; ———, maximal-ratio combination.

(e) The probability of fade outage per call minute is almost equal, for small n, to the number of fade outages per minute:

$$n = \frac{60}{v(t_1, t_2)} \tag{9.9}$$

(f) For the probability that two to five fade outages occur per call minute

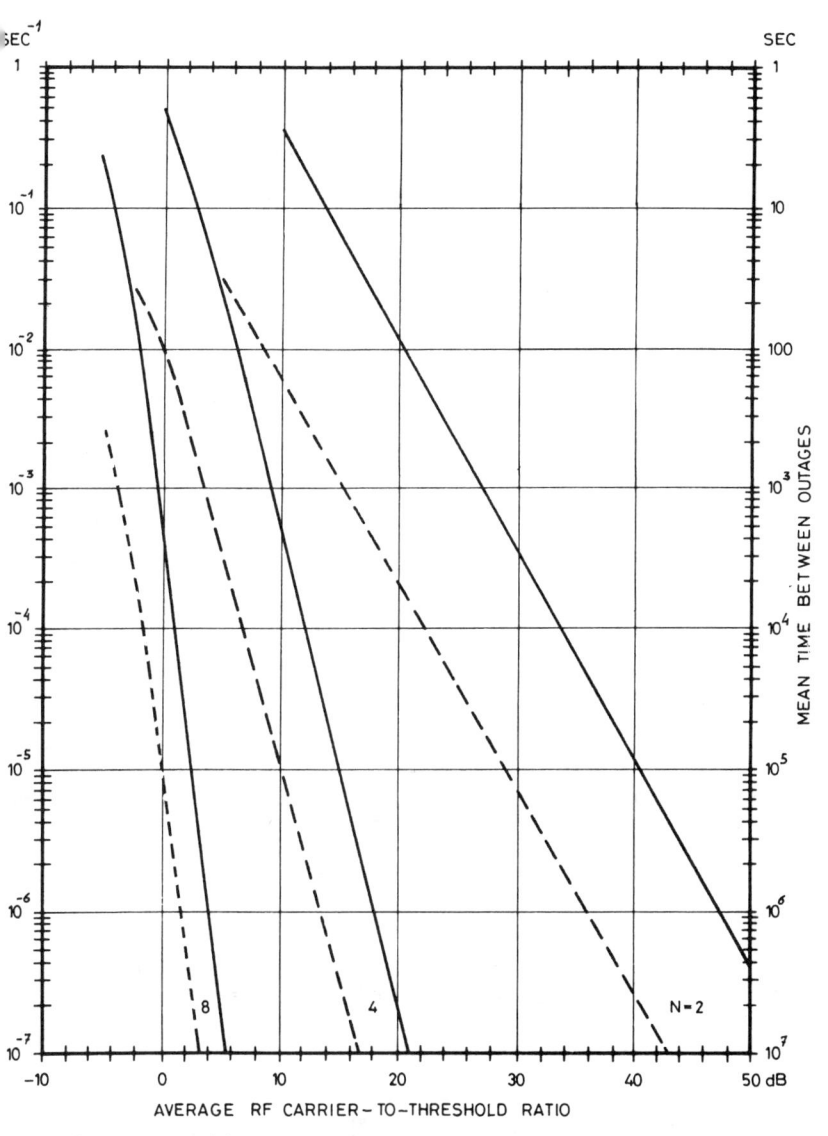

[Figure 9.4 plot with axes: x-axis "AVERAGE RF CARRIER-TO-THRESHOLD RATIO" from -10 to 50 dB; left y-axis in SEC^{-1} from 10^{-7} to 1; right y-axis "MEAN TIME BETWEEN OUTAGES" in SEC from 1 to 10^7. Curves labeled 8, 4, N=2.]

9.4 Outage rate with maximal-ratio combination for mean fade rates of 5 Hz (———) and 0.1 Hz (- - -): N, order of total diversity.

the values for s are calculated using

$$2/60 \leqslant J \leqslant 5/60 \tag{9.10}$$

(g) For the probability that more than five fade outages occur per call minute the values for s are calculated using

$$J > 5/60 \tag{9.11}$$

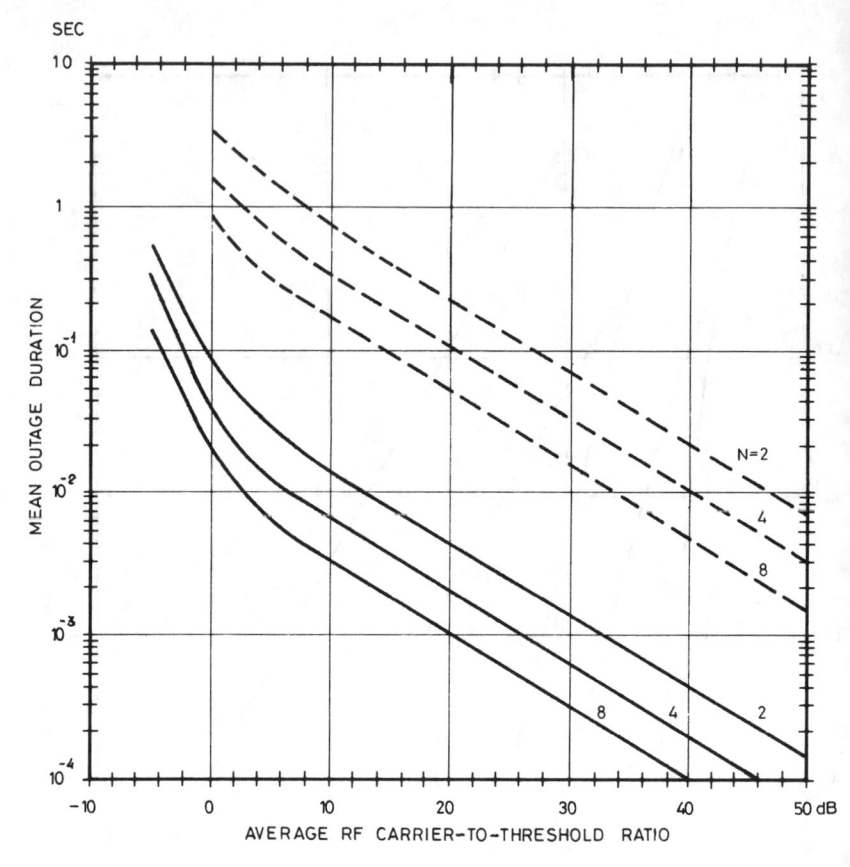

Fig. 9.5 Outage duration with maximal-ratio combination for mean fade rates of 5 Hz (———) and 0.1 Hz (- - -): N, order of total diversity.

(h) The probability of an n-bit block in error, i.e. the probability of at least one error in a block of n bits, is obtained from eqn (5.9) in which $P_o = 1/n$.

The above parameters appear in the performance criteria.

9.2 Quality of digital transmission

The quality of a digital radio link is defined in principle by the achievement of a specified maximum BER which is not exceeded for a given high percentage of time in the received data stream.

However, when the radio link is of the troposcatter type, this criterion is unsatisfactory because troposcatter transmission does not present errors randomly distributed in time but is characterized by periods of error-free transmission interleaved with periods of error bursts depending on the

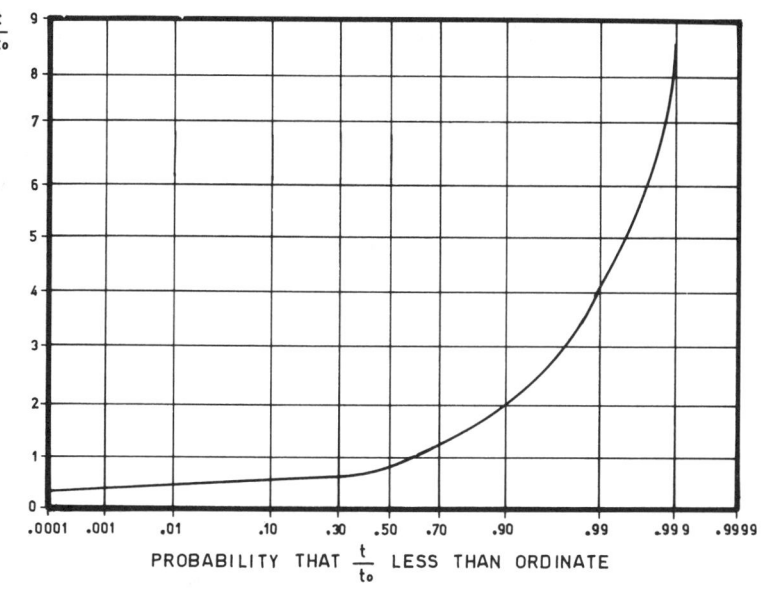

Fig. 9.6 Distribution of outage duration: t_o, mean outage duration.

occurrence of deep fades. In practice an error burst represents an interruption of transmission. The quality criteria for troposcatter radio links are therefore more logically based on the presence or absence of the received data stream.

As in the case of analog troposcatter radio links, the stringency of the performance requirements for digital links depends on the desired transmission quality which in turn is related to the cost. We shall therefore consider both high quality and conventional (low cost) links.

The only large organization which has so far issued detailed performance specifications for high quality digital troposcatter radio links is the American DCS Center, which controls very large communications networks including many troposcatter hops. After theoretical studies and examination of the results of many field tests, this Center proposed recommendations and path design objectives in a suitable form for the designer [9.1]. We shall follow their procedures in subsequent sections.

The quality criteria proposed by the DCS for telephony are based on the percentage of outages and their duration during an average telephone call in the worst month. The criteria for data transmission are based on the percentage of error-free blocks received.

Before specifying the performance objectives it is necessary to examine in more detail some characteristics of the transmission outages due to fast fading. It is assumed that the problem of intersymbol interference has been solved, as stated at the end of Section 9.1.4. The analysis will mainly refer to high quality troposcatter radio links, as no particular problems arise for conventional links.

9.2.1 Outages and quality parameters

The troposcatter channel is affected by frequent fades at a rate of $0.1-5 \text{ s}^{-1}$ and with a mean duration ranging from milliseconds to seconds. In order to classify the types of fades and corresponding outages according to the importance of the disturbance that they introduce in the transmission, the following main types of outages can be considered [9.1] :

(i) outages lasting less than 200 ms which normally occur on troposcatter links when the long-term mean r.f. received signal is high (isolated fades yield negligible disturbance but recurrent sets of fades, which normally occur with an average rate of $2-5 \text{ min}^{-1}$ when the long-term mean r.f. signal is low, can cause substantial problems in the transmission);

(ii) outages lasting from 0.2 to 5 s which occur from time to time on a troposcatter link when the r.f. long-term mean received signal is high or with a high frequency of occurrence if the r.f. signal is lower;

(iii) outages lasting more than 2 min or recurrent outages occurring at a rate greater than 5 min^{-1}; these outages normally occur when the long-term mean r.f. received level is near or below the threshold level.

It is assumed that calls with outages longer than 5 s will be abandoned by the subscriber.

The DCS proposed the following parameters as a measure of the transmission quality [9.1].

(a) *For voice transmission*: the probability that a 5 min voice call is disturbed by an error burst with a specified duration. The probability for 5 min is approximately five times greater than that for 1 min.

(b) *For data transmission*: the probability of error-free transmission of data blocks.

In order to achieve an acceptable transmission quality the DCS proposes to adopt as far as possible systems and equipment with the following minimum characteristics:

(1) PCM voice analog-to-digital conversion at 64 kbit s^{-1};

(2) troposcatter radio equipment using explicit quadruple diversity as far as frequency allocations permit;

(3) the design of troposcatter paths based on 95% service probability;

(4) the use of forward error-correction techniques for voice and/or data transmission.

DESIGN OF DIGITAL TROPOSCATTER LINKS   245

The loss of synchronization or bit count integrity (BCI) can cause longer outages (Chapter 7, Section 7.8.2). However, for clear voice the resynchronization time is of the order of 100 ms, and when using encryption equipment it is of the order of 1 s. Therefore the loss of BCI is important only when crypto equipment is used and can be neglected in the other cases. The DCS performance objective is that not more than 1% of troposcatter fade outages should cause a loss of BCI.

9.2.2 Hypothetical reference circuit

For digital transmission the American DCS Standards propose a *global reference circuit* composed of the following four segments in cascade:

one terrestrial segment 3862 km (2400 miles) long
two satellite segments
one terrestrial segment 3862 km long

The terrestrial segments are covered for 70% of the mileage by LOS links and for 30% of the mileage by troposcatter links. Each terrestrial segment is composed of four cascaded sections 965 km (600 miles) long. Each section includes 14 LOS hops each 48 km (30 miles) long and one troposcatter hop 290 km (180 miles) long. These lengths are considered as representative of such types of links. The global reference circuit includes five tandem analog-to-digital conversions.

This configuration has been studied in order to represent a moderately worst case; it is more pessimistic than the average case but less pessimistic than the worst case. Design criteria for a single hop derived from the performance requirements specified for the global reference circuit would generally be on the conservative side.

9.2.3 Performance objectives

9.2.3.1 General
In order to design a radio link properly it is necessary that suitable criteria are made available. When the radio link to be designed may become part of an important international or even intercontinental connection it must present a suitable quality, normally much better than that required for a link which is part of an independent small system not intended to be interconnected with other systems.

Design criteria for a single radio link are then derived by establishing that the hypothetical reference circuit, which is built up from a combination of individual links of average characteristics, has a certain specified transmission quality from end to end. It is possible to obtain the quality required for a single link and the corresponding design criteria from the overall quality of the reference circuit.

9.2.3.2 Objectives for the global reference circuit

The performance objectives specified by the DCS for end-to-end digital transmission on the global reference circuit are as follows.

(a) The overall *unavailability* (including equipment failure) shall not exceed 1% of the time (unavailability is any loss of continuity or excessive degradation occurring for more than 1 min).

(b) The voice channel quality shall be such that not more than 10% of all basic 5 min calls sustain any noticeable disturbance and not more than 1% of all such calls sustain a disturbance lasting longer than 5 s.

(c) The data channel quality shall be such that 99% of all 1000 bit data blocks transmitted shall be error free.

These percentages are subdivided among the LOS, troposcatter and satellite paths composing the reference circuit in order to find the requirements for a single hop.

Table 9.1 synthesizes the results of the calculations yielding the carrier-to-threshold ratios to be exceeded for the times indicated for a troposcatter hop in order to meet the various clauses. The requirements for voice transmission are found to be more stringent than those for data transmission, and so it is sufficient to design the hop for voice only.

9.2.3.3 Requirements for a single troposcatter hop

The objectives and requirements analyzed so far refer to 1 min values where the fast fading has a Rayleigh distribution about its median value. The cumulative distribution of these medians during the worst month (log-normal distribution) is now considered.

An analysis of the values given in Table 9.1, as applied to many existing troposcatter links and assuming that typical paths present a standard deviation of the log-normal distribution from 3 to 6 dB, shows which of the four clauses is the most stringent for the design of typical paths. It is then sufficient to satisfy this clause. After such an analysis the DCS have simplified Table 9.1 and replaced it with the set of requirements given in Table 9.2. The values in decibels in this table represent the average carrier-to-threshold ratio at the input to the individual receiver which must be exceeded for the percentage time availability shown in the final column in order to meet all DCS requirements.

Corresponding requirements have been calculated by the DCS for data transmission, but it was immediately found that the most stringent requirements are those for voice transmission and no additional clause need be added for data transmission.

Table 9.1 Allocation of fade outage probabilities and minimum allowable levels for a single troposcatter hop

Clause	Criteria	Objectives for			Minimum carrier-to-threshold ratio					
		Reference circuit	Terrestrial section	Single troposcatter hop	755–2400 MHz			4400–5000 MHz		
					\multicolumn{6}{c}{Order of explicit (total) diversity}					
					2(4)	4(8)		2(4)	4(8)	
	Voice channel									
(a)	Probability per call minute of a fade outage lasting 0.2–5 s	0.02	2.5×10^{-3}	7.5×10^{-4}	+3	+1		−2	−4	
(b)	Probability per call minute of a fade outage lasting 5–60 s	0.002	2.5×10^{-4}	7.5×10^{-5}	−3	−4		−6	N/A	
(c)	Probability of recurrent outages shorter than 0.2 s with a rate in the range 2–5 min^{-1}	0.01	2.5×10^{-3}	2.5×10^{-3}	−1	−5		+7	0	
(d)	Probability of recurrent outages with a rate greater than 5 outages per minute	—	—	1×10^{-4}	N/A	N/A		+6	−1	
	Data channel									
(e)	Probability of 1000 bit data blocks in error	0.01	2.5×10^{-3}	2.5×10^{-4}	—	—		—	—	

N/A, not available.

Table 9.2 DCS design requirements for a digital troposcatter hop

Frequency band (MHz)	755–2400	4400–5000
5 min average carrier-to-threshold ratio R at demodulator input (dB)		
Hop with low dispersion, quadruple (explicit) diversity	6	4
Hop with high dispersion and implicit dual diversity		
Dual (explicit) diversity	3	6
Quadruple (explicit) diversity	1	−1
Percentage of time related to the value of R (%)		
Unavailability	0.075	0.01
Availability	99.925	99.99

Low multipath dispersion hops, generally those with a symbol rate not exceeding 2 megasymbols s^{-1}.
High multipath dispersion hops, generally those with a symbol rate exceeding 2 megasymbols s^{-1}.

The probability of equipment failure should not exceed 8×10^{-6}, and loss of synchronization or BCI should not occur for more than 1% of all fade outages. It is also stated that the connection in cascade of digital troposcatter hops will not cause additional transmission quality problems if each hop is designed according to DCS standards. The performance requirements for conventional radio links may be simpler and refer only to a given outage time not to be exceeded for a large percentage (e.g. 99.9%) of the worst month (see Figs 9.3–9.6).

9.3 Design of a digital troposcatter link

The principles for the design of a digital troposcatter path are basically the same as those used for the analog path. The system capability must not be lower than the maximum expected path loss. However, consideration should also be given to the multipath delay which should not exceed certain limits related to the duration of the transmitted symbols in order to avoid excessive intersymbol interference.

9.3.1 Path calculation

In general the path can be calculated in three steps.

(a) Check that the multipath delay is not excessive. Owing to the difficulty of predicting this delay, only rough estimates can be made (see Table 9.3 and Fig. 6.11) if direct path tests are not performed. Also the best calculations available [9.2] are complex and not very reliable (Chapter 6, Section 6.10).

DESIGN OF DIGITAL TROPOSCATTER LINKS 249

Table 9.3 Evaluation of pulse distortion

1. Path length d (km)	185	185	185
2. Path frequency f (MHz)	900	1500	1500
3. Antenna heights			
At site A h_A (m)	64	64	64
At site B h_B (m)	23	23	23
4. Elevation angles			
At site A θ_A (mrad)	−3.9	−3.9	−3.9
At site B θ_B (mrad)	−2.3	−2.3	−2.3
5. Scatter angle θ (mrad)	15.5	15.5	15.5
6. Horizon–chord angle			
At site A α (mrad)	7.2	7.2	7.2
At site B β (mrad)	8.3	8.3	8.3
7. Antenna diameter D (m)	6.4	6.4	12.0
8. Antenna 3 dB beamwidth Ω_0 (mrad)	68.0	40.8	21.8
9. Path antenna beamwidth (Fig. 3.4) Ω (mrad)	51.0	30.6	16.3
10. Path length difference $\Delta d_{12} = 0.5\{(\alpha+\Omega)(\beta+\Omega)-\alpha\beta\}d \times 10^{-3}$ (m)	314	130	48
11. Time delay $\tau = \Delta d_{12}/3 \times 10^2$ (μs)	1.05	0.43	0.16
12. Gross bit rate (kbit s^{-1})	2.176	512	1.024
13. Modulation type	4PSK	2FSK	4FSK
14. Symbol rate (kilosymbols s^{-1})	1.088	512	512
15. Symbol duration T (μs)	0.92	1.95	1.95
16. Ratio τ/T	1.14	0.22	0.08
17. Distortion is negligible	No	?	Yes

(b) Compare the capability of the system to overcome attenuation with the maximum expected loss.

(c) Find the most convenient technical and economical solution by varying the different parameters of the system.

As already stated the multipath delay dispersion can generally be ignored for bit rates of up to 2 or 3 Mbit s^{-1} unless low r.f. frequencies and small antennas are used (e.g. in mobile systems). Critical paths are discussed in Section 9.1.4. Special adaptive modems must be used for high dispersion paths.

Once the problem of intersymbol interference is solved, a path equation can be written as for the analog path.

9.3.2 Path equation

The path equation can be written as follows:

$$P_t + G_p - l_i - T_d - R - i = L(50\%) - Y(q) + u \tag{9.12}$$

where P_t is the transmitted power (in dBm) at the antenna port output of the branching system, G_p is the path antenna gain in decibels (see Chapter 7, Section 7.7.3), l_i is the loss of feeders (transmit plus receive) and receive branching systems (in decibels), T_d is the r.f. threshold level (in dBm), R is a factor (in decibels), i is the implementation margin which accounts for various degradations, $L(50\%)$ is the median path loss (in decibels) in the worst month, $Y(q)$ is the fading margin (in decibels) corresponding to $q\%$ of the time and u is the prediction uncertainty (in decibels) (see Chapter 6, Section 6.4.1). The actual values used depend on the system quality required as discussed below.

The implementation margin takes into account miscellaneous losses due to receiver noise figure aging, waveguide aging, mounting inaccuracies, redundant equipment switching, unexplained transients, minor antenna misalignments, effects of r.f. interference etc. DCS takes $i = 4$ dB.

9.3.3 High quality troposcatter links

If the DCS systems are taken as a reference a high quality radio link normally requires a quadruple explicit diversity, at least for decreasing the multipath dispersion and increasing the equipment availability. The associated multiplex should correspondingly effect the PCM analog-to-digital conversion at 64 kbit s^{-1} for optimizing the quality of voice transmission.

The following values are chosen for some of the parameters in the path equation (eqn (9.12)): T_d is the r.f. threshold level (in dBm) corresponding to a BER of 1×10^{-4} in the presence of steady Gaussian noise at the input to a single receiver; R is the factor (in decibels) given in Table 9.2; $Y(q)$ is the fading margin (in decibels) corresponding to $q\%$ of the time, where q is given in the final column of Table 9.2; u is the prediction uncertainty (in decibels) calculated with 95% confidence (Chapter 6, Section 6.4.1). In this way the path attenuation is calculated with a 95% service probability, and the overall path performance should meet the DCS quality requirements with a high degree of probability. Examples of path calculation are given in Table 9.4.

9.3.4 Conventional troposcatter links

When a high quality is not required some characteristics may be less stringent, so that we have the following.

(a) The path attenuation can be evaluated for a 50% or 84% service probability as in the case of analog radio links.

(b) The diversity may become dual.

(c) The analog-to-digital conversion in the associated multiplex may be made at 32 or even 16 kbit s^{-1}, thus permitting a narrower r.f. bandwidth owing to the reduced bit rate or a greater number of channels over the same r.f. bandwidth at the same bit rate.

Table 9.4 Examples of calculations for high quality paths

1.	Path frequency f (MHz)	900	1500	1500
2.	Path length d (km)	185	258	213
3.	Scatter angle θ (mrad)	15.5	17.8	16.0
4.	Median path loss $L(50\%)$ (dB)	195.1	195.6	192.8
5.	Availability q (Table 9.2) (%)	99.925	99.925	99.925
6.	Fading (long term) $Y(q)$ (dB)	-21.0	-17.5	-19.5
7.	Allowance $u(95\%)$ (dB)	13.4	11.6	12.6
8.	Maximum path attenuation $L(q) = L(50\%) - Y(q) + u(95\%)$ (dB)	229.5	224.7	224.9
9.	Antenna diameter D (m)	18	18	12
10.	Transmitted power P_t (kW)	1	1	1
11.	Order of diversity (explicit) N	4	4	4
12.	Nominal bit rate (kbit s^{-1})	2048	2048	1024
13.	Threshold T_d for BER $= 10^{-4}$ (dBm)	-94.0	-94.0	-97.0
14.	Factor R (Table 9.2) (dB)	$+6.0$	$+6.0$	$+6.0$
15.	Minimum received r.f. level $T_r = T_d + R$ (dBm)	-88.0	-88.0	-91.0
16.	Transmit level P_t (dBm)	60.0	60.0	60.0
17.	Path antenna gain G_p (dB)	77.5	81.2	78.0
18.	Feeder loss l (dB)	3.0	3.0	3.0
19.	Implementation margin i (dB)	1.0	1.0	1.0
20.	Minimum received r.f. level T_r (dBm)	-88.0	-88.0	-91.0
21.	Maximum attenuation overcome $E_q = P_t + G_p - l - i - T_r$ (dB)	221.5	225.2	225.0
22.	Margin $E_q - L(q)$ (dB) (if $\geqslant 0$ DCS is met)	-8.0	$+0.5$	$+0.1$

In the path equation (9.12) the values of some of the parameters are as follows: T_d is the r.f. threshold (in dBm) corresponding to a BER of 1×10^{-3} or even 1×10^{-2}, R is the factor (in dB) taken from Fig. 9.3, $Y(q)$ is the fading margin (in dB) corresponding for example to 99% or 99.9% of the worst month as requested, and $u = 0$ or $u = \{13 + 0.12 Y^2(q)\}^{1/2}$ for a service probability of 50% or 84% respectively (see Chapter 6, Section 6.4.1).

Table 9.5 shows examples of design calculations for conventional troposcatter links. The hops considered are the same as those used in Table 9.4 to facilitate comparison of parameters and results for two different transmission qualities.

The performance requirements of conventional links are specified as the availability (long-term) for more than 99.9% of time in the worst month with an overall outage time of less than 1% (short-term). This means that only during $100 - 99.9 = 0.1\%$ of the 43.200 min of the worst month (i.e. 43 min) does the overall outage time last more than 1% of time, i.e. 26 s.

Comparison of Tables 9.4 and 9.5 shows that the maximum path attenuation considered is something like 10 dB less for conventional paths. Furthermore the capability of the system to overcome losses is increased by the lower threshold and the lower value of R. Much smaller antennas and a

Table 9.5 Examples of calculations for conventional paths

1.	Path frequency f (MHz)	900	1500	1500
2.	Path length d (km)	185	258	213
3.	Scatter angle θ (mrad)	15.5	17.8	16.0
4.	Median path loss $L(50\%)$ (dB)	195.1	195.6	192.8
5.	Required availability q (%)	99.9	99.9	99.9
6.	Fading (long term) $Y(q)$ (dB)	−18.5	−16.5	−18.0
7.	Allowance u (dB)	3.6	3.6	3.6
8.	Maximum path attenuation $L(q) = L(50\%) - Y(q) + u$ (dB)	217.2	215.7	214.4
9.	Antenna diameter D (m)	10	8	8
10.	Transmitted power P (kW)	1	0.5	0.3
11.	Order of diversity (explicit) N	4	4	4
12.	Nominal bit rate (kbit s^{-1})	2048	2048	1024
13.	Threshold T_d for BER = 10^{-3} (dBm)	−96.0	−96.0	−99.0
14.	Factor R (Fig. 9.3) (dB)	3.0	3.0	3.0
15.	Minimum received r.f. level $T_r = T_d + R$ (dBm)	−93.0	−93.0	−96.0
16.	Transmit level P_t (dBm)	60.0	57.0	54.8
17.	Path antenna gain G_p (dB)	70.5	73.5	73.5
18.	Feeder loss l (dB)	3.0	3.0	4.0
19.	Implementation margin i (dB)	3.0	3.0	4.0
20.	Minimum received r.f. level T_r (dBm)	−93.0	−93.0	−96.0
21.	Maximum attenuation overcome $E_q = P_t + G_p - l - i - T_r$ (dB)	217.5	217.5	216.3
22.	Margin $E_q - L(q)$ (dB) (if $\geqslant 0$ path is feasible)	+0.3	+1.8	+1.9

lower r.f. transmitted power are therefore required. The factor R, which is fixed in Table 9.4, is taken from Fig. 9.3 in Table 9.5. For example, in quadruple diversity with maximal-ratio combination the hourly average (1.6 dB above the median) received r.f. signal should be 1.5 dB above threshold for an outage time less than 1% of the hour. Therefore the median r.f. signal level received during the worst hour of the worst month is $R = 1.5 - 1.6 = -0.1$ dB above the threshold. However, in Table 9.5 a value $R = 3$ dB was introduced to take into account the fact that the theoretical curves in Fig. 9.3 are perhaps slightly optimistic with respect to the actual performance. The 1.6 dB difference between average and median values for R is ignored in Table 9.4.

9.4 Comments and suggestions for further reading

For deeper insight into this topic it is advisable to read ref. 9.1 or at least ref. 9.3.

References

9.1 Kirk, F. W. and Osterholz, J. L. DCS digital transmission system performance, *Tech. Rep. 12-76*, Defense Communications Center, Reston, VA, November 1976.
9.2 Bello, P. A. A troposcatter channel model, *IEEE Trans. Commun. Technol.*, **17** (2), 130–137, April 1969.
9.3 Osterholz, J. L. Design considerations for digital troposcatter communications systems, *AGARD Conf. Proc.*, **244**, 22-1–22-15, October 1977.

Chapter 10

Topographic and Survey Problems

Although we have occasionally mentioned that a parameter can be measured on site, so far we have dealt with problems that required mostly office work. However, at some stage of the work, the engineer will have to face problems that can only be solved on the spot after some preliminary work both in the home office and in a local office, such as an agency, a ministry, the PTT or similar.

The detailed maps required may not be available at the home office as they may be considered secret by the local authority, and the engineer may only be allowed to work on them at the office of this authority. In some parts of the world such maps may not even exist or they may be accurate only in some regions, e.g. near roads or inhabited areas, and very poor elsewhere. When on site the engineer should be able to determine the exact direction of the distant station(s), both for positioning the antenna foundations correctly, particularly for billboard antennas which are fixed on the ground, and for orienting tower-mounted parabolic antennas before or during system line-up. Sometimes the site coordinates are unknown and must be obtained by measurement, together with the skyline profile of the horizon.

In order to deal with problems that arise during surveys and installation of equipment the engineer must be able to extract appropriate information from the available maps and to obtain any other data required from on-site measurements. Therefore he should be capable of using small portable instruments (Abney level, gradienter, magnetic compass etc.) or sophisticated instruments (theodolite, gyrocompass etc.) depending on the problem to be solved and the accuracy required.

These and similar problems that we can describe as topographical or survey problems are discussed in this chapter. Some of them call for the specialist skills of a surveyor, and therefore we shall not discuss them in detail, but we believe that the engineer or designer should have at least a working knowledge of the techniques required in order to be able to follow the work of others or even to do it himself.

10.1 Preliminary work using maps

The initial design stage (which could be called a feasibility study) starts with the choice of a route on a map, which could be a large-scale map or even a road map which enables sites with easy access to be selected. At this stage the elevation angles of the horizons may be estimated in order to provide a rough evaluation of the feasibility of the path and the suitability of the design. This approximate evaluation may also result in some modifications of the route originally chosen.

Detailed maps, i.e. maps with a scale of 1:100 000 or 1:25 000 provided with geographic coordinates and contour lines at least every 50 m, are essential for the next stage. These maps are composed of separate sheets which are laid edge to edge on a suitable flat surface (table, wall, floor etc.) such that the whole region of interest is covered. A single path normally crosses several sheets. The line connecting two sites is not strictly a straight line on this set of maps, but the difference is negligible for the distances encountered in practice. However, spherical trigonometry formulae permit a better approximation to be made from a set of straight lines, one for each sheet of the map, as we shall see in Section 10.1.2.

When possible sites have been chosen and their geographic coordinates have been determined, it is necessary to calculate for each site the azimuth (direction with respect to the north) of the other site so that the connecting line can be drawn on the maps (see Section 10.1.1). Furthermore the profile of the horizon for each site can be derived from the maps (Section 10.1.3).

The above data are used for the on-site measurements. The elevation angles and the skyline profile of the horizon obtained in the office can be checked on the spot or directly measured if maps are unavailable. The azimuth of the other site must be known in advance, and it can be determined from astronomical measurements *in situ* (see Section 10.2.4) and then marked on the terrain.

10.1.1 Calculation of the antenna azimuth

Consider the spherical triangle shown in Fig. 10.1, where A and B are the two sites and B has a greater north or south latitude (it is nearer to the respective pole). We can write the following spherical trigonometry formulae:

$$\tan \alpha = \cot\left(\frac{\Delta\lambda}{2}\right)\frac{\cos\frac{1}{2}(\varphi_B - \varphi_A)}{\sin\frac{1}{2}(\varphi_B + \varphi_A)} \quad (10.1)$$

$$\tan \beta = \cot\left(\frac{\Delta\lambda}{2}\right)\frac{\sin\frac{1}{2}(\varphi_B - \varphi_A)}{\cos\frac{1}{2}(\varphi_B + \varphi_A)} \quad (10.2)$$

$$\tan\frac{\gamma}{2} = \tan\left(\frac{\varphi_B - \varphi_A}{2}\right)\frac{\sin \alpha}{\sin \beta} \quad (10.3)$$

where φ_A is the latitude of site A, φ_B is the latitude of site B ($\varphi_B > \varphi_A$), $\Delta\lambda$ is the difference in longitude between sites A and B,

$$\begin{aligned}\alpha &= \tfrac{1}{2}(Y+X) \\ \beta &= \tfrac{1}{2}(Y-X) \\ X &= \alpha - \beta \\ Y &= \alpha + \beta\end{aligned} \quad (10.4)$$

(see Fig. 10.1) and γ is the angular distance between A and B. We can derive α, β, γ, X and Y, in that order, from these formulae.

The geographic azimuth (direction with respect to true north) of each site from the other site for the triangle in Fig. 10.1 (angles in degrees) is as follows:

northern hemisphere

X	azimuth of site B from site A
$360° - Y$	azimuth of site A from site B

(10.5)

southern hemisphere

$180° + X$	azimuth of site B from site A
$180° - Y$	azimuth of site A from site B

(10.6)

In these formulae X and Y are the left-hand and right-hand angles of the triangle. In the calculations X and Y can be inverted (e.g. when AB has the opposite tilt). The distance AB can be calculated by assuming that for a mean Earth radius of 6370 km the length of the arc subtended by 1° is 111.18 km. Therefore, if γ is expressed in degrees, the path length d is given by

$$d = 111.18\gamma \text{ km} \quad (10.7)$$

The sum of the angles in the spherical triangle is slightly greater than 180° and this difference is known as the *spherical excess*. This can be used as a check on the results.

The Earth is assumed to be an ideal sphere in the above calculations, whereas in fact it is approximately ellipsoidal. This implies that the vertical line through the site does not pass through the Earth's center, although it does cross the axis of rotation. The *geographic latitude* (altitude of the celestial upper pole as measured from the site) is a little higher than that which would be measured on a sphere at the same point. For a more accurate calculation using the above formulae we should replace the geographic latitudes by the corresponding *geocentric latitudes*

$$\varphi' = \varphi - 696'' \times \sin(2\varphi) \tag{10.8}$$

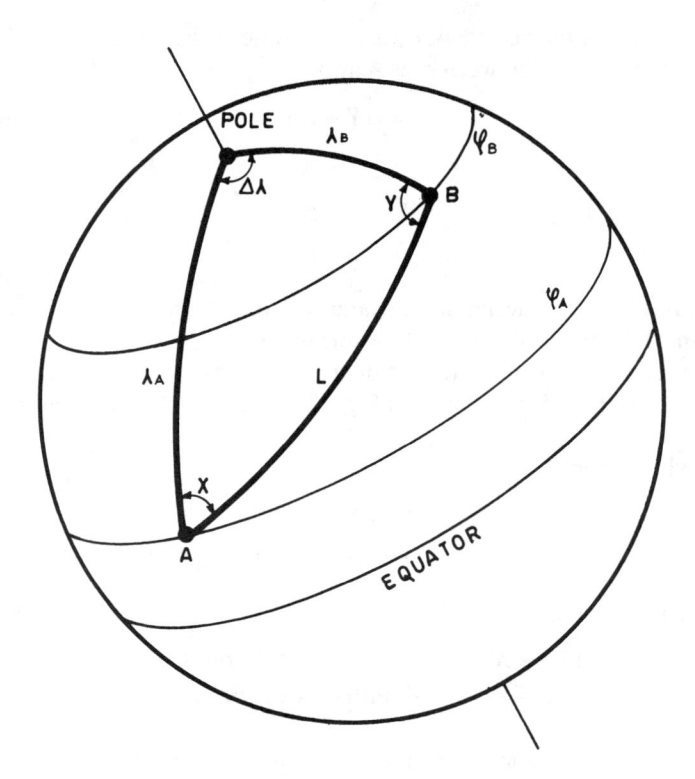

Fig. 10.1 The geographic triangle.

The maximum difference between the two latitudes occurs at a latitude of 45° and is 696″ or about 12′ or 0.2°. If no correction is applied the error in the azimuth could reach 6′ or 0.1°.

The above formulae can easily be programmed into a microcomputer so that the azimuths can be obtained immediately from the site coordinates. An example of azimuth calculation including the preliminary results is shown in Table 10.1, which refers to Fig. 10.2.

Table 10.1 Calculation of antenna azimuths from geographic data

Station	Geographic longitude λ	Geographic latitude φ	Correction $\varphi - \varphi' = 696'' \times \sin(2\varphi)$	Geocentric latitude φ'
A	26°48'01"	37°37'08"	0°11'13"	37°25'55"
B	22°29'21"	39°21'55"	0°11'23"	39°10'32"

Preliminary results
$\alpha = 88°39'48''$ $\beta = 27°15'15''$ $\gamma = 3°48'19''$
$X = 61°24'19''$ $Y = 115°55'17''$ $\Delta\lambda = 4°18'40''$
$X + Y + \Delta\lambda = 181°38'16''$
Spherical excess $= 1°38'16''$
$\gamma = 3°48'19'' = 3.8054°$

Final results
Azimuth from station A $360° - X = 298°35'40''$
Azimuth from station B $Y = 115°55'17''$
Distance $d = 111.18 \times 3.8054 = 423.1$ km

10.1.2 Drawing the connection line on the maps

As the detailed (small-scale) map is composed of several sheets, a segment of the line connecting the two end sites appears on each sheet. It is possible to draw these segments independently using the formulae given below. If the coordinates of site A and the azimuth X are known, these formulae can be used to calculate for each latitude (or longitude) the corresponding longitude (or latitude) of the points of the connecting line. Therefore, given the latitude or the longitude of the edge of a sheet, it is possible to calculate the other coordinate on the same edge. We can obtain two points for each sheet and thus draw the line from edge to edge (or edge to site).

The formulae for calculating the latitude φ_C when the longitude difference $\Delta\lambda$ is given are

$$\cos y = \sin X \sin \varphi_A \sin \Delta\lambda - \cos X \cos \Delta\lambda \qquad (10.9)$$

$$\cos \varphi_C = \frac{\sin X \cos \varphi_A}{\sin y} \qquad (10.10)$$

and the formulae for calculating the longitude difference $\Delta\lambda$ when the latitude φ_C is given are

$$\sin y = \frac{\sin X \cos \varphi_A}{\cos \varphi_C} \qquad (10.11)$$

$$\cot\left(\frac{\Delta\lambda}{2}\right) = \tan\left(\frac{y-X}{2}\right) \frac{\cos \frac{1}{2}(\varphi_C + \varphi_A)}{\sin \frac{1}{2}(\varphi_C - \varphi_A)} \qquad (10.12)$$

It is not necessary to draw the lines along all the path but only from each site to its horizon using the method explained in the next section.

NORTH

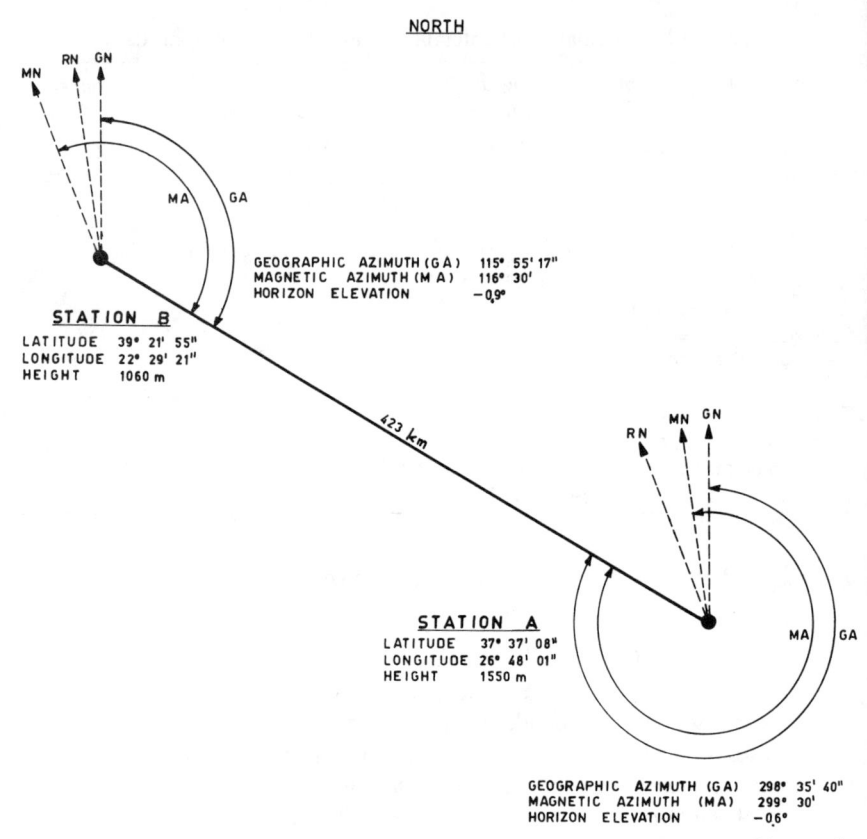

GEOGRAPHIC AZIMUTH (GA)   115° 55' 17"
MAGNETIC AZIMUTH (MA)   116° 30'
HORIZON ELEVATION   −0,9°

STATION B
LATITUDE   39° 21' 55"
LONGITUDE 22° 29' 21"
HEIGHT   1060 m

423 km

STATION A
LATITUDE   37° 37' 08"
LONGITUDE 26° 48' 01"
HEIGHT   1550 m

GEOGRAPHIC AZIMUTH (GA)   298° 35' 40"
MAGNETIC AZIMUTH (MA)   299° 30'
HORIZON ELEVATION   −0,6°

Fig. 10.2 Geographic parameters of a troposcatter path: GN, geographic north; MN, magnetic north; RN, north of reticule; GA, geographic azimuth; MA, magnetic azimuth.

10.1.3 Plotting the horizon skyline profile

We know that for the design of a troposcatter path it is sufficient to derive the elevation angle of the two horizons so that the scatter angle and the path loss can be calculated. The elevation angles can be calculated using eqn (3.2) when the horizon is on land and eqns (3.3) and (3.8) when it is on the sea. It is good practice to perform these calculations not only for the direction of the other station but also for symmetrical directions within an angle of say $\pm 4°$ in order to obtain the horizon skyline profile which is also useful for checks during surveys.

The method to be used is explained with reference to Table 10.2 and Figs 10.3 and 10.4. We draw on the map the connecting line and pairs of symmetrical lines starting from the site in angular steps of 1°. We do not draw

Table 10.2 Determination of the horizon skyline

Station	Relative azimuth (deg)	Peaks of the path profile: height/distance (m/km)			Calculated elevation angles (deg)			Far horizon[a] Elevation (deg)	Far horizon[a] Distance (km)
A									
Height: 1578 m a.s.l.	−4	1240/12.5	1237/24		−1.6	−0.9		−0.9	24
Antenna azimuth:	−3	1196/12.5	1404/24		−1.8	−0.5		−0.5	24
298°35′40″	−2	1152/13.2	1488/24		−1.9	−0.3		−0.3	24
	−1	1096/13.5	1156/23.8	1086/55	−2.1	−1.1	−0.7	−0.7	55
	0	1190/13.5	990/23.8	1086/55	−1.7	−1.5	−0.7	−0.6	54.8
	+1	1332/13.5	833/23.5	1182/54.8	−1.1	−1.9	−0.6	−0.6	54.6
	+2	1395/14.1		1183/54.6	−0.8		−0.6	−0.8	14.1
	+3	1666/14.2		712/54.2	+0.3		−1.1	+0.3	14.2
	+4	1567/14.2			−0.1			−0.1	14.2
B									
Height: 1070 m a.s.l.	−4	600/9	743/13.5		−3.0	−1.4		−1.4	13.5
Antenna azimuth:	−3	600/8	743/13.5		−3.4	−1.4		−1.4	13.5
115°55′17″	−2	650/8	650/13.5		−3.0	−1.8		−1.8	13.5
	−1	620/8	650/13.5		−3.2	−1.8		−1.8	13.5
	0	500/9	730/13.5	450/14.5	−3.6	−1.5		−1.5	13.5
	+1	712/8.3	510/11	660/14.5	−2.4	−2.9		−1.5	16
	+2	680/8.3	570/11	880/14.7	−2.7	−2.6	−2.4	−1.1	14.7
				400/16					
	+3	680/8	603/11	850/14.6	−2.8	−2.5	−1.9	−0.9	14.6
				550/16			−2.5/−1.7		
	+4	680/8	550/11	930/14.5	−2.8	−2.7	−1.7	−0.6	14.5
				600/18			−1.1/−1.5		
				671/16			−0.9		
				650/16.2			−0.6/−1.5		

[a] Sea horizon from station A: elevation, −0.9°; distance, 135 km.

262   TROPOSCATTER RADIO LINKS

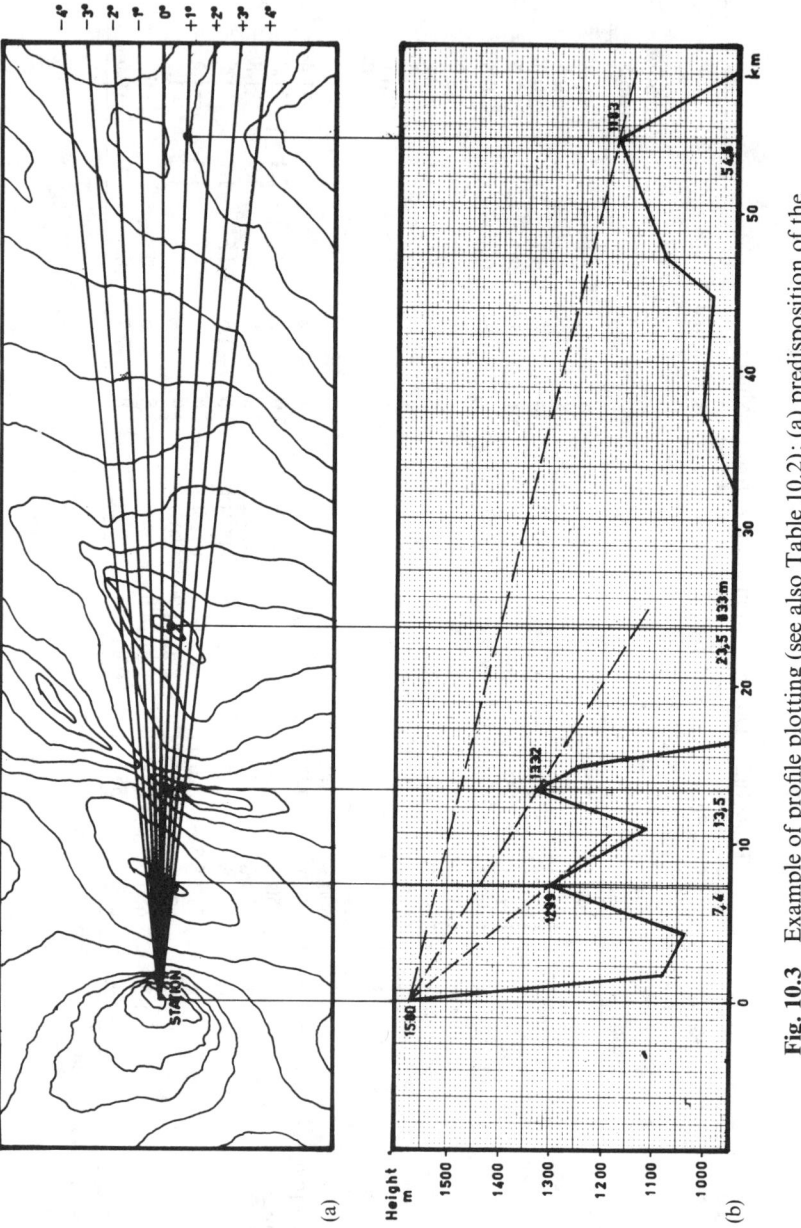

Fig. 10.3 Example of profile plotting (see also Table 10.2): (a) predisposition of the map for profile plotting; (b) profile plotting along the line of azimuth $+1°$.

the profiles along these lines but only mark the height and distance of the highest points, as in Table 10.2. We then use the formulae referred to above to calculate the elevation angle of each peak (and of the sea if necessary). The highest elevation for each azimuth line is the elevation of the horizon. The radio horizon skyline of Fig. 10.4 is obtained using the data given in Table 10.2. The other elevations yield the profile of lower terrain.

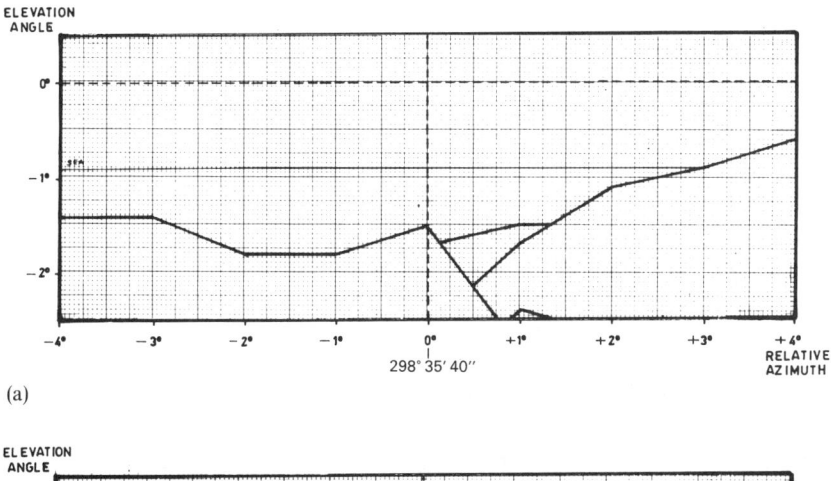

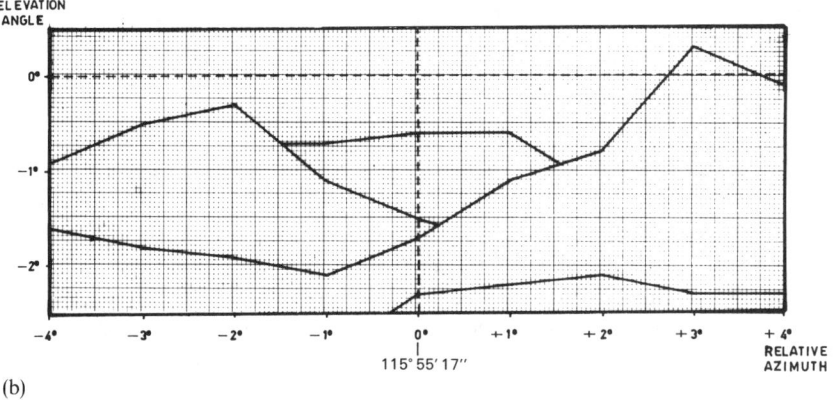

(a)

(b)

Fig. 10.4 Radio horizon as seen from (a) station A and (b) Station B (correction factor k for the Earth's radius is taken as 4/3). Calculated using data from Table 10.2.

10.2 Surveying problems on site

After the above preliminary work, the engineer should have the main data available in a form like that shown in Figs 10.2 and 10.4. These data include magnetic azimuths, which are derived from the map or estimated and can be used for quick orientation. Using a magnetic compass and the calculated skyline profile he can recognize, on a clear day, the exact direction of the path on site. However, he may need to obtain the exact azimuth on the terrain and in some cases may have to determine the geographical coordinates of the site. Furthermore, he may have to determine the horizon skyline from local measurements.

Apart from the horizon profile, these data can be calculated from theodolite measurements using classical spherical astronomy methods programmed into a microcomputer. The following equipment designed specifically for this purpose is also available.

(a) The azimuth can be found by using a north-seeking gyroscope in conjunction with a theodolite. This is a precision instrument, but direct-reading gyrocompasses which are easier to use and are not subject to the errors inherent in magnetic compasses are also available.

(b) The site coordinates can be measured with sufficient accuracy (error of the order of hundreds of meters in position) by using simple satellite navigation equipment which automatically makes triangulations from synchronized signals received from satellites.

If this equipment is available the methods described in most of Section 10.2 are superseded. However, it is useful to have some knowledge of the classical methods, and therefore we proceed by first reviewing some general concepts of geographical astronomy which are required to understand these methods.

10.2.1 Basic principles

Geographical astronomy calculations are based on various coordinate systems and their interrelations. These systems are based on spherical trigonometry and deal only with angles. Distances are not considered.

At each site (of given geographical coordinates) on the surface of the Earth we can establish a fixed system of two polar coordinates created by the theodolite, which locates any object on the ground or in the sky by a horizontal angle (azimuth) and a vertical angle (elevation or zenith distance). Terrestrial objects are stationary with respect to this local fixed system, but celestial objects (the Sun, planets, stars) are moving and their location with the theodolite requires a third coordinate, i.e. the exact time of observation in hours, minutes and seconds of local civil time. Celestial objects are referred to a system which is fixed with respect to the stars and revolves with them at a constant angular velocity (one turn in 23h 56m 4s). This system, which is called the uranographic equatorial system, can be obtained by projecting the system of geographical coordinates on the celestial sphere from the center of the Earth. The system is shown schematically in cylindrical projection in Fig. 10.5,

TOPOGRAPHIC AND SURVEY PROBLEMS   265

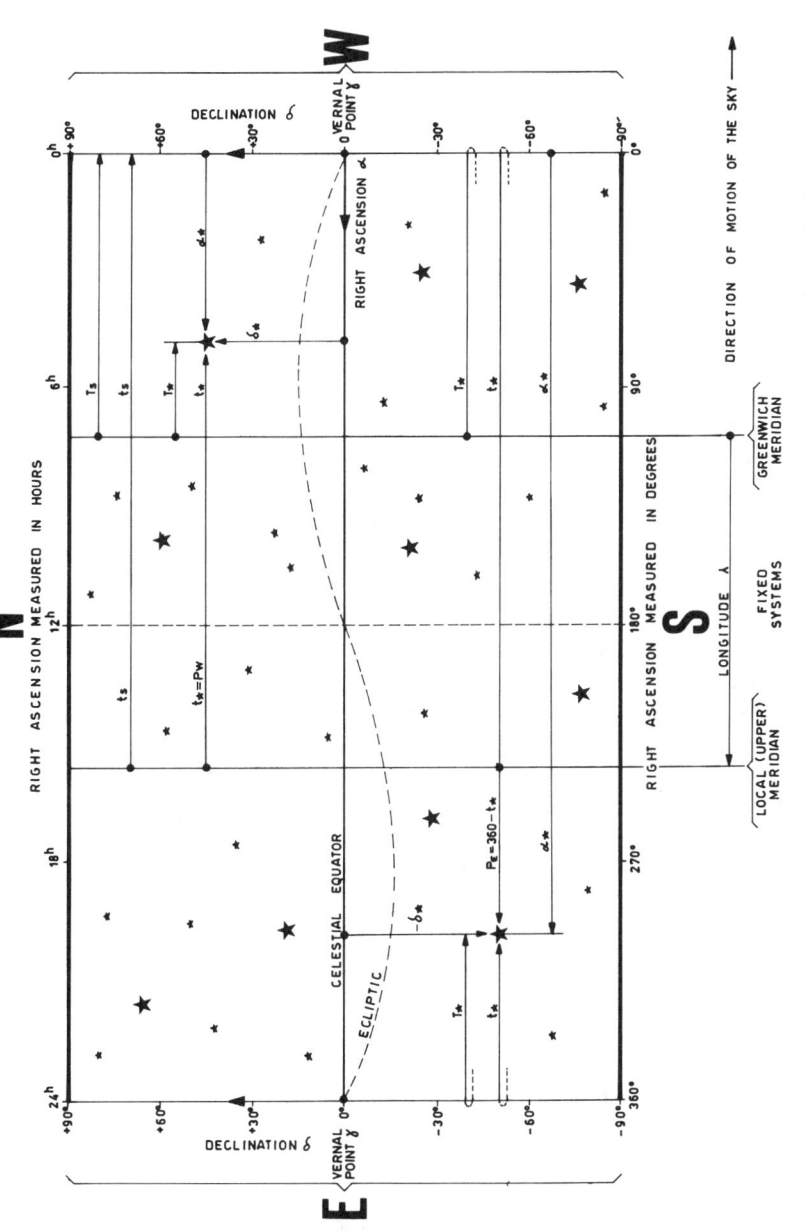

Fig. 10.5 Idealized map of the sky: uranographic equatorial system of spherical coordinates and its relations with the local hourly coordinates system.

in which we can see the corresponding coordinates, i.e. the right ascension α (measured in degrees or time) and the declination δ, their origin at the vernal point γ and several other parameters used in this field, which are defined in the next section.

The solution of problems in geographical astronomy is based on the interrelation between the fixed terrestrial system and the rotating celestial system at any given time. The interrelation is obtained by replacing the right ascension by a more suitable coordinate—the local hour angle—which implies another system of coordinates. These systems are defined in Section 10.2.2.

Some celestial bodies are effectively fixed in the sky (stars), while others are moving (Sun, Moon, planets). The position of the main bodies can be obtained directly or derived from the ephemeris or from nautical almanacs, which also include explanations and examples of calculations together with any corrections required (e.g. for refraction by the atmosphere). The coordinates are the hour angle at Greenwich and the declination for the Sun, the Moon and the planets, and the right ascension and declination for stars. The time is Universal time.

We note here that it is not advisable to refer to the Moon, as it requires more complex calculations because of the effects of parallax and rapid motion. The nautical almanacs also contain simple maps of the sky which permit easy recognition of the main stars. Explanations of the use of the tables, with simple examples, are also included.

The data in the ephemeris and the almanacs (e.g. position and time coordinates at any hour) are referred to the Greenwich meridian, but they can be transferred to the local system in any part of the world at any time by a simple transformation. Thus if the position of a celestial body in the mobile system is known at a given time, we can find the same position in the fixed local system of the theodolite by means of an appropriate transformation. As the positions of the celestial bodies are given with great accuracy, it is possible for example to use the theodolite and the chronometer to find their precise azimuth at a given time in the fixed system, which corresponds to finding true north or any other fixed azimuth with the same accuracy.

At this point we should note that the trigonometric expressions appearing in the formulae have been tabulated, and tables are commercially available which enable the azimuth to be determined quickly with a minimum of calculations when an accuracy of about $0.5°$ is acceptable. We should also note that many radio stations in the world broadcast time signals with an accuracy which is more than sufficient for our purposes. A quartz clock or watch can be set exactly with reference to one of these stations so that the error in time could be of the order of 1 s.

Before proceeding further, we review the most important parameters of geographical astronomy.

10.2.2 Fundamental parameters of geographical astronomy

The parameters defined below should be examined carefully with reference to Figs 10.5 and 10.6. The spherical trigonometry calculations are based on the astronomical triangle (Fig. 10.6) formed on the celestial sphere by the upper pole, the zenith (the vertical at the site) and the star or other celestial body under examination. The meridian crossing the zenith is the local upper meridian. We also need to define the star's hour angle, astronomical azimuth, declination etc. The local and Greenwich meridians (the longitude is taken as positive eastwards, but sometimes the opposite convention is used), the local and Greenwich hour angles of a star (t_*, T_*, P_E, P_W etc.), the local and Greenwich sidereal time etc. are shown in Fig. 10.5. These figures and the explanations below should be studied carefully. By paying careful attention to the figures and looking at the sky, the reader will find that long explanations are unnecessary and that these parameters and the fundamental principles are much simpler than they may appear in our rapid synthesis.

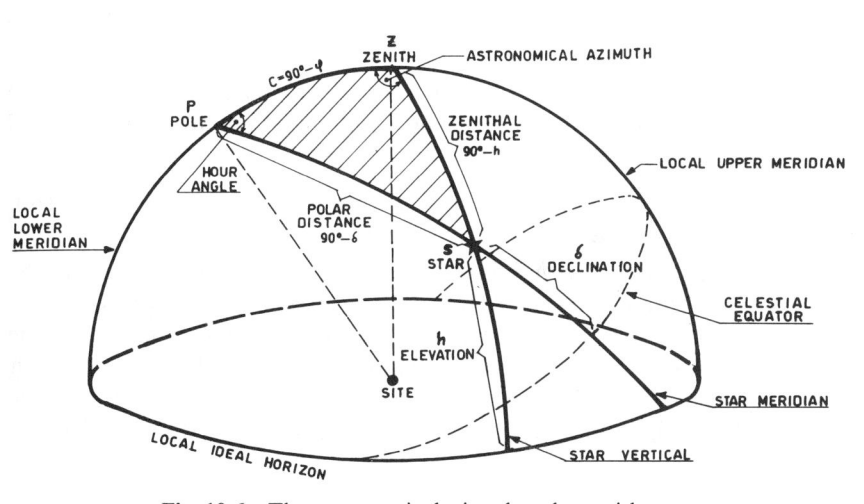

Fig. 10.6 The astronomical triangle pole–zenith–star.

The fundamental parameters to be considered are as follows.

(a) *Zone time correction* Δ is the difference between the zone time under consideration and the first zone time (Greenwich). Attention must be paid to changes due to the use of summer time or local time in some countries at some periods of the year.

(b) *Universal time* T_u, which is also known as the Greenwich mean time T_m (GMT) is the time at the Greenwich meridian.

(c) *Sidereal time* T_s is the hour angle of point γ (the origin of astronomical coordinates) at Greenwich. This angle is given by the ephemeris as a function of Universal time.

(d) *Local zone time* t_m is the time given by the clock at the site and corresponding to the local zone. It is related to Universal time by $t_m = T_m + \Delta$.

(e) *Local sidereal time* $t_s = T_s + \lambda$ is the local hour angle of point γ.

(f) The time T_* (at Greenwich) or t_* (locally) of a celestial body ($t_* = T_* + \lambda$) is the hour angle, measured from the upper meridian (corresponding to noon) in the west direction from 0° to 360°. It is calculated from $T_* = T_s - \alpha$ or from $t_* = t_s - \alpha$. The values for the Sun, the Moon and the main planets are given in the ephemeris and in nautical almanacs as a function of Universal time.

(g) *Geographic longitude* λ is the difference between the hour angles of a celestial body (e.g. the Sun) as measured simultaneously from Greenwich and from the site ($\lambda = t_* - T_*$). It is considered as positive east of Greenwich and negative west of Greenwich (sometimes the opposite).

(h) *Geographic latitude* φ is the altitude of the celestial upper pole as measured (ideally) from the site.

(i) The *altitude* h (or alternatively the *zenithal distance* $w = 90° - h$) is the vertical angle between the celestial body under consideration and the local horizontal plane (ideal).

(j) The *elevation* of the horizon is the vertical angle between the horizon and the ideal horizontal plane.

(k) *Geographic azimuth* Z is the angle between the north pole and the horizontal direction under consideration measured in the clockwise direction from the pole ($Z = 0°$) and varying between 0° and 360°. The geographic azimuth Z of a star is equal to z if the star is on the east side of the meridian. If it is on the west side of the meridian, $Z = 360° - z$. If the observer is in the southern hemisphere, $Z = 180° - z$ for an east star and $Z = 180° + z$ for a west star.

(l) *Astronomical azimuth* z and *altitude* h (alternatively Z, h or Z, w) are coordinates of the celestial bodies in the local system of horizontal coordinates. Note that z is measured from the lower meridian (or from the upper pole) and varies between 0° and 180° either east or west.

(m) *Right ascension* α and *declination* δ are coordinates of the celestial bodies in the uranographic equatorial system (they are given by the ephemeris).

(n) *Hour angle* t_* and declination δ (alternatively hour angle P and declination δ) are coordinates of the celestial bodies in the local system of hourly coordinates. Note that P is measured from the upper meridian and can be west ($P_W = t_*$) or east ($P_E = 360° - t_*$).

10.2.3 Coordinate systems

Measurements of (geographic) azimuths, elevation of horizons etc. at a site on the Earth's surface are made using a theodolite. This instrument establishes (at the site) a local system of spherical coordinates related to an ideal horizontal plane and to the vertical of the site which has given latitude and longitude. The two coordinates are as follows:

(1) geographic azimuth, which is the horizontal angle between 0° and 360° measured clockwise starting from the north direction (also for the southern hemisphere);

(2) altitude (elevation), which is the vertical angle between the observed object and the horizontal plane.

This system is called the altazimuthal or horizontal system.

However, the sky revolves around the polar axis which, together with the equatorial plane, is related to another local system of spherical coordinates, which are as follows:

(i) hour angle t_*, which is the angle between the meridian crossing the star and the local upper meridian (south direction in the northern hemisphere and vice versa);

(ii) declination δ, which is the angle between the star and the equatorial plane (it is given in the ephemeris).

This is the system of hourly coordinates.

The transformation formulae between the two systems at a site of latitude φ are as follows, where M is an auxiliary angle which is always positive:

$$\tan M = \cot \delta \cos P \qquad 0 \leq M \leq 180°$$

$$\sin h = \frac{\sin \delta \sin(\varphi + M)}{\cos M} \qquad (10.13)$$

$$\tan z = \frac{\tan P \sin M}{\cos(\varphi + M)} \qquad (10.14)$$

The inverse transformation is given by the same formulae with the following exchanges: $h \longleftrightarrow \delta$, $P \longleftrightarrow z$. Normally P and δ are known for a given star and h and z have to be calculated. Note that the astronomical azimuth angle z has to be transformed into the geographic azimuth Z. Actually P and δ are derived from the ephemeris or from nautical almanacs, which give the following data for the celestial bodies:

(1) the hourly coordinates T, δ of the Sun, the Moon, the main planets and the γ point $(T_s, 0)$ at Greenwich as a function of Universal time; the

transformation to the local system is

$$t_* = T_* + \lambda \qquad \delta = \delta \text{ (no variation for } \delta\text{)}$$

(2) the astronomical coordinates α, δ of the most important stars; the transformation to the local hourly coordinates is

$$t_* = T_s + \lambda - \alpha \qquad \delta = \delta \text{ (no variation for } \delta\text{)}$$

Therefore it is possible to derive from the ephemeris, nautical almanacs etc. the exact position of the celestial bodies in local hourly coordinates for any place in the world and any instant of time and to change to horizontal coordinates using the above transformation.

The results are compared with measurements made using the theodolite so that unknown variables can be deduced.

10.2.4 Determination of the direction of the distant station

On site the direction of the distant station is first found approximately using a magnetic compass. Good compasses are available with the needle immersed in oil for damping and with adjustable scale and transit facilities. The compass error can be several degrees and in some cases can be very high. The correct azimuth is determined accurately using a theodolite as described below.

(a) Set the theodolite in a suitable position on site, e.g. on the spot on which the antenna will be installed.

(b) With an arbitrary azimuth origin, measure the azimuth of some fixed far points (steeples, mountain peaks or even stakes or bench marks provided for this purpose in the terrain). This will serve for future reference.

(c) With the same arbitrary azimuth origin measure the azimuth of the Sun, a planet or a star (it is better to avoid the Moon which moves rapidly and requires more troublesome and accurate calculations) and take note of the exact time of observation. It is advisable to make this measurement several times and/or with various celestial bodies.

(d) Using the data of the ephemeris or the nautical almanac, calculate the true azimuth of the celestial body at the exact time of observation and compare it with the provisional azimuth measured. An example of this calculation is shown in Table 10.3. When an accuracy within about 0.5° is acceptable, tables for the determination of azimuths can be used. The relevant trigonometric expressions are tabulated and the azimuth is easily obtained.

(e) Correct the azimuths measured in (b) to obtain the true azimuths which will serve as references at the time that the foundations and the antenna are oriented.

(f) Use stakes to mark the direction of the distant station on the terrain.

Table 10.3 On-site determination of the antenna azimuth with reference to the azimuth of a star

Date: Monday 20 July 1970		Station:	Aden
Zone time correction: $+3^h$		Longitude λ	44°59′55″E
		Latitude φ	12°47′32″N
Selected star		Sun	Venus
Declination of star	δ (ephemeris)	+20°40′30″	+9°9′0″
	α (ephemeris)		
Relative azimuth observed[a]	Z_o (theodolite)	231°08′12″	215°16′00″
Time of observation	t (clock)	$17^h 34^m 00^s$	$18^h 50^m 00^s$
Zone time correction	Δ	3^h	3^h
Greenwich mean time	$T_m = t - \Delta$	$14^h 34^m 00^s$	$15^h 50^m 00^s$
Greenwich sidereal time	T_s (ephemeris)		
Right ascension	α		
Hour angle at Greenwich	$T_* = T_s - \alpha$	36°56′12″	14°14′00″
Longitude of station	λ	44°59′55″	44°59′55″
Local hour angle	$t_* = T_* + \lambda$	81°56′07″	59°13′55″
Star hour angle	$P_* = \begin{cases} P_W = t_* \\ P_E = 360° - t_* \end{cases}$	81°56′07″	59°13′55″
Auxiliary angle $\tan M = \dot{\cos} P_*/\tan \delta$	$M (\geqslant 0 \text{ and } \leqslant 180°)$	20°23′40″	72°31′00″
Astronomical azimuth of star: $\tan z = \tan P^* \times \sin M/\cos(M + \varphi)$	z	71°19′00″	87°05′00″
True geographic azimuth[b]	$Z = \begin{cases} Z = z_e \\ Z = 360° - z_w \end{cases}$	288°41′00″	272°55′00″
Relative azimuth observed	Z_o	231°08′12″	215°16′00″
Relative azimuth of north (or correction of relative azimuths)	$N = Z_o - Z$	302°27′12″	302°21′00″
Precalculated antenna azimuth	A	307°54′00″	307°54′00″
Relative antenna azimuth Assumed value 250°18′00″	$A_o = A + N$	250°21′12″	250°15′00″

[a] Relative to a provisional arbitrary origin.
[b] For southern hemisphere $Z = 180° - z_e$, $Z = 180° + z_w$.

(g) Make a small map of the site which also indicates the position of the stakes.

10.2.5 Determination of the horizon skyline

When the maps available are not accurate enough for deriving the horizon skyline, it is necessary to determine it by direct measurement on site on a clear day. An accurate measurement of the elevations is essential for a reliable calculation of the path loss. An error of 6' in the elevation will give an error of about 1 dB in the median propagation loss.

The measurement is made easily using a theodolite. However, neither light nor radio waves propagate in a straight line along the Earth's surface; normally they bend downward slightly and the bending is not the same for light and radio waves. Therefore, with reference to a diagram such as that in Fig. 10.4, we should consider the following:

(a) the geometrical profile, which is the true profile derived from the maps using the true radius of the Earth but which has no practical value;

(b) a radio profile, which is seen by the radio waves and can be derived from maps using the Earth's radius corrected using the k factor (usually $k = 4/3 = 1.33$);

(c) a visual profile, which is obtained using a theodolite and which corresponds on the average to a k factor of 1.18.

If the extreme variations of atmospheric refractivity over the world are taken into account, it can be seen that the difference between the optical and radio elevations is less than 6' when the distance to the horizon is less than 70 km. The error may become intolerable for farther horizons. When the horizon is the sea, the elevation should be calculated from eqns (3.3) and (3.8).

10.2.6 Determination of latitude and longitude

The simplest methods for finding the latitude and the longitude independently are given first, followed by a method for finding both together.

(a) *Latitude from culmination of stars*: with a theodolite follow a star with declination δ as it approaches the upper meridian and measure its altitude. In this way find the maximum altitude h which is reached at the culmination, i.e. when the star crosses the meridian. After correction for the atmospheric refraction using the tables given in the almanac, the maximum true altitude h_v is found and the latitude is derived from

$$\varphi = \delta - h_v + 90° \qquad (10.15)$$

It is advisable to repeat this measurement for several stars in order to compensate for errors.

(b) *Latitude from the altitude of Polaris*: measure the altitude of the Polaris at a certain time using a theodolite, correct it for atmospheric refraction and obtain its true altitude. Then, using the tables given in the ephemeris or in the nautical almanac, find the latitude of the site.

TOPOGRAPHIC AND SURVEY PROBLEMS 273

(c) *Longitude from hour angles*: if the latitude and the direction of the meridian are known, measure the altazimuthal coordinates of a star at a given time. Then using the inverse transformation (eqns (10.13) and (10.14)) find the hour angle of the star and compare it with the hour angle at Greenwich taken from the ephemeris for the same time. The longitude is the difference between the two hour angles (recall the definition of longitude). Note that a clock error of 1 min or 1 s gives a longitude error of 15' or 15" respectively. If the direction

Table 10.4 Determination of the geographic position

Date: Monday 13 July 1970 Zone time correction: $+3^h$		Station: Sana'a Estimated position $\begin{cases} \lambda_s & 44°11'24''E \\ \varphi_s & 15°21'00''N \end{cases}$	
Observed star	Right ascension	Declination	Measured altitude h	Time of observation t
Jupiter	—	$-9°10'06''$	$18°18'21''$	$22^h51^m05^s$
τ Scorpii	$248°34'41''$	$-28°09'56''$	$37°20'20''$	$23^h15^m05^s$

Observed star		Jupiter	τ Scorpii
Measured altitude	h (theodolite)	$18°18'21''$	$37°20'20''$
Refraction correction	c (diagrams)	$1'54''$	$1'18''$
Semidiameter correction	d (tables)	—	—
True altitude	$h_u = h - c + d$	$18°16'27''$	$37°19'02''$
Time of observation	t (clock)	$22^h51^m05^s$	$23^h15^m05^s$
Zone time correction	Δ	3^h	3^h
Greenwich mean time	$T_m = t - \Delta$	$19^h51^m05^s$	$20^h15^m05^s$
Greenwich sidereal time	T_s (ephemeris)	—	$235°03'54''$
Right ascension	α	—	$248°34'41''$
Hour angle at Greenwich	$T_* = T_s - \alpha$	$23°49'01''$	$346°29'13''$
Estimated longitude	λ_s	$44°11'24''$	$44°11'24''$
Local hour angle	$t_* = T_* + \lambda$	$68°00'25''$	$30°40'37''$
Star hour angle	$P_* \begin{cases} P_W = t_* \\ P_E = 360° - t_* \end{cases}$	$68°00'25''$	$30°40'37''$
Points on the altitude circle First approximation	$\begin{cases} \lambda_1 \\ \varphi_1 \\ P_2 \\ \lambda_2 \\ \varphi_2 \end{cases}$	$44°11'24''$ $15°31'37''$ $68°10'25''$ $44°21'24''$ $14°56'20''$	$44°11'24''$ $15°20'57''$ $30°50'37''$ $44°21'24''$ $15°14'18''$
Second approximation	$\begin{cases} P_3 \\ \lambda_3 \\ \varphi_3 \\ P_4 \\ \lambda_4 \\ \varphi_4 \end{cases}$	$68°01'25''$ $44°12'24''$ $15°28'07''$ $68°05'25''$ $44°16'24''$ $15°14'03''$	$30°41'37''$ $44°12'24''$ $15°20'17''$ $30°45'37''$ $44°16'24''$ $15°17'38''$
True calculated position	$\begin{cases} \lambda = 44°15'10'' \\ \varphi = 15°18'30'' \end{cases}$		

of the meridian is not known, it is found by following the Sun or a star with a theodolite for some hours when they are near to crossing the estimated meridian direction. By taking periodic note of time, altitude and azimuth, it is possible to find by interpolation the azimuth corresponding to the maximum altitude and therefore to the meridian.

(d) *Simultaneous latitude and longitude determination using altitude circles*: the locus of the points on the Earth's surface from which at a given instant of time a star with declination δ is seen at a certain altitude h is a circle with the equation

$$\sin \varphi \sin \delta + \cos \varphi \cos \delta \cos P - \sin h = 0 \qquad (10.16)$$

where $P_W = T_* + \lambda$ and $P_E = 360° - T_* - \lambda$. In this equation the altitude and the time are known, as they were measured at the same instant. The only unknown variables are the two coordinates. If the measurements are repeated for a second star, there are two equations and two unknowns, and the system can easily be solved using a microcomputer. There are two crossing points for the circles, but the second point is normally far away from the first and there is no ambiguity. The calculation can proceed by successive approximation starting from an estimated position.

For maximum accuracy the above method is normally applied to a dozen stars in all possible positions in the sky in order to compensate for various errors. The crossing points of the pairs of circles will never coincide, but we shall obtain a set of points close together from which we will take the average position as the most probable.

An example of a calculation is shown in Table 10.4 and should be self-explanatory. Data on the stars and the corrections are taken from the ephemeris; other parameters are calculated using the simple formulae given. The final parameters are entered in the microcomputer and the result is obtained by successive approximation. Programs for calculating the geographic coordinates from observed data for stars are also available commercially.

10.3 Topographic problems in antenna installation and line-up

The method used to find the direction of the distant station(s) and to mark it on the terrain was discussed in Section 10.2.4. The inaccuracy in the direction should be a small fraction of the antenna beamwidth. When the method described here are used the error in the direction should be much less than 0.1°

TOPOGRAPHIC AND SURVEY PROBLEMS 275

The antennas may be installed several months after the surveys have been made. We now briefly discuss the positioning of the antenna foundations, which is not a problem for parabolic antennas mounted on self-supporting towers which can easily be oriented, but requires great accuracy for billboard antennas which are fixed on the terrain and should not need orientation adjustments after installation.

In the remainder of this section we shall discuss the orientation of tower-mounted parabolic antennas in the horizontal and vertical planes, and the way in which the reflector profile is checked in large billboard antennas.

10.3.1 Orienting the antenna in the horizontal plane

If the antenna is of the billboard type, so that it is directly fixed to the terrain, the problem of orientation is solved as soon as the position of the foundations has been determined. After that, it is only possible to make small corrections (no more than about $\pm 2°$ in azimuth) by moving the illuminator, but such a solution should be avoided because it affects the radiation pattern, particularly with regard to sidelobe intensity.

If the antenna is a paraboloid mounted on a tower, it usually has provision for azimuth adjustments of the order of $\pm 5°$ if it is mounted on a face of the tower and of the order of $\pm 20°$ if it is mounted on a corner of the tower.

The final orientation of the antenna is made using one of the following methods.

(a) *By finding the direction of maximum field strength*: the use of a field strength recorder connected to one of the radio receivers is mandatory because the "instantaneous" field is very variable owing to the unavoidable presence of fast fading. Only the average field strength is considered. First the radio connection between the two stations is found by tentatively moving both antennas. Then the antenna is moved horizontally in angular steps of a quarter of the beamwidth (or in larger steps when far from the final azimuth) and it is kept for some minutes in each position. The direction of the maximum average field is determined from the field strength records. The procedure should be followed alternately for the two terminal stations in order to maximize the fields. This method takes time and is not accurate. There are variations in the average path loss during measurements owing to slow fading, which cause errors in the evaluation.

(b) *By geometrical methods*: the direction of the distant station is marked on the terrain or, on clear days, found from the horizon skyline. The antenna is normally made of spikes which define a vertical line on its surface, and the illuminator with its supporting tripod or tubular beam may yield other reference lines. On a clear day a person on top of the antenna can easily transit the antenna top center, the axis of the illuminator and the horizon skyline with a plumb line. The antenna is moved horizontally until the correct direction is reached. Alternatively, a point (marked by a stake) 50 or 100 m in front of the antenna in the exact direction of the distant station, the antenna vertical center line and the illuminator center can be transited by a plumb line and will be in line only when the antenna is moved to the required direction. If combined electrical and geometrical methods are used, this last procedure allows the position of the antenna sidelobes to be checked. An azimuth displacement of 1° 100 m from the antenna is equivalent to a linear displacement of 1.75 m perpendicular to the antenna direction. A set of stakes can be placed on this perpendicular line, thus determining the different directions and permitting a check of the radiation pattern within, say, ±5° or even more. Instead of standing in front of the antenna the person can stand on one side and transit the two edges of the parabola. The error in orienting the antenna by these simple methods will be less than 0.1° (6').

10.3.2 Orienting the antenna in the vertical plane

The vertical orientation can be carried out using the electrical or the geometrical method, but it is perhaps advisable to use both methods together. Also, the refractivity of the air is slightly different for light and for radio waves, and therefore the optical elevation is not exactly equal to the radio elevation. However, as stated in Section 10.2.5 the difference is on average less than 0.1° if the horizon is at a distance of less than 70 km.

According to the type of construction of the antenna, the elevation can normally be varied by ±1° or ±2° using one of the following methods:

(a) moving the complete antenna, reflector with illuminator;

(b) moving only the illuminator with respect to the reflector, which is kept vertical or has been set to an initial inclination (which prevents degradation of the radiation pattern).

This movement is generally obtained using a lead screw device controlled through a steering wheel. It is necessary to predetermine the number of turns of the wheel per degree of elevation, as well as the central position, so that the elevation of the beam is known for each wheel position. If only the illuminator moves, the beam direction is opposite to the illuminator direction with respect to the paraboloid axis!

The elevation to be provided is known from the preliminary calculations. However, it is advisable to check the maximum average field strength using a recorder connected to one of the receivers and base the final orientation on the results of both methods. The vertical angular steps should be of the order of a quarter of the beamwidth or less.

10.3.3 Assessment of the profile of large billboard antennas

The gain of a parabolic antenna is related to the accuracy of the profile of the reflector surface. In practice it is impossible to measure the gain of an installed antenna. Correct mounting should guarantee that the full gain is obtained but in practice, owing to unavoidable inaccuracies during installation, the tolerance may be exceeded at some parts of the surface. It may therefore be advisable to check the profile.

This can be done using a microgeodesic method called *double intersection from three stations*. Figure 10.7 shows the geometry of the method. The horizontal coordinates X and Y and the vertical coordinate Z are established from the vertex of the ideal paraboloidal surface. The physical paraboloid may be both horizontally and vertically symmetrical (center-fed type) or asymmetrical (offset type), as in Fig. 10.7, depending on which portion of the ideal surface is utilized by the reflector.

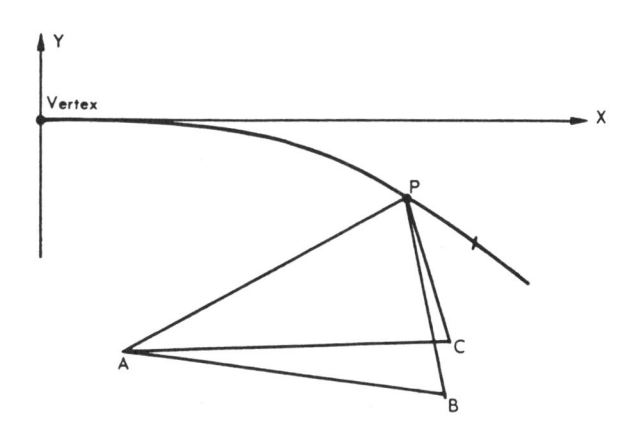

Fig. 10.7 Geometry of the method of double intersection from three stations.

Three positions A, B and C are chosen for the theodolite, and α and β (Fig. 10.8) indicate the positions of the extreme feet of the structure. Stations A, B and C are chosen such that the measured zenithal angles are always greater than 45°. Billboard antennas normally have diameters of 18 m or 27 m, so that the distances AB and AC are chosen to be about 40 m, and the distances between the stations and the antenna are chosen to be about 35 m. The stations are marked by stakes with a nail in them. The distances AB, AC, Aα, Aβ, Bα, Bβ, Cα and Cβ are measured twice using a steel tape and then corrected for temperature and tilt. Horizontal and vertical angles in the direction of the other stations and of α and β are measured from each station. The resolution of triangles ABα, ABβ, ACα and ACβ enables the coordinates of A, B and C to be calculated twice in the same system as α and β and also enables the azimuth of AB and AC to be calculated. The results can be checked by recalculating the coordinates of α and β from the coordinates obtained for A, B and C.

The reflector surface is made of panels (Fig. 10.9) which can be identified by numbering them vertically. The position of each panel is adjustable along the Y axis, and the panels are identified by a small label in the center or by a suitable rivet.

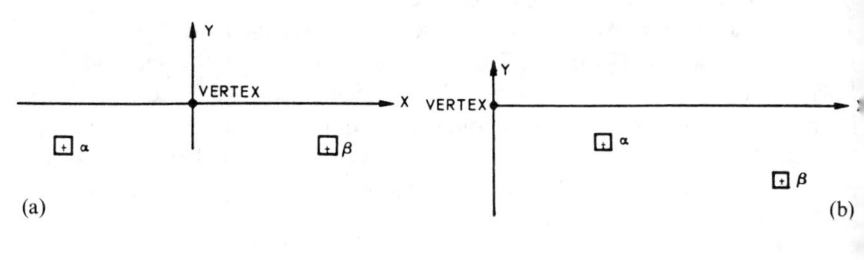

	X	Y	Z	X	Y	Z
α	−9.1440	−5.2470	−10.5680	4.5720	−2.0115	−15.0500
β	+9.1440	−5.2470	−10.5680	32.0200	−10.2495	−15.0500
f		−9.1440			−30.4800	
	Distance α–β		18.2880 m	Distance α–β		28.6576 m
	$\theta_{\alpha\beta}$		100.0000 grad	$\theta_{\alpha\beta}$		118.5624 grad

Fig. 10.8 Position of the reflector with respect to the coordinate axes (lengths in meters): (a) billboard center-feed type (diameter, 18 m); (b) billboard offset-fed type (diameter, 27 m).

The azimuthal and zenithal angles of each point P are measured from stations A, B and C. The coordinates of the points are given by

$$X_P = X_A + (Y_P - Y_A) \tan \theta_{AP} \tag{10.17}$$

$$Y_P = Y_A + \frac{(X_A - X_B) - (Y_A - Y_B) \tan \theta_{BP}}{\tan \theta_{BP} - \tan \theta_{AP}} \tag{10.18}$$

$$Z_P = Z_A + AP \cot \varphi_{AP} \tag{10.19}$$

where θ_{AP} is the azimuth of the segment AP referred to the Y axis, θ_{BP} is the azimuth of the segment BP referred to the Y axis, φ_{AP} is the zenith distance of P measured from A and

$$AP = \{(X_P - X_A)^2 + (Y_P - Y_A)^2\}^{1/2}$$

The coordinates X_P and Y_P are calculated twice (once from X_B, Y_B and θ_{BP}, and once from X_C, Y_C and θ_{CP}) and the mean is taken. Similarly, Z_P is calculated three times and then the mean is taken.

TOPOGRAPHIC AND SURVEY PROBLEMS   279

The intersection of a paraboloid with a plane parallel to the axis is the parabola

$$Y = \frac{X^2 + Z^2}{4f} \qquad (10.20)$$

where f is the focal distance. If the values of X_P and Z_P obtained above are substituted in (10.20) and the values of Y_P calculated in this way are compared with those obtained above, the error ΔY in the position of the panel is found.

The following set of errors is obtained:

(1) the average unwanted tilt of the antenna;

(2) the deformation diagram which is represented using contour lines as on a geographic map (Fig. 10.9);

(3) the r.m.s. error which is compared with the specified error;

(4) the plan of the panels which need adjusting and the amount by which they need to be adjusted.

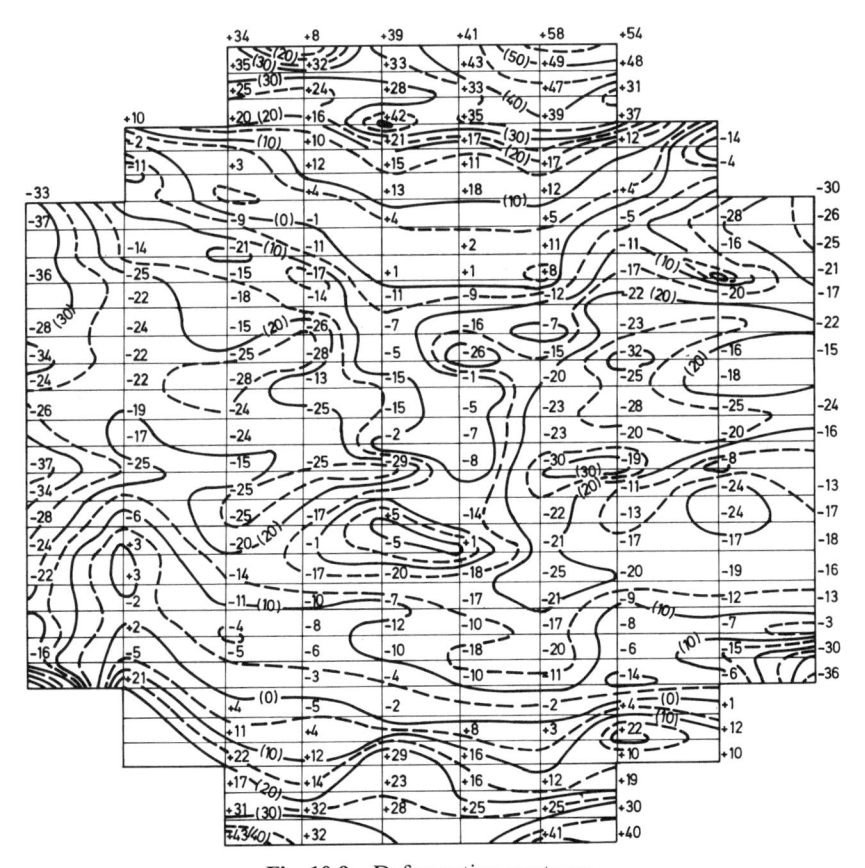

Fig. 10.9   Deformation contours.

In an actual installation with six center-fed billboard antennas of diameter 27 m and six offset-fed billboard antennas of diameter 18 m, working at 900 MHz, 244–266 points were observed for each antenna. The unwanted tilt was -3 to -9 mrad, the r.m.s. error was 8–22 mm, 35–176 points were out of tolerance and the error in the height of the illuminator ranged from 30 to 90 mm. It was therefore necessary to introduce corrections in order to remain within the allowed tolerances.

An analysis of the accuracy of the above method indicates that the error in the calculated coordinates can be kept well within ± 3 mm. Measurements made at night and repeated in full sunlight yielded differences which were still well below 3 mm, thus indicating that the measurements can be made in the sun without risk of error and that the antenna gain does not vary from day to night.

10.4 Comments and suggestions for further reading

Equations (10.1)–(10.3) and (10.9)–(10.12) were taken from ref. 10.1 where, however, they are not proved. The principles, concepts and formulae of geographical astronomy are given in ref. 10.2. Books on spherical astronomy, geographical astronomy, nautical astronomy and geodesy also provide useful information.

References

10.1 Transmission loss predictions for tropospheric communication circuits, *NBS Tech. Note 101*, Vols 1 and 2, January 1967 (National Bureau of Standards, U.S. Department of Commerce).

10.2 Mackie, J. B. *The Elements of Astronomy for Surveyors*, Griffin, London, 1978.

Chapter 11

From Planning to Implementation and Maintenance

In preceding chapters we have examined the theoretical and practical aspects of the design of troposcatter systems. The various topics, including path geometry, propagation problems, evaluation of path loss and distortion, equipment, performance prediction, topographic problems arising before and after a site survey, installation etc., were considered separately. Since we now have the necessary technical background, we can look at the whole process in the round. This chapter is wide ranging: it gives a panoramic view of the subject without going into excessive detail. In practice a detailed description is unnecessary, and a brief mention is generally adequate.

We shall start with planning, which involves office work and, in most cases, site surveys. Site surveys are also necessary for the final siting of stations and before installation. The frequency band and the individual working frequencies are then chosen, unless they are allocated by the local authorities on the basis of crowding of the radio spectrum. The design proceeds with calculation of the path and choice of equipment. The next step is installation and its associated problems. Radiation hazards are dealt with in a separate section. Finally, the installed system is tested. However, path tests may also be made during the design stage.

We give a brief description of the problems of operation and maintenance and discuss the cost of a troposcatter system.

11.1 Planning the system

Planning a troposcatter system is generally one step in the planning of a communication system connecting a number of locations according to the expected traffic requirements. The organization needing the system generally issues an enquiry or a request for bid on the international market. This enquiry contains information on the requirements ranging from a minimum, in which only the stations to be connected and the number of circuits required are indicated and the designers are asked to propose a complete system, possibly with alternatives, to a maximum in which the system is defined in terms of a preliminary study which is to be completed by detailed surveys and final design work. The complexity of the system ranges from a network including troposcatter hops, LOS hops and other types of communications to a single troposcatter path. Our analysis is limited to the troposcatter portion of the system.

Planning proceeds as follows.

(a) A route plan is drawn, indicating the route of each required telephone channel connecting the different terminals and showing the traffic capacity necessary for each radio path.

(b) Using a large-scale map, on which roads are also indicated, the designer chooses possible sites for repeater stations, trying to site them at the highest points but at the same time considering the problem of accessibility through existing roads or through roads to be constructed. The same consideration is given to the problem of energy supply. When the terminals are in towns or lowland areas it is often necessary to site the nearest troposcatter station on top of a nearby mountain and to connect it to the terminal station by means of an LOS link.

(c) When a tentative choice of sites has been made on a large-scale map, it must be checked on a smaller-scale map (generally 1:100 000). The geographical parameters can then be derived.

(d) The geographical and geometrical data obtained are used to perform a feasibility study for each hop: the path loss is calculated in one or more frequency bands and the main parameters of the equipment required for realizing the system (transmitted power, diameter of parabolic antennas etc.) are evaluated.

(e) When preliminary proposals have been obtained from the feasibility study, it is necessary to survey the sites in order to solve local problems and finalize the design. It may be found that there are local obstructions, such as trees, buildings etc., so that the antennas must be installed on high towers. Furthermore, sites could be affected by local radio noise or interference, which must be avoided if full exploitation of the low noise characteristics of troposcatter receivers is to be obtained.

Table 11.1 Example of a list of materials required for a troposcatter terminal station

Item	Description	Quantity Dual diversity	Quantity Quadruple diversity
1.	Troposcatter radio terminal equipment composed of		
1.1.	Radio transmitter (exciter)	2	2
1.2.	Radio power amplifier	2	2
1.3.	Radio receiver	2	4
1.4.	Diversity combiner	1	1
1.5.	Service channel	1	1
1.6.	Set of r.f. and antenna filters	1	1
1.7.	Duplicated power supply	1	1
1.8.	Ancillaries (local and remote alarms, built-in test instruments etc.)		
2.	Parabolic antenna with dual polarized illuminator and supporting tower (plus obstruction lights if required)	1	2
3.	Coaxial feeder cable with connectors	2	4
4.	Pressurizer–dehydrator for feeders and antennas	1	1
5.	Grounding plant and protection against surges and lightning	1	1
6.	Set of spare parts for 2 years' maintenance	1	1
7.	Set of normal and special tools	1	1
8.	Set of measuring instruments	1	1
9.	Set of instruction manuals	1	1
10.	Electric power generating station (if electricity not available) composed of		
10.1.	Diesel engine	3	3
10.2.	Alternator	3	3
10.3.	Control board	1	1
10.4.	Daily tank with automatic refilling	1	1
10.5.	Large underground fuel tank	1	1

(f) After the feasibility study and the site survey have been completed, all evidence for finalizing the design and evaluating the costs should be available. A complete technical proposal is then prepared: it includes all data and calculations justifying the proposed solution and terminates with a complete and detailed list of the equipment and materials necessary for each station. For guidance a list of materials is given in Table 11.1. This should be completed with the main characteristics: traffic capacity, frequency band, r.f. power, supply voltage etc. for the radio equipment; diameter, wind speed etc. for the antennas; lengths of feeders, list of spares, instruments etc. Detailed characteristics should be included in other parts of the proposal.

Allowance should be made for possible future expansion of the system.

Although the main planning procedure involves the steps listed above, other points should be taken into consideration for various types of system. The system may be for commercial or military use. In the first case it is generally of the fixed type, and in the second case it may be of the transportable or mobile type. The following additional points should be considered for systems of the transportable or mobile type.

(g) If total mobility is required, the terminals should be capable of being completely mounted and dismounted (both radio and antennas) at any site chosen. In this case the preliminary planning of the main network is performed at the office so that the operators know in advance which are the best positions for the terminal stations in the region considered. Graphs giving the maximum allowable elevation angle for the antennas versus the distance to be covered by each hop should also be prepared. The operators should then have enough information to choose the sites.

(h) If partial mobility is sufficient, predetermined sites, i.e. concrete platforms and basements on which the station can be set up when required, can be provided by the commissioning authority.

11.2 Site surveys

Site surveys are necessary for both the preparation of a complete technical proposal and the detailed planning of the installation. Sometimes, particularly in regions where radiometeorological data are scarce, additional direct propagation tests from site to site are made to obtain a more reliable value of the path attenuation than that predicted by calculations. We shall discuss this subject further in Section 11.9.

During the survey possible station sites are inspected and all available information is gathered. As a guide to the collection of information we have listed in Table 11.2 all possible questions related to the design, siting, installation, costs, existing facilities, logistics etc. If all this information is available, the designers will be able to prepare a plan of the work which is complete in both technical and economic detail. Careful examination of Table 11.2 gives an idea of the various problems which require consideration and solution, and will avoid the need for long explanations. We only draw attention to minor, but important, practical problems arising from point 1 of this table.

Table 11.2 Fundamental requirements for a detailed site survey

1. *Materials and instruments recommended for survey and site inspection*
   (a) Maps, road maps, detailed small-scale maps
   (b) Guides, dictionary, information and tourist manuals etc.
   (c) Slide rule, graduated rule, angle-measuring device, scale meter
   (d) Geologist's compass
   (e) Abney level, tachometer, theodolite
   (f) Data on ephemeris and on trigonometric points
   (g) Radio and watch
   (h) Metric strip
   (i) Electric torch, knife, lighter
   (j) Photographic camera, films, flash
   (k) Small multimeter
   (l) Binoculars, telescope
   (m) Forms to be filled in
   (n) Block-notes, agendas, pencils, erasers
   (o) Firm's paper and envelopes (normal and airmail)
   (p) Portable typewriter, carbon paper
   (q) First-aid kit
   (r) Radio telephones
   (s) Geologist's hammer
   (t) Altimeter
   (u) Nautical tables (azimuths)
   (v) Programmable calculator

2. *Inspection of site*
   (a) Take photographs
   (b) Obtain maps, coordinates, elevation
   (c) Obtain information about real estate and boundaries
   (d) Examine soil type (for foundations and drainage)
   (e) Estimate soil bearing capacity
   (f) Examine soil shifting and earthquake probability
   (g) Height of water table
   (h) Nearest potable water
   (i) Nearest water supply for concrete
   (j) Distance from nearest town or road
   (k) Characteristics of access roads
   (l) Unloading and storing areas
   (m) Terrain features (wooded, marsh etc., sloping)
   (n) Major obstructions
   (o) Horizon elevation angles and clearances

3. *Inspection of existing buildings*
   (a) Size, shape, ceiling height; record or obtain plans
   (b) Sketch all openings, relative to north, landmark or direction of propagation, dimensions of openings, height from floor etc.
   (c) Type of building construction, type of floor
   (d) Permissible roof loading, floor loading, wall loading
   (e) Type of heating or air conditioning
   (f) Sketch all obstructions within building—radiators, pipes, existing panels etc.
   (g) Distance of building from other structures
   (h) Accessibility of building—public transportation available, roads etc.
   (i) Water availability in or near building (potable or not); toilet facilities
   (j) Parking availability
   (k) Take photographs

4. *Availability of power*
   (a) Type and capacity of power source—voltage, currents, phase, frequency, continuity, stability tolerance back-up, wire size, reliability etc.
   (b) Future plans for additional power—same data as above
   (c) Distance from power takeoff, type of transformers if used
   (d) Data on standby power—diesel, or commercial—name of commercial companies

involved, or diesel manufacturer with type and model of diesel; fuel capacity and consumption at full load; all other information from the diesel generator's name plate
    (e) Wiring employed

5. *Existing communications equipment*
    (a) Type, manufacturer and characteristics of existing equipment
    (b) Record existing installation on plans (considering future modifications)
    (c) Interface characteristics of existing equipment
    (d) Location and frequencies of nearby electronic equipment or facilities
    (e) Equipment and facilities that can be utilized in the new plant
    (f) Interference possibilities to existing or new service

6. *Technical information from administrations*
    Communications
    (a) Type, characteristics, description of existing equipment
    (b) Technical ideas about the new plant—its characteristics (discuss the specifications), operation, maintenance, use of personnel
    (c) Ideas about future developments
    (d) Local regulations on frequency, power, antenna size etc.
    (e) Painting, lighting or other distinguished marking required
    (f) Are antennas considered an air hazard; how high; warning lights

    Climate
    (a) Maximum, minimum, average temperature throughout the year
    (b) Maximum, minimum, average humidity throughout the year
    (c) Maximum, minimum, average wind speed throughout the year
    (d) Maximum, minimum, average rainfall throughout the year
    (e) Maximum, minimum, average snowfall throughout the year
    (f) Direction of winds at each site
    (g) Data on icing
    (h) Number of stormy days in the year
    (i) Data on lightning, lightning areas
    (j) Data on corrosion problems
    (k) Data on sandstorms
    (l) Available data on atmospheric refractive index and radioclimatology

7. *Logistics*
    Roads, conditions, availability, frequency of traffic interruptions (for snow, sandstorm, ice etc.)
    (a) Average and maximum grade
    (b) Minimum width
    (c) Type and conditions of surface
    (d) Minimum radius of turns
    (e) Weight limitation
    (f) Capacity of bridges
    (g) Clearance, height of bridges, underpasses, tunnels etc.
    (h) Possibility of flooding, icing, drifting snow, sandstorm
    (i) Railroads, siding distance, rail gauge etc.
    (j) Airfields, name, distance, location, size of plane handled, airlines using etc.
    (k) Location of areas suitable for helicopter landing and access

    Transportation
    (a) Availability of cars, trucks, planes, helicopters for hire
    (b) Unloading equipment available and type
    (c) Unloading areas and types—rail, water, plane etc.

(d) Water transport (if any): waterways, usable period, distance to site, shipping schedules or cycles, unloading and storing facilities
(e) Unloading and storage facilities (cranes, power shovels, bulldozers, trucks, trenchers, concrete mixers)

8. *Fuel supplies*
   (a) Nearest source of diesel fuel
   (b) Nearest source of gasoline
   (c) Grade of diesel fuel
   (d) Amount available
   (e) Delivery problems

9. *Accommodation*
   (a) Living conditions, housing, stores
   (b) Hotels, laundry services
   (c) Entertainment facilities available
   (d) Facilities for family living
   (e) Available water; any unusual qualities of available water
   (f) Sewage—type and handling capability
   (g) Availability of medical facilities

10. *Others*
    (a) Land area involved around site—ownership
    (b) Fences and size of openings
    (c) Approximate number of personnel attached or assigned to area
    (d) Construction material availability
    (e) Availability of telephone, telegraph, mail facilities
    (f) Repair and electronics test shop available
    (g) Local construction practices

11. *Local labor, contractors, subcontractors*
    (a) Local hire labor in nearest market: general class available, guards, common labor, semiskilled, clerks, stenographers, draughtsmen, warehousemen or property men, technicians, engineers, heavy equipment operators, mechanics, riggers, carpenters, masons, laborers
    (b) Contractors: building and construction, civil engineers and surveyors, electrical, estimate size, competence, responsibility
    (c) Subcontractors: for transportation, tower erection etc.

12. *Administrative*
    (a) Local population, type and language
    (b) Civil development and facilities
    (c) Political situation
    (d) Major local authorities
    (e) Eventual code name of sites
    (f) Type of station and function
    (g) Proximity to military installations
    (h) Proximity to radio, TV, HF etc. installations
    (i) Proximity to airports, ports, stations etc.
    (j) Problems of security, vandalism, sabotage
    (k) Local legal problems
    (l) Local working days and holidays
    (m) Military restrictions

13. *Commercial and financial*
    (a) Monetary restrictions
    (b) Employment statutes
    (c) Licenses or registration requirements
    (d) Income and other taxes
    (e) Cost of living
    (f) Cost of hotels and restaurants
    (g) Cost of travel (train, airplane, taxi etc.)
    (h) Cost of hiring a car, a truck, a small plane, a helicopter, a motorboat
    (i) Cost of transportation (by truck, train, plane) per ton km and per cubic meter

(j) Cost of construction materials	(n) Cost of labor
(k) Cost of a cubic meter of concrete (with and without steel) in place	(o) Customs duty
	(p) Cost of insurance
	(q) Cost of local installation materials
(l) Cost of excavation	
(m) Cost of loading and unloading	(r) Cost of storage

It is always advisable to take two pocket instruments for roughly measuring the horizontal coordinates (Section 10.2.3) of the relevant sections of the horizon or of other objects of interest. These instruments are as follows:

(1) a magnetic compass with an oil-damped needle, which can measure azimuths with an error ranging from zero to a few degrees, and which can measure relative angles (between a known point and the required direction) to a much better approximation, provided that all measurements are made from the same point;

(2) a gradienter or an Abney level, which measures vertical angles with an error of $0.1°$.

These instruments can also be used in subsequent surveys, when the major references, directions and elevations are known from previous theodolite measurements and only relative checks have to be made. A mountain peak of known azimuth and elevation can be used for reference. Another important use is for siting mobile troposcatter stations after the azimuth and elevation of the horizon have been measured.

If these pocket instruments are not available, it is possible to measure horizontal and vertical relative angles from a reference point using a simple ruler or even a strip of paper held between the thumb and index finger in front of the eyes by the outstretched arm. If the distance between the eye and the strip is, say, 50 cm, each degree is equivalent to about 8 mm on the strip. If this ruler or strip is not available, the angular width of the thumb 50 cm from the eye is about $2°$, the angular width of the closed fist is about $12°$ and the angular width of the open hand with all fingers spread is about $25°$.

11.3 Siting the stations

The sites for troposcatter radio links are generally chosen in elevated locations so that the takeoff angle of the antennas toward the horizon is if possible negative with respect to the horizontal, i.e. the antennas should look slightly downwards. The aim is to obtain the minimum possible scatter angle and therefore the minimum path loss. It is convenient to check the elevation angle and the horizon profile by means of a theodolite. It should be noted that a variation of 1° in the scatter angle means a variation of about 12 dB in the path loss. Of course on flat terrain (e.g. in the desert) the antenna will be pointing horizontally or, when unavoidable, upwards.

The space in front of the antennas must be clear of all obstructions and the horizon must be at a distance of some kilometers. If there are trees or other obstructions that cannot be removed, the antenna should be mounted on a tower high enough for there to be a clearance equal to at least the radius of the antenna. Care must be taken that there is no manmade radio noise or possibility of interference in the area of the station. There should be no motor roads in front of the antenna, as motor ignition systems generate radio noise. The site requirements are summarized in Table 11.3. The chosen site should provide a sufficient area for the station. Leveling of the terrain could be advisable but it is not often necessary.

Table 11.3 Site requirements

1. Elevated position, the higher the better
2. No obstructions in front of the antennas
3. Horizon distant at least some kilometers
4. No mountains or hills in the direction of the distant station, obliging the antennas to point upwards
5. Horizon possibly depressed with respect to the horizontal plane from the site so that the antennas can be pointed slightly downward
6. No motor roads or other facilities generating radio noise particularly in front of the antennas

An idea of the required area is given in Fig. 11.1 which shows a typical layout of a dual space diversity (or a quadruple space-frequency diversity) station. The distance between the antennas is typically 100 wavelengths. The hut for the radio equipment is centered between the two antennas in order to minimize the feeder length and to equalize the delays. Of course a larger area is required for a troposcatter nodal or repeater station, which contains more antennas. There could also be a separate power building for the electric generators, and a parking area for motor vehicles. The water supply for the station could present a problem.

Figure 11.1 also gives an idea of the clearance in the near region required for the antenna beam, which can be assumed to be cylindrical and to have the same diameter as the parabolic reflector out to a distance $D^2/2\lambda$ from the antenna (Fresnel region) and to be conical with the vertex in the image of the illuminator thereafter (Fraunhöfer region). In the far region the clearance should be the same as for the conventional radio links (the first Fresnel zone is free), but in practice the beam is normally free from obstruction after leaving the site until it reaches the horizon.

The site requires an access road for the transportation of materials and for the maintenance crews. Sometimes this road does not exist and must be purpose built, thus adding to the cost of the system.

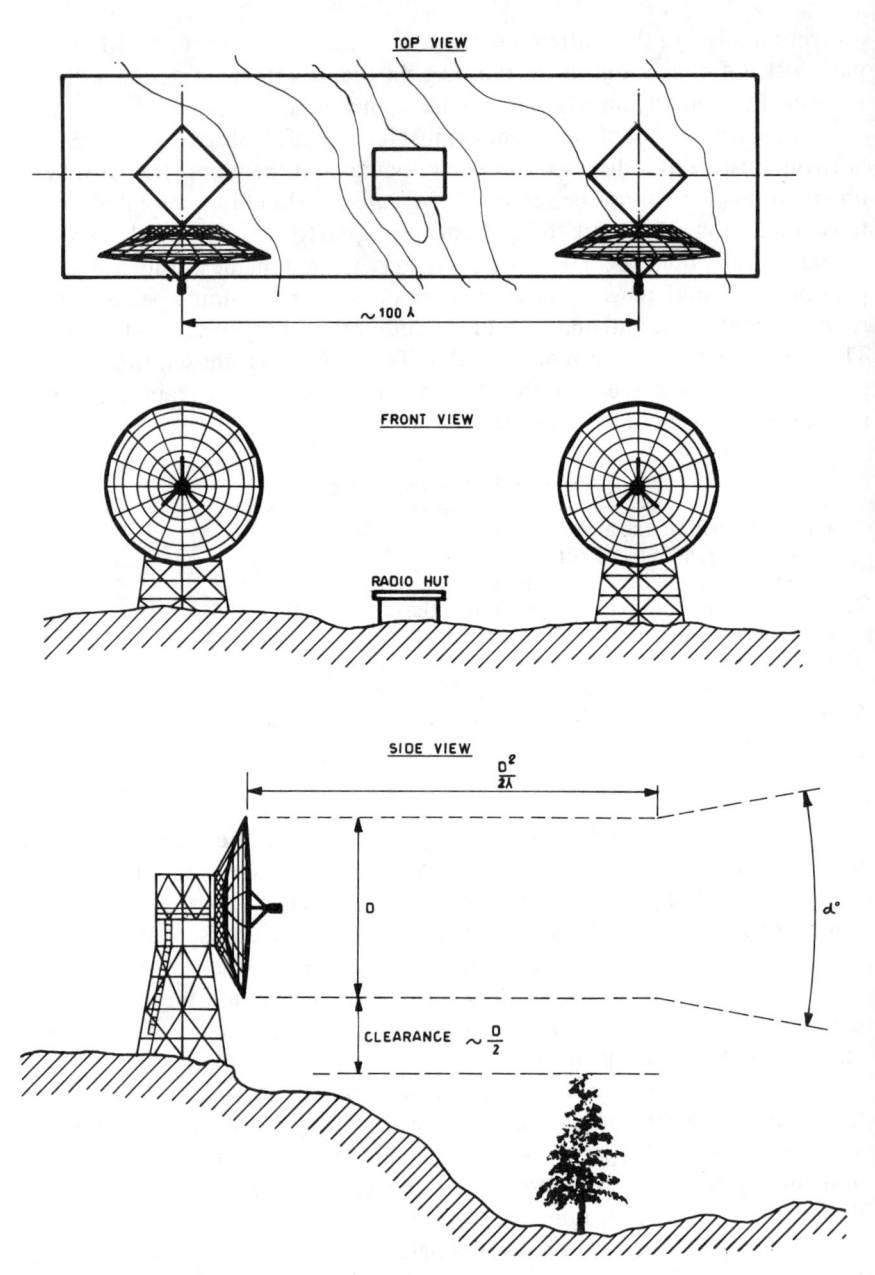

Fig. 11.1 Typical layout of a space diversity fixed station.

An example of a fixed station is shown in Fig. 11.2.

Before definitely accepting a site it is advisable to check the direction of the antennas with respect to the position of the Sun at certain hours of the day.

Fig. 11.2 Example of a fixed station.

Sometimes it is found that an antenna points directly into the Sun (near sunrise or sunset) and this creates a lot of noise in the circuits and overheating of the illuminator which, as it is at the focus of the reflector, is heated by all the thermal energy collected by the reflector.

11.4 Choice of frequency band

In many countries the radio spectrum is rather crowded, and the frequency band for new radio links is often determined by local radio regulations. However, it is sometimes possible to choose between two or more frequency bands. In this case the following points should be taken into consideration.

(a) It is easier and less expensive to obtain high r.f. power in lower frequency bands. The power amplifiers use triodes or tetrodes below 1 GHz and klystrons or traveling-wave tubes at higher bands. These tubes are more complex in operation and much more expensive than triodes or tetrodes.

(b) The lower the band, the easier it is to obtain good noise figures for the receivers by using low noise mixers or preamplifiers. It is unnecessary to use sophisticated circuits or to add parametric amplifiers or bandwidth compression devices.

(c) The antennas are comparatively less expensive in the lower bands because the tolerance on the surface profile is less stringent (± 4 cm at 400 MHz) and the reflecting surface may be lighter (it may be of the mesh or the perforated type).

(d) Coaxial cables can be used for antenna feeders below 1 GHz, but expensive and fragile waveguides are required in the higher bands.

(e) The peaks of fast fading are less steep and less numerous in the lower bands. Thus a narrower band is required for the control of the baseband diversity combiner and a physical service channel can be used instead of a translated service channel.

(f) The upper bands allow shorter hops but also lower path intermodulation, thus improving the quality of high capacity links.

(g) In space diversity stations, where the two antennas must be displaced by 100 wavelengths, the upper bands require a smaller area and shorter feeders.

In order to give a simple reference it is assumed that the transmitter power and the receiver threshold are independent of the frequency band; thus the

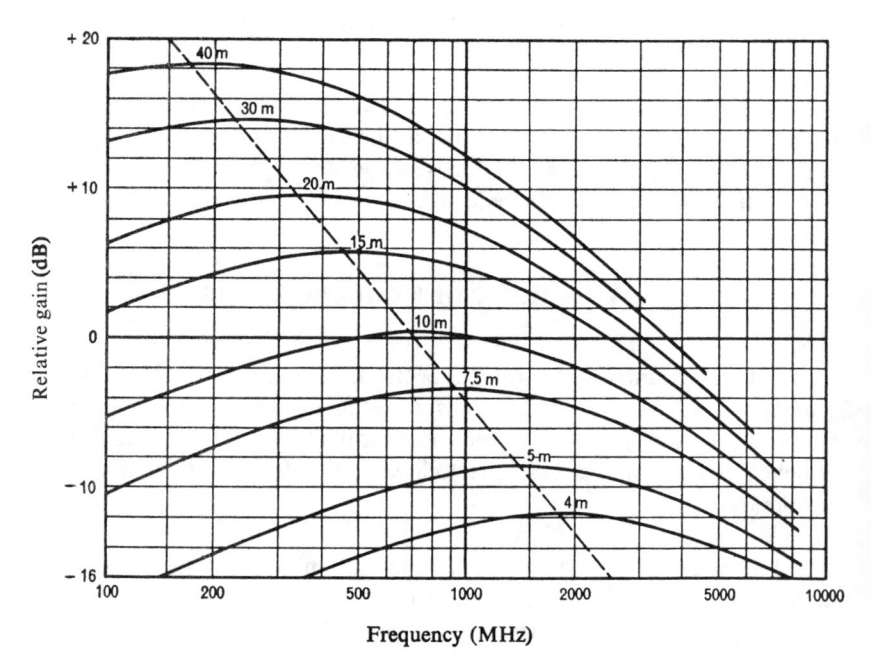

Fig. 11.3 Relative gain versus frequency in a troposcatter path with antennas of given diameters.

only significant parameters remaining are the path loss, which varies as the cube of the frequency, the antenna gain, which varies as the square of the frequency, and the degradation of antenna gain. The curves in Fig. 11.3, which is taken from *CCIR Report 285-6* and shows the relative gain between two troposcatter terminals with antennas of the same diameter when the frequency is varied, were obtained by combining these parameters. The reference gain (0 dB) is relative to a antenna of 10 m diameter at 1000 MHz. The path length is assumed to be between 150 and 500 km. It is apparent that, for each antenna diameter, there is a maximum relative gain at a particular frequency. This gain decreases at lower frequencies because the free-space antenna gain decreases, and it decreases at higher frequencies because of the increase in antenna gain degradation.

The frequency corresponding to the maximum relative gain decreases when the diameter of the antenna is increased. In practice, frequency bands below 1 GHz seem to be more suitable for antenna diameters greater than 7 m. However, since the maximum is rather flat, a 100% shift in frequency away from the maximum has a very small effect.

11.5 Frequency plan

When the geographical configuration of the system is defined and the frequency band is allocated, a frequency plan has to be prepared for the paths so that no interference occurs within the system or with other existing systems within a radius of, say, 1000 km. The CCIR gives only general and generic indications for frequency allocations of troposcatter systems. Frequency diversity should be avoided in order to be economical with the spectrum. We give here the main principles of frequency planning in order to aid the designer in the choice of the actual frequencies.

When a signal composed of two or more frequencies enters a nonlinear circuit the output signal contains the original frequencies plus a large number of spurious frequencies or intermodulation products which are linear combinations of the original frequencies. If the input spectrum is composed of frequencies $F_1, F_2, \ldots, F_i, \ldots, F_n$, the output spectrum contains an infinite number of frequencies, for which the general term is

$$f = |a_1 F_1 + a_2 F_2 + \ldots + a_i F_i + \ldots + a_n F_n| \tag{11.1}$$

$$= \left| \sum_{i=1}^{n} a_i F_i \right| \qquad a_i = 0, \pm 1, \pm 2, \pm 3, \ldots$$

where the a_is can appear in any combination. The sum $\Sigma |a_i|$ is called the order of the term. The number of terms increases very rapidly with n, but their level generally decreases with increasing order.

A nonlinear circuit may be required, as in the case of a mixer at the receiver input. Here the receive frequency R and the local oscillator frequency H yield the required output intermediate frequency $|H-R|$, and the other unwanted products are filtered out by the i.f. filter. There may also be unwanted nonlinear elements, such as dirty contacts in coaxial cables, rusty bolts on the surface of the parabolic antenna, rusty contacts in metallic fences near the antenna etc., which can generate unwanted and dangerous spurious frequencies. These can be eliminated by accurate mounting.

In practice the origin of most harmful spurious frequencies is the receiver mixer. It is necessary to ensure that signals different from the required receive frequency reach the mixer input at a sufficiently high level after filtering. If an intermodulation product at the mixer output falls into the i.f. band, it cannot be eliminated and will result in spurious tones in the baseband and an increase in the noise or the BER. There may be many spurious frequencies at the mixer output, but they are normally filtered out by the i.f. filter.

In practice, in order to generate harmful products the unwanted signals must reach the mixer input (and the receiver input) with a sufficiently high level. Interference curves should be obtained from the manufacturer or measured directly. If, for example, we take the antenna connector of the radio equipment as the reference point, these curves represent the level and frequency of an unwanted signal which generates the maximum acceptable degradation of performance when the receive signal is low (say 3 dB above threshold at the same reference point).

The action of the r.f. and i.f. filters shapes the interference curve in the form of a V centered on the receive frequency with a width approximately the same as or slightly larger than the radio channel. The safe zone is below the curve, but if the unwanted signal is above the curve (input level higher than, say, -40 or -30 dBm) it may be harmful at any frequency as the receive r.f. filters cannot attenuate it sufficiently. This is the case for signals from other troposcatter transmitters at the same station which may reach the input with levels which are too high as a result of coupling between antennas, insufficient filtering etc.

For example, the antenna of our receiver is decoupled by 35 dB from another antenna radiating 1 kW ($+60$ dBm). The unwanted signal at the input connector has a level of $60-35 = +25$ dBm. If the frequency difference allows a filtering attenuation of 60 dB, the unwanted level at the receiver input remains at $25-60 = -35$ dBm. This level is dangerous if an intermodulation product (11.1) falls into the i.f. band. In this case the frequencies F_1, F_2 etc. are the unwanted signals which intermodulate with the receive frequency, the corresponding transmit frequency and the heterodyne frequency (local oscillator). The frequencies equal to the difference between any two of these frequencies and the multiples of these differences are particularly dangerous.

The analysis of all possible cases is troublesome and may overlook or overestimate possible situations. However, the problem can be simplified by observing that in general interference can be avoided by preventing the following signals from reaching the antenna connector of the radio cabinet:

(1) signals of any frequency with a level higher than, say, $-40\,\text{dBm}$ (although only specific frequencies are dangerous);
(2) signals which are too near the receive frequency (less than the adjacent channel).

This corresponds to remaining in the safe zone below the interference curve. However, if strong signals cannot be avoided, the following criteria can be adopted at each station:

(a) the frequency distance between any transmitter and any receiver should not be equal to or a submultiple of the intermediate frequency;

(b) in a conventional channeling plan the distance (a) should be an odd multiple of half the channeling step, so that at most the spurious products fall midway between two receive frequencies;

(c) the channeling step itself should not be a submultiple of the intermediate frequency;

(d) the frequency distance between any transmitter and the heterodyne of any receiver should not be equal to or a submultiple of the intermediate frequency.

If the above criteria are satisfied there is a reasonable probability of avoiding interference at the station.

11.6 Procedure for design calculations

The procedure used for the design calculations for a troposcatter system, based on the topics discussed in previous chapters, is summarized here.

We assume that the system under study is a cascade of troposcatter hops and that the end-to-end performance requirements are given. First we derive the performance of the single hops from that of the complete link. If the system is analog and the required performance is as specified by the CCIR, it is sufficient to design each hop independently according to CCIR Recommendations. If a lower performance is required, the end-to-end noise distribution should be split into a sum of distributions, one for each hop. In general a subdivision in which all distributions are equal is acceptable, and the curves $G(\sigma;z;q)$ of Fig. 8.1 can be used for the log-normal noise, remembering that we can assume that deep fades occur independently and one at a time in the various hops. If the system is digital and of high quality it is sufficient to design each hop according to DCS standards; otherwise lower requirements can be derived for the single hops with the assumption of the independence of deep fades as before. We now have the requirements for the single hops, whose exact ends and routing are drawn on topographic maps. We can therefore proceed to perform the geometrical calculations as in Table 6.1.

The path loss can now be calculated using one (or more, for comparison) of the methods given in Chapter 6. If we use CCIR method I we first decide which type of climate is applicable to our case and then derive the value of the surface refractivity from Fig. 4.1 as explained in Chapter 4, Section 4.1.1. We then calculate the median path loss as described in Table 6.2. We obtain the elements for drawing a diagram of path loss distribution like that in Fig. 6.5 from Fig. 6.3 and eqn (6.12).

If the system is analog we now determine the intermodulation noise as explained in Chapter 8, Section 8.2.2, and perform the calculations as in Tables 8.1 and 8.2. The path is then calculated as described in Section 8.3. If the system is digital it is necessary to evaluate the multipath distortion by performing a calculation of the type shown in Table 9.3, using Fig. 6.11 or, better, calculating the delay power spectrum using (6.28). If the distortion problems are overcome, the path calculation can proceed as described in Tables 9.4 and 9.5.

11.7 Choice of equipment

The main considerations to be made in choosing the equipment necessary for a troposcatter system are given below.

11.7.1 Radio equipment

This equipment should have the following main characteristics.

(1) Design should be simple, avoiding complex and/or sophisticated circuits or units which need skilled personnel for commissioning and maintenance. This simplicity should also decrease the possibility of faults, operating errors, instability to variations in environmental conditions and supply voltage etc.

(2) Operation should be simple: after switch-on, the station should require a minimum of operations to obtain the required output r.f. power. During normal operation (24 h per day) no attention should be required, so that the equipment can be installed in an unattended station.

(3) Maintenance should be simple so that unskilled personnel can make routine checks by reading built-in instruments and observing indication lamps. This should enable a faulty unit to be identified immediately. Semi-skilled personnel should also be able to replace units.

(4) The output r.f. power should be adjustable in steps, so that full power

Table 11.4 Example of a specification for a troposcatter radio terminal

1	Frequency range	790–960 MHz
2	Traffic capacity	12, 24, 60, 120 channels selectable by substitution of i.f. filter
3	Output power	Three steps, 100 W, 300 W, 1 kW nominal
4	Order of diversity	Dual or quadruple diversity
5	Type of combination	Maximal-ratio baseband combination, with possibility of excluding diversity branches
6	Noise figure of receivers	5 dB maximum
7	Baseband levels, impedances and frequency limits	According to *CCIR Recommendation 380-2*
8	Output–input r.f. impedances	50 Ω coaxial
9	Baseband response	Within ±0.5 dB
10	R.m.s. deviation per channel	According to *CCIR Recommendation 404-2* and *Report 446*
11	Adjustment of deviation	±4 dB with respect to nominal
12	Emphasis–de-emphasis	According to *CCIR Recommendation 275-2*
13	Pilot frequency and deviation	According to *CCIR Recommendation 401-2*
14	Receiver i.f. frequency	According to *CCIR Report 285-2* and *Recommendation 403-2*
15	Service channel	Included
16	Background noise	150 pW0p maximum
17	Total noise	300 pW0p maximum with CCIR load
18	Muting circuit	Included
19	Alarms	Local optical and acoustical alarms; contacts available for remote alarms
20	Routine measurements	Test points and built-in meters
21	Field strength measurement	Socket for external recorder
22	Solid state	Fully solid state except for power amplifier tubes
23	Ambient conditions	Temperature −10 to +50 °C, humidity up to 95%
24	Tropicalization	On request
25	Cooling	By forced air
26	Power supply	220 V ±5%, 50 Hz ±5%, three-phase

is used only during the months of bad propagation, while during the other periods the same performance is obtained with lower power, lower consumption, fewer interference problems and longer amplifier tube life.

(5) FM receivers should have a low r.f. threshold which is obtained by using low noise mixers or preamplifiers and avoiding the utilization of sophisticated and delicate parametric amplifiers and/or bandwidth compression (threshold extension) techniques.

(6) A maximal-ratio diversity combiner, which is the best type and the most common, should be used.

(7) It should be possible to adjust the frequency deviation and, for the highest capacities, the i.f. bandwidth (by substituting an i.f. filter) in order to optimize the performance of the system in terms of total noise.

(8) CCIR Recommendations should be complied with as far as possible.

(9) A simple and reliable cooling system should be provided for the power amplifiers.

(10) Other electrical characteristics should be as for conventional radio equipment.

A typical specification for troposcatter radio equipment is given in Table 11.4.

11.7.2 Multiplex equipment

The multiplex equipment is highly standardized. The description of analog and digital multiplex in Chapter 7, Section 7.8, should provide sufficient information for a choice to be made.

In order to increase the availability of the system, the analog multiplex is often provided with a duplicate carrier generator with automatic switchover in case there is a fault in the primary generator. In conventional radio links, where the analog telephone channels are noisier, the apparent quality can be improved by adding a compandor to the channels.

11.7.3 Antennas

The antennas were described in Section 7.7. The diameter of the reflector is chosen according to the path calculations in order to obtain either the necessary gain or a narrow beam which limits the path distortion. The reflector surface can be chosen from a number of materials (e.g. steel, aluminum, fiberglass etc.) and can be solid, perforated or mesh, depending on the environmental conditions. Radomes are never used in troposcatter antennas, except for the illuminator itself, and special heating facilities for de-icing are not normally required. Attention should be paid to the sidelobes, which should be as small as possible to decrease the risk of interference.

An important point to consider is that the structural design of the antenna must be such that the shape of the reflecting surface is kept within the specified tolerances with respect to the ideal paraboloid, particularly for higher frequency bands. Both the inaccuracies in mounting the reflecting surface and the deformation of the installed structure in the presence of wind, ice etc. should be taken into account. Modern technology allows antennas with diameters of up to 15 m (working at 5 GHz and remaining within tolerance under the actual mounting conditions) to be designed and manufactured.

In order to realize the full antenna gain, the displacement of the reflector from the ideal surface should not exceed $\pm 1/16$ of the wavelength, which means ± 4 cm at 450 MHz and ± 4 mm at 4.5 GHz. We shall return to this subject in Section 11.8.1.4 when we discuss antenna mounting.

For fixed antennas the height of the supporting tower is determined during the site survey and should provide the necessary clearance with respect to local obstructions. The antenna–supporting tower structure must be able to withstand the maximum winds and a coat of ice. It must be designed so that its maximum deformation has a negligible effect on the elevation angle of the antenna beam. For example, the structure should withstand, without permanent deformation, gusts of say 220 km h^{-1}, and at a permanent wind speed of say 180 km h^{-1} the deformation angle should not exceed a quarter of the antenna beamwidth. When the tower is being provided, the elevation angle of the antenna beam is known. It may be appropriate to design the attachments to the tower so that the mounted antenna has the required tilt. This may avoid the necessity of moving the illuminator (causing a possible degradation of the sidelobe pattern) to set the elevation angle. The illuminator should be moved only for minor adjustments. An example of a specification for a fixed antenna is given in Table 11.5.

Table 11.5 Example of a specification for a troposcatter antenna

Electrical characteristics	
Frequency band	900 MHz
Gain with respect to isotropic radiator	37.2 dB
Half-power beamwidth	2.4° maximum
Polarization	Two, horizontal and vertical
Polarization decoupling	Higher than 35 dB
SWR within working band	Less than 1.2 in a band 50 MHz wide
Power-handling capacity	10 kW
First sidelobe level	At least 20 dB below main lobe
Front-to-back ratio	Higher than 45 dB
Connection to illuminator	By EIA 1⅝ in flange (coaxial)
Ratio of focal length to diameter	0.43
Mechanical characteristics	
Diameter of the parabolic reflector	10 m
Azimuth adjustment	$\pm 5°$
Elevation adjustment	$\pm 1.5°$
Tolerance of reflecting surface	$\pm \lambda/16$ in static conditions
Allowed wind velocity	150 km h^{-1} from any direction without permanent deformation
Ice load	Layer 2.5 cm thick
Ambient condition	Thermal excursion of metal from -20 to $+70$ °C; humidity up to 100%
Surface protection	Hot dip galvanization
Pressurization of illuminator	Possible through the coaxial feeders with nominal 40 g cm^{-2}, maximum 80 g cm^{-2}

In the case of mobile antennas, ease of transportation, simplicity and rapidity of deployment in the field and the possibility of operation in various types of terrain (sandy, rocky, unleveled etc.) and environment (wind, snow etc.) should be taken into consideration.

11.7.4 Power equipment

Electric power may not be available at some sites and so a power station must be provided. This is generally composed of three diesel–alternator sets, two of which work in turn for a certain number of hours, while the third is on standby. Fully automatic power stations are available commercially. In some cases an automatic power station is kept on standby when the reliability of commercial power is poor. Transportable or mobile power stations may be provided for transportable or mobile troposcatter stations.

The primary power requirements of a troposcatter terminal with quadruple diversity are typically as follows:

output r.f. power of 100 W, 1–1.3 kVA
output r.f. power of 1 kW, 12–18 kVA
output r.f. power of 10 kW, 80–100 kVA

Additional power should be included for other services such as lighting, air conditioning or heating, power supply for test equipment, obstruction lights, automatic fuel pumps etc. The rating of the power station should include a margin above the expected steady consumption because some equipment, e.g. air conditioners, produces a strong current surge when it is switched on, which may become a problem for a fully loaded generator (e.g. in mobile systems). An example of the evaluation of power consumption is given in Table 11.6.

The main considerations in the choice of power equipment are as follows.

(a) The power-generating station can be of either the break type or the no-break type, depending on whether the electricity supply is interrupted during intervention or switchover of a diesel–alternator set.

(b) The engine should be air cooled, rather than water cooled, with cold starting. The velocity should be low (e.g. 1500 rev min^{-1}) to ensure a longer life. A bath of heavy duty lubricating oil should be provided.

(c) The alternator should be of the brushless type and the exciter should also be brushless.

(d) The station should have the capacity for continuous unattended full-load operation for a long period (e.g. 500 h).

In the case of low power consumption when commercial power is

Table 11.6 Power consumption of a troposcatter station equipped with a single 1 kW quadruple diversity terminal

Equipment	Power
1 kW HPA no. 1	6700 W
1 kW HPA no. 2	6700 W
Low power radio	100 W
Multiplex	200 W
H.f. TX–RX radio set, 100 W r.f.	300 W
Internal lighting	200 W
Obstruction lights on towers	1000 W
Air conditioning (two conditioners)	3300 W
Allowance for test instruments	1000 W
Total	19500 W
Apparent power at $\cos \varphi = 0.8$	24400 V A
Minimum required rating for power station	25000 V A

available but a reserve is required, a static no-break power station may be provided. In this station the mains, or a reserve diesel–alternator set, feeds a floating battery through a rectifier and the battery feeds the equipment without interruption through a static d.c.–a.c. converter. If the mains fails, the reserve generator is automatically started, after some delay, and keeps the floating battery charged.

11.7.5 Air conditioning

An air conditioning plant, with a reserve if necessary, has to be provided in certain areas or for shelter-mounted equipment. Table 11.7 shows an example of the calculations for the dimensioning of a cooling system. Note that the power amplifiers are cooled by air circulation from outside to outside, which limits the internal dissipation. In the example in the table an air conditioner capable of 3000 kcal h^{-1} is necessary. In other areas a heating system could be required.

11.8 Installation of the system

We consider the installation of fixed, transportable and mobile systems separately in order to emphasize the specific problems of each type of installation. We shall draw attention only to the main points of our interest, to avoid overloading the text with details which are inappropriate to our present purposes. A detailed treatment of the problems of installation would require several chapters and is not within the scope of this book.

Table 11.7 Dimensioning of a cooling system for a 1 kW sheltered troposcatter station with quadruple diversity

Internal power dissipation	
1 kW HPA no. 1	250 W
1 kW HPA no. 2	250 W
Low power radio	100 W
Multiplex	200 W
H.f. TX–RX radio set, 100 W r.f.	200 W
Lighting	200 W
Miscellaneous	600 W
Total[a]	$\begin{cases} 1800 \text{ W} \\ 1550 \text{ kcal h}^{-1} \end{cases}$
Heat from outside	
Outside temperature	50 °C
Inside temperature	22 °C
Temperature difference ΔT	28 °C
Heat transfer coefficient q	1 kcal m^{-2} °C^{-1} h^{-1}
Surface area S of shelter	42 m^2
Heat $qS\Delta T$ introduced	1176 kcal h^{-1}
Heat to be removed (1550 + 1176)	2726 kcal h^{-1}
Rating of air conditioner	3000 kcal h^{-1}

[a] 1 kW = 860 kcal h^{-1}.

11.8.1 Fixed systems

The installation of fixed systems is first studied at the office using the results and measurements obtained during site surveys. An installation plan is prepared, complete with site maps, building maps, layout of equipment and cables, calculation of antenna foundations, indication of azimuths etc.

11.8.1.1 Station layout

In studying the station layout care must be taken to minimize the routing of feeders (coaxial cables and waveguides) in order to ensure minimum attenuation between the radio equipment and the antennas. Furthermore the disposition of the antennas and the feeder lengths should be such that the transmission delay of the signal is the same for all diversity branches. In the rare cases in which this does not happen it may be necessary (depending on the value of the delay and of the transmitted bandwidth) to add a delay equalizer, which is normally made of a length of thin coaxial cable, to the i.f. stage of some receiver.

We have already shown the general layout of a quadruple diversity station with conventional parabolic antennas (Fig. 11.1). The feeder length appears to be minimized in this arrangement. However, for large antennas the feeder length can be reduced further by using, instead of conventional parabolic antennas, two offset-fed billboard antennas. Their reflecting surfaces can be designed as two symmetrical portions of the same ideal paraboloid. In this case both reflectors have the same focus and both illuminators can be mounted together on a small tower near the building or the hut. The feeder length can thus reach an absolute minimum, but the area of the station is extended by a triangular portion on which the building and the illuminator tower are located. This solution also increases the efficiency of the antenna, since the illuminator tower is out of the antenna beam and does not obstruct it as in the case of center-fed antennas.

The feeders can also be made very short in cases in which a single antenna is required (e.g. for dual frequency or dual angle diversity), unless provision is made for extension of the diversity order.

When more than one radio terminal is to be installed at a site, the antennas should be located in such a way as to minimize, under given conditions, coupling between themselves. This is to ensure that intermodulation and auto-interference are avoided in the system. For example, antenna beams should not be allowed to cross each other.

The station area is often protected by a fence, which may run in front of or near the antennas and be immersed in a strong electromagnetic field. Metallic fences may be dangerous, as rusty contacts act as mixers and may generate unwanted intermodulation frequencies which are harmful for very sensitive receivers. Leveling of the station area, particularly on mountain tops, is not always necessary. Small level areas at the end of the access road, in front of the equipment building and in front of the power building are often sufficient. In the layout of a station with a very high r.f. power, care must be taken with radiation hazards, which could result in restriction of access to some areas during operation. This problem is discussed further in Section 11.10.

11.8.1.2 Installation of equipment and feeders

An example of the internal layout of a repeater building is shown in Fig. 11.4. The most important problems of installation are encountered with the high power amplifiers (HPAs) and the feeders.

The HPAs require a cooling system. In the most common case of air-cooled 1 kW klystron amplifiers the bay is mounted near an external wall. The cooling air is taken from either outside or inside the building through a dust filter which must be cleaned periodically. The hot air is discharged to the outside through a suitable pipe protected from the entrance of insects.

The feeder runs are protected from the environment by installing them in either elevated cable ducts (which can contribute to minimization of length) or underground trenches. It is possible to have a single run connecting the radio equipment to the antenna, thus avoiding multiple connectors, cable tails and

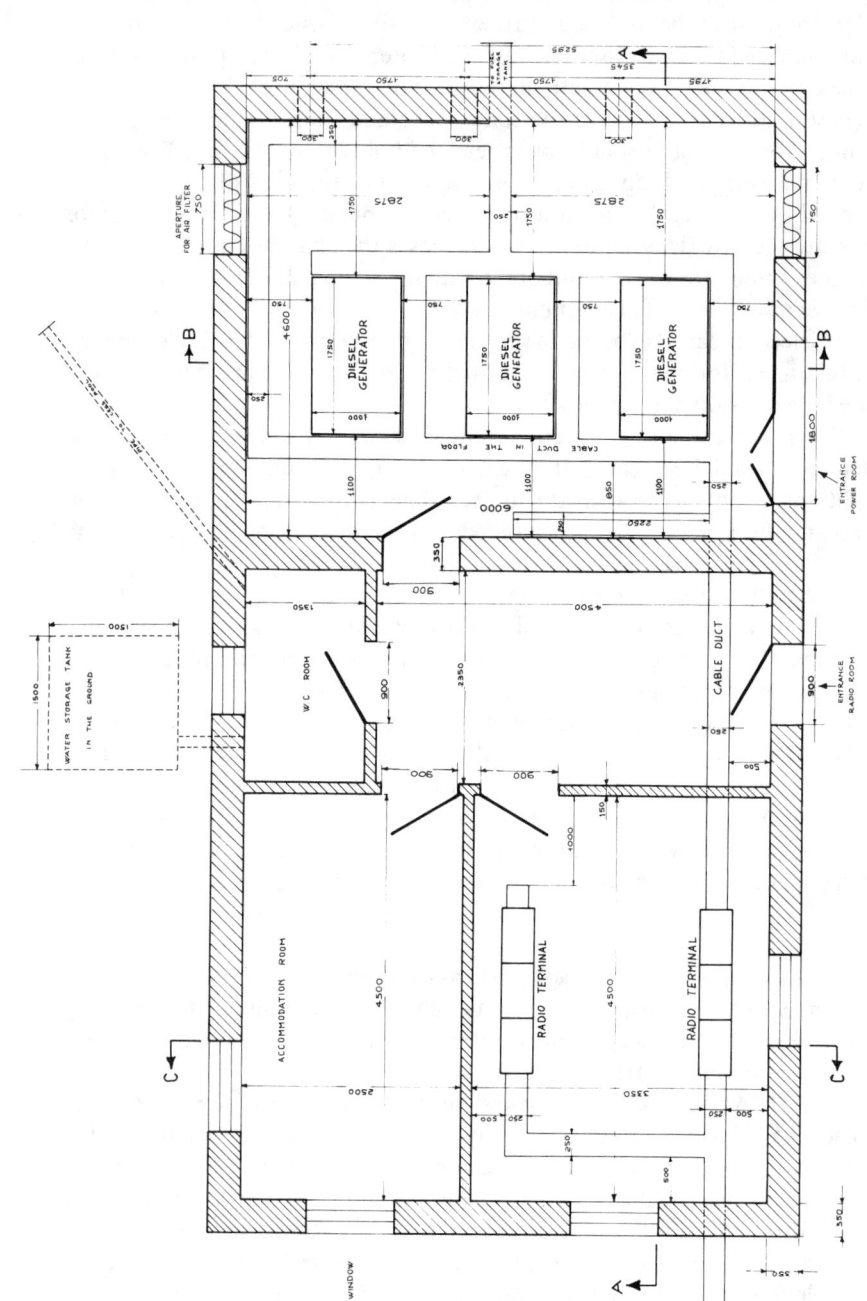

Fig. 11.4 Building layout for a fixed station (lengths in millimeters)

similar components which increase the attenuation and generate reflections (and intermodulation in the case of poor contacts).

A dehydrator–pressurizer is often installed to protect the air dielectric feeders and the antenna illuminators from moisture. However, it is always better to allow for the possibility of removing water from inside the feeders. Therefore the horizontal sections should be slightly tilted, particularly if the feeder is composed of a single run, so that there is an accessible lowest point where all the water can collect. There should be a small (normally closed) outlet in the outer conductor at this point.

When air dielectric coaxial cables terminated with connectors are installed, care should be taken that the inner conductor is not shifted with respect to the outer conductor as the cable is lifted onto the tower. Otherwise the inner conductor of the cable may be disconnected from the inner conductor of the terminating connector, producing an additional strong attenuation which is difficult to locate and to eliminate. Feeders should not be squeezed during installation so that unwanted reflections are avoided.

11.8.1.3 Foundations of the antenna tower or structure

The foundations of billboard antennas consist of a set of plinths or of a basement with foundation bolts, and are generally designed by the manufacturer of the antenna for various types of terrain. They must be prepared before the installation is started.

The correct position and the azimuth of the antenna axis should be found and marked on the terrain as described in Chapter 10, Sections 10.2.3 and 10.2.4. The exact positions of the plinths and the foundation bolts can then be determined, and the foundation can be constructed.

The self-supporting towers for conventional parabolic antennas are erected on four concrete foundation plinths which are also generally designed by the antenna or tower manufacturer. The maximum positive or negative vertical and horizontal forces acting on each plinth for various wind conditions (speed and direction) are calculated. The vertical force is the resultant of the weights of the antenna, the tower and the ice, if any, and of the vertical forces due to the moments at the base generated by the pressure of the wind. The maximum vertical forces under the worst conditions of wind speed and direction are calculated for each plinth. The weight of the plinth must be greater than the uplifting force (vertical and directed upwards), with a safety margin of say 2 (this weight also includes a mound of earth covering the plinth). The downward-directed forces plus the weight of the plinth (and of the covering ground) must produce a pressure which does not exceed the allowed load on the terrain (e.g. 0.5 kg cm^{-2} for poor terrain or 2 kg cm^{-2} for average terrain). The horizontal forces on the plinths are weaker than the vertical forces and can often be disregarded.

The dimensions of plinths can be determined using the above procedure (Fig. 11.5). An example of such a calculation is given in Table 11.8 in a

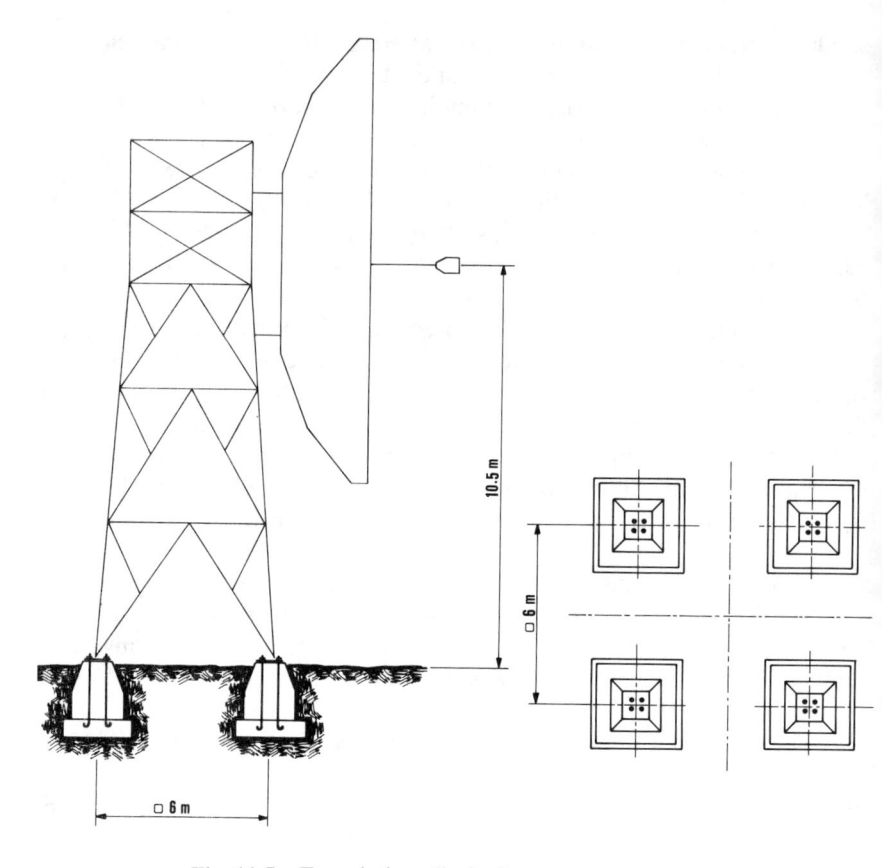

Fig. 11.5 Foundation plinths for an antenna tower.

simplified form. Table 11.9 provides some data on the foundations and on the installation of antennas.

11.8.1.4 Mounting the antennas

The gain of large antennas is almost never measured in practice, either at the factory or after installation, because of the difficulty and cost of this procedure. However, it can be determined with a negligible error if the radiation pattern of the illuminator is correct and the parabolic profile of the reflector is within its tolerances. Large antennas are normally manufactured and delivered in pieces and must therefore be mounted and installed on site with the necessary accuracy to ensure that the gain is not degraded.

The antenna gain decreases rapidly with increasing r.m.s. error in the profile. A statistical theory [11.1] states that the gain decreases by 1 dB for an r.m.s. error equal to 4% of the wavelength, by 3 dB for an error of 7% and by 6 dB for an error of 9%, which is the same as reducing the antenna diameter by

FROM PLANNING TO IMPLEMENTATION AND MAINTENANCE 307

Table 11.8 Example of the calculation of foundation plinths for an antenna supporting tower

1. Starting data	
1. 1 Maximum wind velocity (km h^{-1})	150
1. 2 Allowed load on terrain (kg cm^{-2})	5
1. 3 Thickness of ice coat (cm)	3
1. 4 Antenna diameter (m)	13
1. 5 Antenna height above ground (m)	10.5
1. 6 Side of square tower base (m)	6
1. 7 Weight of antenna (kg)	6800
1. 8 Weight of tower (kg)	7500
1. 9 Specific weight of terrain (kg m^{-3})	2200
1.10 Specific weight of concrete (kg m^{-3})	2500
1.11 Friction angle of terrain (deg)	30
1.12 Aerodynamic coefficients	
Antenna C_p	1.50
Tower C_t	1.75
1.13 Eccentricity of parabola (m)	4
1.14 Security factor for foundations	1.5
2 Calculation of forces	
2. 1 Wind pressure P at 150 km h^{-1} (kg m^{-2})	160
2. 2 Front area A of antenna (m^2)	133
2. 3 Front area A_t of tower + ice (m^2)	9
2. 4 Horizontal forces	
Parabola PAC_p (kg)	31920
Tower PA_tC_t (kg)	2520
2. 5 Sum of horizontal forces (kg)	34440
2. 6 Weight of antenna and tower (6800 + 7500) (kg)	14300
2. 7 Weight of ice: 2000 + 1000 (antenna + tower) (kg)	3000
2. 8 Sum of weights (kg)	17300
3 Calculation of moments at the base	
3. 1 Antenna: 31 920 × 10.5 (kg m)	335160
3. 2 Tower: 2520 × 10/2 (kg m)	12600
3. 3 Eccentricity: 4 × (6800 + 2000) (kg m)	35200
3. 4 Total maximum moment M_t (kg m)	382960
3. 5 Corresponding force on one leg: $(M_t/6)/2$ (kg)	±31913
4 Maximum forces on one plinth	
4. 1 Compression force	
Due to moments (kg)	31913
Due to weights: 17 300/4 (kg)	4325
	36238
4. 2 Uplifting force	
Due to moments (kg)	−31913
Due to weights (kg)	+4325
	−27588

Table 11.8 (continued)

4. 3 Shear force: 34 440/4 (kg)	8610
5 Assumptions for plinth calculation	
5. 1 Compression force: $1.5 \times 36\,238$ (kg)	+54000
5. 2 Uplifting force: $1.5 \times 27\,588$ (kg)	−42000
5. 3 Shear force: 1.5×8610 (kg)	12900
6 Dimensions of anchor bolts (four bolts per plinth)	
6. 1 Maximum acting force: 42 000/4 (kg)	−10500
6. 2 Minimum section (cm^2)	9.2
6. 3 Corresponding minimum external diameter (cm)	4.0
7 Calculation of plinths	
7. 1 Weight of plinth (kg)	14000
7. 2 Weight of terrain within 30° from plinth base (kg)	28000
7. 3 Force against uplifting (kg)	42000
7. 4 Minimum base area of plinth: $(42\,000 + 54\,000)/5$ (cm^2)	19200
7. 5 Minimum volume of plinth: $14\,000/2500$ (m^3)	5.6

The shape of the plinth follows from the above conditions.

half! It is therefore important to check the profile of the parabolic reflector after mounting and installation.

The profiles of antennas with diameters of up to about 12 m can be checked during mounting on the terrain at the foot of the tower, with the paraboloid axis directed upwards. The check is performed using a plane steel template with one parabolic edge. It is set along a diameter of the reflector with its edge on the parabolic surface, thus revealing the profile errors. It is then rotated about the axis of the paraboloid and is set along other diameters until the whole surface is covered. The errors can be corrected before the structural bolts are tightened. The structural deformation that occurs when the paraboloid is installed leaves the surface within the tolerance.

Checking and correcting the profile and the elevation angle of a billboard antenna is a necessary and long job. The method used is described in Chapter 10, Section 10.3.3. Measurements, calculations and final adjustments for each billboard antenna may take several days. Heating due to exposure to the sun does not create problems.

In mounting the antenna and the supporting structure care should be taken to avoid dirty contacts between metallic parts, especially in the reflector. These contacts act as rectifiers and may generate harmful intermodulation frequencies in the presence of high r.f. power at different frequencies.

In some installations (e.g. in the vicinity of airports) it may be necessary for antennas and their supporting towers to be painted white and orange and identified by red warning lights according to ICAO Regulations [11.2]. These

Table 11.9 Installation data for troposcatter antennas

Antenna		Tower height (m)	Foundation data			Antenna and tower installation data	
Type	Diameter (m)		Excavation (m^3)	Materials[a]	Holes[b]	Optimum crew	Working time
Parabolic	12	12	40	12–20 m^3 concrete	—	4 Riggers 5 Local laborers	4 days for tower 6 days for antenna = 10 days total
Billboard	18 (= 60')	—	370	150 m^3 concrete 10 tons iron 250 m^2 formwork	—	3 Riggers 7 Local laborers	2400 h = 1 month
Billboard	27 (= 90')	—	600	255 m^3 concrete 12 tons iron 600 m^2 formwork	120	4 Riggers 16 Local laborers	4500–6000 h = 1 month

Flat sandy soil; load, 1.6–2 kg cm^{-2}; wind speed, 200 km h^{-1}; safety coefficient, 2.
[a] Gravity type foundations.
[b] Rockbolt type foundations.

require a specially designed lighting plant. Figure 11.6 shows phases of the erection of troposcatter antennas.

(a)

(b)

Fig. 11.6 Mounting and erection of troposcatter antennas: (a) parabolic antenna on a tower; (b) billboard antenna.

11.8.1.5 Grounding plant and protection

Particular attention should be paid to grounding the plant, especially in regions where lightning strikes are expected. All metallic parts of the station, including antennas and towers, electronic equipment bays, outer conductors of feeders, power generators, oil tanks, power transformers, lightning arresters, Faraday cages etc., must be connected together and to a common ground consisting of one or more copper strands or tapes buried in the earth around the building, the towers or even the entire station area, and also grounded through metal rods set vertically in the terrain. It may also be convenient to install an isolation transformer and lightning arresters at the entry point of the external power line, as many faults caused by lightning arise at this point. Incoming telephone lines should also be protected in a similar way.

CCIR Report 932 gives all the necessary information (see Appendix 1).

11.8.2 Installation of transportable systems

The main difference from fixed systems is that the electronic equipment (and the power source if any) is installed at the factory in a suitable shelter or van (Fig. 11.7), which is then transported to the site as a complete unit and connected to the antennas, the power and communication lines, and the grounding system.

The connections are made through special windows provided in the shelter. In some regions of the world the shelter may require a single or dual air-conditioning system or a heating system. To avoid damage during transportation, the power tubes and other delicate or heavy items of equipment are normally dismounted and packed separately.

11.8.3 Deployment of mobile systems

Mobile systems are transported with the antennas and the power generators towed by jeeps or trucks and with the electronic equipment on the truck. They can also be designed to be transportable by aircraft, train and helicopter. A complete station with two antennas, a radio container and a power generator can be transported in a single C-130 aircraft.

The weight of the system varies depending on the type and dimensions. The following figures give a general idea:

radio equipment shelter or van	1500–2000 kg
complete mobile antenna	2800–5000 kg
mobile power generator	1000–1500 kg

All the equipment should be able to withstand shock and vibration without

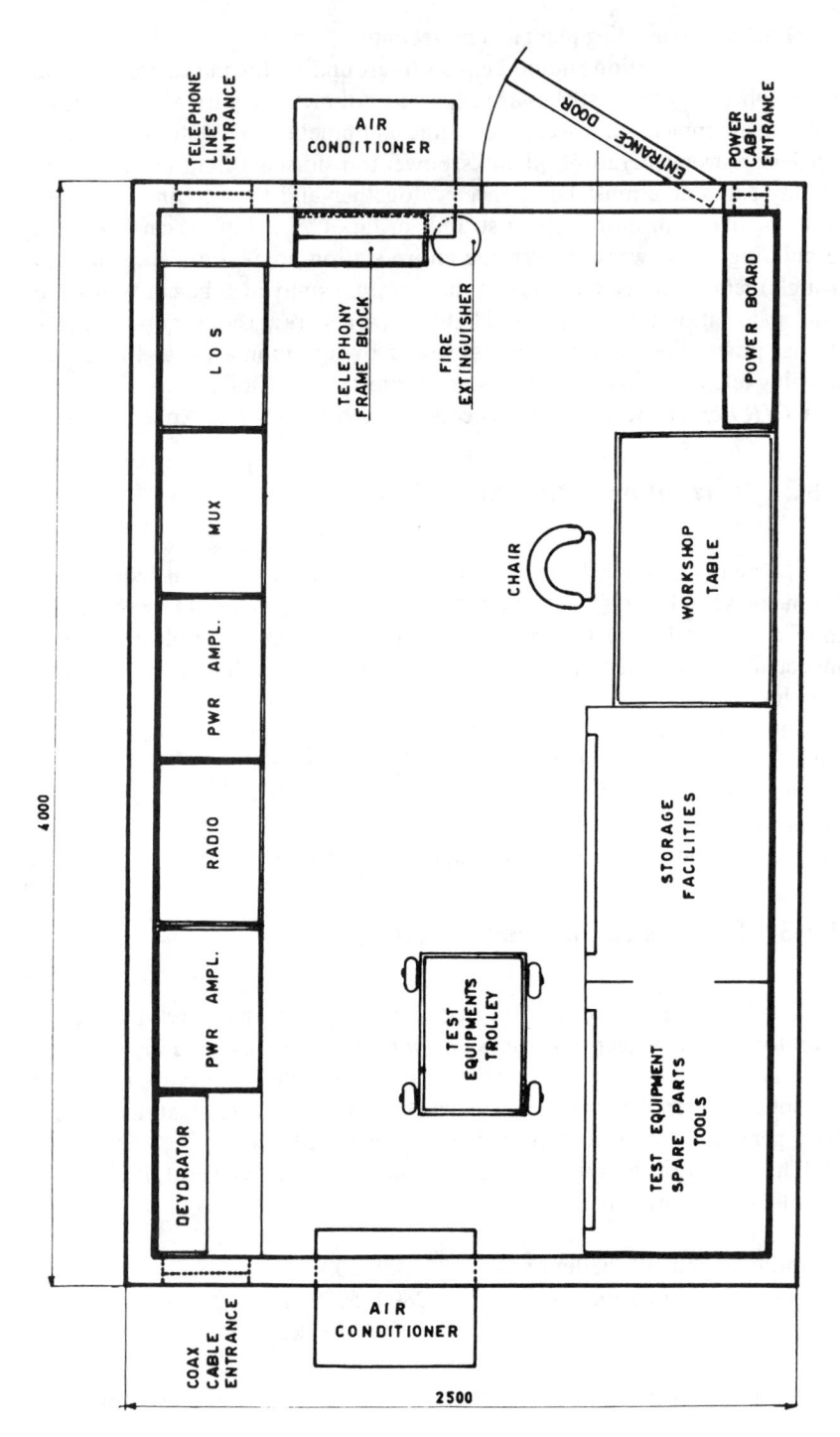

Fig. 11.7 Layout of a sheltered station.

damage and without the necessity of dismounting any delicate parts. Shock absorbers should be used throughout. The military specifications of these characteristics are very stringent, as the mobile station should be capable of moving at an acceptable velocity on almost any type of terrain. An example of a mobile system deployed on site is shown in Fig. 11.8.

Fig. 11.8 Example of a mobile system deployed on site.

The main problems in the deployment of these systems are as follows.

(a) Correct siting (see Table 11.3): it is advisable that the sites are predetermined even if they are not prepared.

(b) The equipment van and the antennas should be correctly positioned for easy interconnection by cables.

(c) The antennas should be oriented correctly, and so the azimuth must be determined with an accuracy to less than 1°.

(d) The operating frequencies, which are set by a synthesizer with tuned r.f. filters and klystron cavities, should be chosen appropriately. If possible the frequencies should be predetermined.

(e) Communications should be established with fine adjustment of antenna orientation.

In a well-designed mobile system in which the above elements are predetermined a crew of four men can establish communications within 30–40 min of arrival on site.

If the sites are not predetermined, their coordinates can be derived from maps, if available, or measured using simple satellite navigation equipment. The antenna azimuth is then calculated using a programmable pocket calculator, and is found on site using a magnetic compass or, better, a gyrocompass. The elevation of the horizon can be measured using a gradienter or an Abney level.

In order that the suitability of the hop can be evaluated rapidly it is advisable for the crew to be provided with precalculated diagrams showing, for example, curves of path length versus elevation angles which should not be exceeded for a given performance or time availability. Before and during the deployment of the station the initial communications with the other site are often obtained using an h.f. radio set.

The maximum permissible tilt of the terrain and the maximum wind velocity acceptable in operation are specified by the manufacturer. The antennas can be deployed by hand or by electric motor, depending on the design, in the presence of a moderate wind. The feeders should be robust and able to withstand an indefinite number of reeling and unreeling operations, and should be in a single piece terminated by special quick weatherproof connectors.

Mobile stations require a grounding system, even if it is simplified. It can be obtained by interconnecting the metal bodies of the container, the antennas and the power generators by a thick copper strand which is then connected to several metal rods implanted in the terrain.

11.9 Tests and measurements

Tests and measurements are performed at various stages of the work. At the design stage it may be necessary to measure the path loss, the path distortion and their statistics because of the difficulty of predicting them with acceptable accuracy, particularly in poorly mapped regions. Each piece of equipment is tested at the factory, and tests are also often performed on a simulated system before delivery to the site. The system is then subjected to further measurement during alignment and commissioning. Thorough tests are made at the time of acceptance of the system before official delivery to the user.

We shall examine the most important problems related to tests and measurements.

11.9.1 Path propagation tests

We know from Chapter 6 that geographical, topographical and climatic conditions have a marked effect on propagation. The prediction methods are derived from empirical results obtained from troposcatter systems installed in some regions of the world. In some areas it can be difficult to make sufficiently reliable predictions. Therefore propagation tests are sometimes performed on a new path under design. However, propagation tests are normally very expensive to carry out. In order to give reliable results, they should last for at least several weeks. It is also advisable to use test antennas of the same diameter as those to be installed in the final system (if test and final frequencies are the same) so that the common volume of the antenna beams and the behavior of the scatterers are unchanged, but this is very difficult to realize in practice. Therefore path tests for the design of a new link are generally avoided, and it is preferred to allow some margin of error or some possibility of expansion of the system.

A more favorable situation exists when path tests are performed to check the possibility of upgrading an existing link, e.g. to expand its traffic capacity or to transform it from analog to digital. In this case the antennas and most of the equipment are already in existence and installed in their final operating conditions, and only the measuring instruments are required for the tests. This is an ideal set-up for obtaining the correct design data.

We now give some information on measurements of the path loss and the multipath delay spectrum.

11.9.1.1 Equipment for measurements of path loss

The equipment used for measuring path loss should be light, transportable, weatherproof and easy to install. The test antennas are of limited size with a diameter of less than 10 m. The frequency band may differ from that assigned to the system, as it is possible to test at one frequency and transform the results to another frequency (Chapter 6, Section 6.6).

In order to overcome the highest path attenuations with reduced r.f. power and antenna gain, the receiver is of a very narrow band type with a very low r.f. threshold. The receiver is connected to a field-strength recorder or, better, to a digital automatic recording device including a computer and capable of yielding a complete statistical analysis of the path loss.

11.9.1.2 Measurement of the multipath delay spectrum

The multipath delay spectrum is measured using a multipath analyzer. This device is based on the rake technique, so called because it "rakes" together the multipath contributions used initially in the h.f. field. It uses the autocorrelation properties of long digital streams described in Chapter 2, Section 2.6. If one such stream is compared with a copy of itself, there is correlation when the time shift between the two streams is less than 1 bit; otherwise, the correlation is zero.

On the basis of this property a transmitted digital stream composed of, for example, 255 bit pseudorandom code sequences is compared at reception with a set of locally generated copies of itself, each of which is delayed by 1 bit with respect to the former. By using a 10 Mbit s^{-1} stream (bit length, 100 ns) and 16 copies shifted in this way it is possible to analyze a received spread pulse for a length of 1.5 µs in steps of 0.1 µs.

The device is capable of isolating the signal energy received with a given delay for each 0.1 µs step and of reconstructing the shape of the received pulse. Of course the equipment used must have the necessary bandwidth. If the integration time of the correlators is 10 ms, a complete measurement can be made in about 90 s [11.3].

11.9.2 Tests on the installed system

When the system has been installed, switched on and aligned with the antennas correctly oriented and all equipment is commissioned, a complete set of acceptance tests is performed before it is delivered to the user. The test results are recorded so that the performance can be evaluated and compliance with the specified characteristics can be checked. The records are also useful for checking the stability of the characteristics with time. In fact the system will be periodically checked or measured according to normal rules of maintenance.

The problems related to test instruments, lists of tests which should be made and performance evaluation are now discussed.

11.9.2.1 Test instruments

Specific instruments are required for the initial tests of the system and for subsequent performance measurements and maintenance tests. The measuring instruments normally recommended for acceptance tests and for different levels of maintenance of an analog system are listed in Table 11.10. A digital system would require fewer instruments, as all the baseband measurements could be made using an oscilloscope and a BER measuring set.

The instruments may be required in more than one unit. For instance, a path is generally tested using a set of instruments at each site. It may be convenient to keep a complete set of instruments at a central station, and to take them to other stations as required. The instruments and their containers should be rugged so that they are not damaged by vibration and shock and can be carried by trucks, jeeps or cars on any type of road.

11.9.2.2 Acceptance tests and routine measurements

A normal test procedure for the acceptance of an analog radio system is given below. Of course all or some of the tests are also performed on other occasions, such as during alignment or maintenance, but the acceptance tests comprise an almost complete set, the results of which remain as a reference for future checks.

Table 11.10 List of recommended measuring instruments (analog equipment)

1 Routine measurements
1.1 Meters incorporated in the equipment
1.2 Portable multimeter — For measuring supply voltages, currents, resistances

2 Ordinary maintenance
2.1 Baseband signal generator — For sending a test tone to the radio or the multiplex
2.2 Baseband level meter — For measuring the baseband test tone
2.3 R.f. generator — For sending a calibrated r.f. carrier to the receiver
2.4 R.f. wattmeter/reflectometer — For measuring forward and reflected r.f. powers
2.5 R.f. dummy load — For terminating a transmitter under test
2.6 Loop connection unit — For looping transmitter and receiver at the r.f. side

3 Further instruments for quality maintenance
3.1 Wave analyzer — For following a single test tone, signalling tone, pilot etc. along the multiplex and the radio circuits
3.2 Oscilloscope — For monitoring the shape of the signals, noise or disturbances
3.3 Frequency counter — For the exact alignment of frequencies

4 Further instruments for sophisticated tests
4.1 Spectrum analyzer — For monitoring the transmitted r.f. signal
4.2 Sweep generator — For alignment of r.f. filters, i.f. circuits, discriminator etc.
4.3 Noise figure meter — For measurements on the receiver
4.4 Deviation meter — For accurate setting of transmitter deviation
4.5 VSWR meter — For reflection measurements on antennas, receive branch etc.

5 Instruments for performance measurement
5.1 Recording microammeter — For recording the received r.f. field strength or the noise in a channel
5.2 Noise measuring set — Including transmitter and receiver for measurements of background and intermodulation noise in baseband slots according to CCIR
5.3 Set of r.f. fixed and variable attenuators — For measurements of r.f. received levels (field strength)
5.4 Psophometer — For monitoring the total psophometric noise in the multiplex channels

(a) Test the power supply voltages both external and internal to the equipment and maintain within their tolerances.

(b) Check that the frequencies are correct.

(c) Make a local loop in the single station between transmit and receive circuits on the r.f. side, possibly also including the duplexer. Test the levels of the complete chain from the transmit side (input to the multiplex channels in a terminal station; input to the radio in a repeater station) to the receive side (output from the multiplex channels in a terminal station; output from the radio in a repeater station). Make the white noise tests recommended by the CCIR (*Recommendations 393* and *399*).

(d) Test the levels and make white noise tests on a single path in both directions. It may also be convenient to loop the receive and the transmit side at baseband level in one station and test the two-way path from the other station.

(e) If the system is composed of more than one path, additional level and noise measurements should be made from terminal to terminal.

The following tests should be made and the results recorded:

(1) factory test results for single panels and units;

(2) the transmitted and received frequencies and their accuracy and stability;

(3) the transmitted power (one or more steps);

(4) the noise figure of the receivers;

(5) the pilot frequency;

(6) the r.m.s. frequency deviation for the test tone, pilot and service channel;

(7) check of the alarms (local and remote);

(8) check of all levels (baseband test tone, pilot, service channel test tone, signaling tone, remote alarm tones);

(9) baseband response curves for the different diversity branches, both individually and together;

(10) check of the transmit and receive r.f. spurious frequencies;

(11) check of the spurious frequencies in the baseband;

(12) white noise tests (according to CCIR methods) for background and total noise (these measurements should also be made for different values of the frequency deviation in order to find the deviation which gives the minimum total noise in the actual hop);

(13) the response curves of the service channel and of the band reserved for remote alarm signals;

(14) check of the dynamic efficiency of the combiner;

(15) check of the operation of the squelch;

(16) calibration of the field strength output connector on the receivers (for the connection of a recorder).

The readings from the built-in instruments should be recorded on test sheets kept in the vicinity of the equipment.

Some of the tests listed above are also valid for digital systems, particularly those related to the r.f. and i.f. sections. Noise measurements in the baseband are replaced by measurements of eye patterns, jitter and BER. Environmental (temperature, humidity) and mechanical (vibrations, bumps) tests may also be specified for some systems, particularly for military applications. The prototypes of mobile systems should undergo mobility tests on standard artificial rough terrain (Fig. 11.9) to demonstrate their ability to withstand vibration and shock.

Fig. 11.9 Test on an artificial difficult path.

11.9.2.3 Evaluation of system performance

We have seen that an analog system has been designed with signal-to-ratios which exceed a given value in the worst telephone channel for a large percentage of time. A digital system has been designed with fade outage probabilities which do not exceed a given outage duration per call minute.

The design was based on a set of parameters of which some are well defined but others are difficult to evaluate. The uncertain parameters in the system design are those related to propagation, and hence there is a predicted path loss distribution as shown in Fig. 6.5. It is of great interest to compare the predicted and actual distributions. This can be done by connecting to the receivers a field recorder or, better, a more complex apparatus including a computer which yields the statistics of the received signal directly.

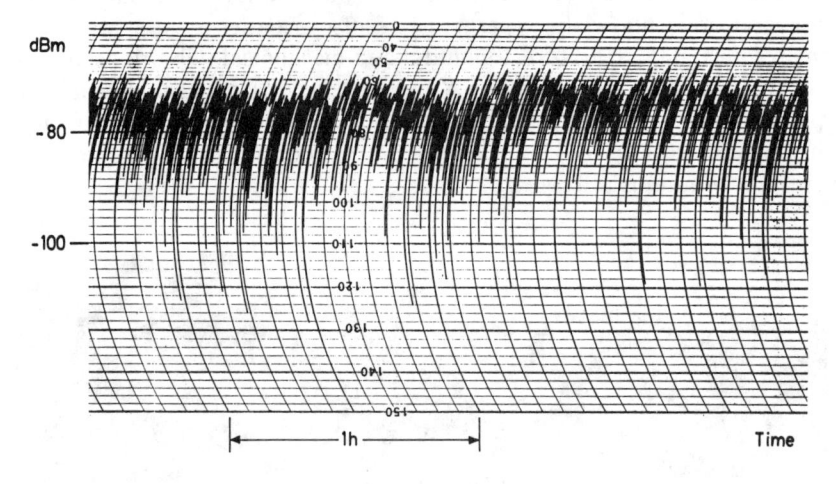

Fig. 11.10 R.f. field strength in one receiver.

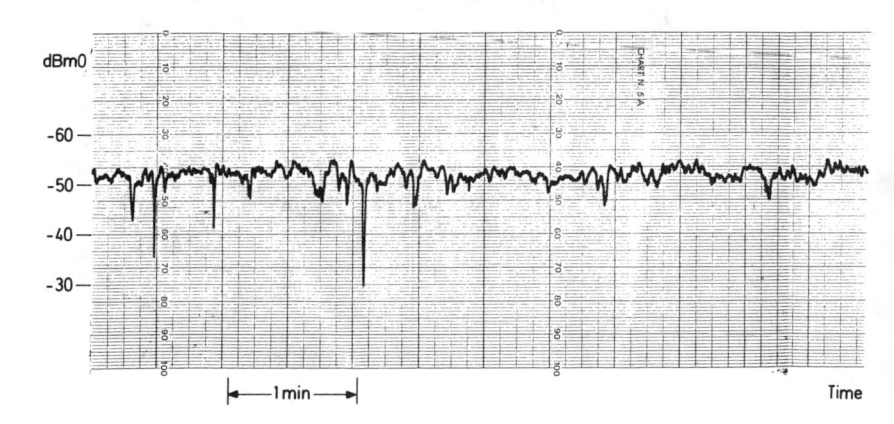

Fig. 11.11 Noise in the upper telephone channel.

Figures 11.10 and 11.11 show respectively the field strength in one receiver and the noise (after combination) in the upper channel of the multiplex for an analog system. The teleprinter error rates can be recorded by making tests at different times of the day, particularly during periods of bad propagation, transmitting the standard CCITT text.

In a digital system it is necessary to record the statistics of the fade outages in various propagation conditions. In order to be reliable such measurements should be made over periods of several weeks.

11.10 Radiation hazards

The possibility of some radiation hazards should be taken into account at the design stage, and it would be wise to make some checks during the test period. This is particularly important for very high power systems, but most ordinary systems do not pose particular problems.

When the radiation density exceeds certain limits it may become biologically hazardous as well as having other unwanted effects, e.g. the generation of sparks between metal parts which may cause fire, activate detonators in military plants, fire rocket motors etc. Some suspect cases have been reported.

In a troposcatter plant the only hazardous components carrying high power are the HPA, the r.f. feeders and the antenna. The feeders should never be opened during operation, and if a fault occurs (e.g. accidental cutting) the protection against reverse power in the HPA should cut the power off. In practice the only dangerous component is likely to be the antenna, which has a radiation field in its environment and which points toward the horizon where there could be an inhabited zone. This problem has been investigated [11.4].

We consider a reference power flux $1.26P/A$, where P is the power radiated by the illuminator and A is the antenna aperture area. Figure 11.12 gives the contours of equal power flux relative to the reference in front of a

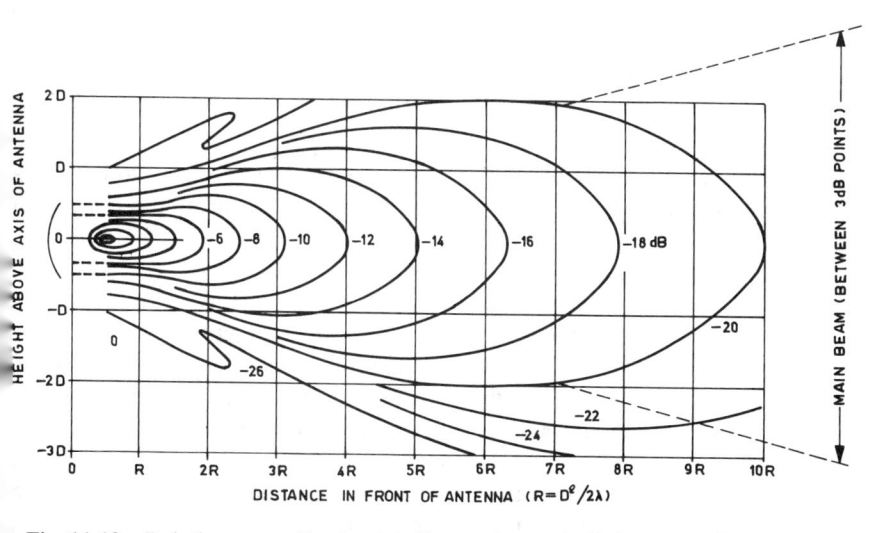

Fig. 11.12 Relative power flux levels in front of a parabolic antenna of diameter D.

parabolic antenna. The distance R is the limit between the near-field region (the Fresnel region in which the radiation beam can be considered to be cylindrical with the diameter of the paraboloid) and the far-field region (the Fraunhöfer region in which the beam becomes conical). Reflections from nearby terrain could enhance the field locally by up to 6 dB.

An analysis of the usual situation found in troposcatter plants, assuming the limits presently accepted by British and American Standards (100 W m^{-2} for 6 min for biological hazard), shows that in general there is no particular hazard in the vicinity of the antenna, except perhaps when using very high power and relatively small antennas which is an uncommon situation. If there is any doubt regarding safety local measurements should be made and access to hazardous areas should be restricted. Of course, in order to avoid risk, personnel should not stay within the main beam in front of the reflector when the power is on, e.g. when working on the illuminator of a center-fed antenna.

11.11 System operation

Many troposcatter systems operate unattended for most of the time. They should be able to keep running indefinitely after switch-on. However, apart from the normal maintenance and repair work, the following operations may be required in practice:

(a) switching the output r.f. power to a different value because of very-long-term variations in the path loss (a few times a year);

(b) periodically changing the power tubes in the HPAs because of deterioration and decrease in efficiency;

(c) changing the lubricating oil in the engines of the electric power generators (at most once or twice a month depending on the duty cycle of the generators);

(d) refuelling the tank of the electric power station (once a year to once a month depending on the power consumption of the station and the capacity of the fuel tank).

The replacement of tubes, particularly the very expensive klystrons, is a most important item of cost in a troposcatter system as there are normally four klystrons per hop. The guaranteed life of a power tube is of the order of a few thousand hours at nominal rating. If operated with care it can last much longer. For example, if the HPA is operated at low output for most of the time, the tube life may become several times longer than the nominal life. In practice, in a continuously running station it should not be necessary to change the tubes more than once a year or, with careful operation, once every few years.

11.12 Maintenance problems

The object of maintenance is to keep the system in continuous operating condition and to reduce to a minimum the number and duration of interruptions, degradations or faults. Maintenance problems are dependent mainly on the geographical siting of the system, the quality and engineering skill of the available personnel, the quality of the equipment and the design of the stations with respect to protection. A potential source of faults may be poorly maintained and inadequately protected external power or communications lines.

The maintenance staff can be subdivided into two types:

(1) unskilled personnel who perform routine measurements using the instruments built into the equipment (this operation is normally sufficient to detect possible sources of faults and more sophisticated measurements are not needed);

(2) skilled technicians, who are necessary for the ordinary maintenance carried out once or twice a year and for repairing the equipment.

Two main types of maintenance should be considered:

(i) preventive maintenance with regular site visits to check equipment in order to anticipate faults;

(ii) urgent maintenance, for immediate elimination of faults (this is generally quality maintenance).

In the first type of maintenance an operation plan is adopted so that, for example, each station is visited, checked and tested once or twice a year. If some test results show evidence of degradation, the personnel should service the equipment and ensure the correct operation before a fault occurs. Part of preventive maintenance is to check using built-in instruments the deterioration of equipment and power tubes in r.f. amplifiers, to anticipate early failure and to effect replacement. These checks may be made more frequently. The second type of maintenance requires a suitable geographical distribution of personnel who must be available to attend to breakdowns in order to minimize the time that the equipment is out of service.

Quality maintenance staff should be provided with test instruments and tool kits. If special tools are required for some of the operations, these should be provided by the manufacturer and kept together with the normal maintenance tool kits. The maintenance staff should if possible avoid repairing equipment on site but should replace faulty units and return them to a central laboratory or to the manufacturer for repair. Thus spare parts should be held in stock and should be available in one or more of the following types.

(a) *Complete panels*: this is the most expensive solution but it permits immediate substitution of faulty panels with a minimum outage time.

(b) *Complete units or subunits*: this is a less expensive solution because the same units could be present in different panels, but more skilled personnel may be required for the substitution.

(c) *Sets of components*: these are necessary for performing repairs in the laboratory or on site if skilled personnel are available and if the fault is not serious. Fuses, lamps etc. must also be provided, and unskilled personnel should be capable of replacing these.

A separate maintenance staff must be available if some of the stations are provided with electric power generators, for which specific maintenance procedures are recommended by manufacturers.

A maintenance procedure for radio relay systems for telephony using frequency division multiplex is outlined in *CCIR Recommendation 290-1* which recommends checks on the transmission quality by maintenance measurements of baseband level stability and total noise.

11.13 Cost considerations

In the economic evaluation of a troposcatter system the cost factors in Table 11.11 should be taken into account.

A troposcatter station is much more expensive than an LOS station. The higher cost of the radio equipment, the additional cost of the power amplifiers and diversity combiners, the enormous cost of the antennas and their supporting structures, the cost of the power station, if required, etc. should all be taken into account in the calculation of the investment costs. The cost of installation is correspondingly higher. Furthermore, the costs of operation and maintenance, specifically the cost of energy which is very high because of the high powers involved, and the costs of the periodic replacement of power tubes (normally klystrons) must be considered.

For a rough comparison, with a great deal of reservation as there are so many parameters to consider, we would expect a single troposcatter hop for a 60 channel analog system to cost the same as six to eight hops in an equivalent LOS system (excluding land, buildings and roads). When all the costs incurred during the life of the system, including the operation and maintenance costs, are considered, a troposcatter system is still more expensive than the corresponding LOS system.

We should also consider the possibility of leasing satellite circuits and installing small satellite stations. This solution is expected to cost much less than implementing a troposcatter system. However, there could be a limitation on traffic capacity for small satellites.

Table 11.11 Cost factors for troposcatter systems

Investment costs	Operation and maintenance

1 Real estate
  (a) Access roads
  (b) Land
  (c) Buildings
2 Equipment
  (a) Electronic
  (b) Antennas
  (c) Power
3 Installation
  (a) Civil engineering
  (b) Foundations
  (c) Equipment installation
4 Miscellaneous
  (a) Engineering, consultants
  (b) Administration, handling
  (c) Damages, insurance
  (d) Purchase, transportation
  (e) Storage
  (f) Legal services
  (g) Taxes
	1 Personnel
	  (a) Technical
	  (b) Administrative
	  (c) Training
	  (d) Travel
	2 Materials
	  (a) Instruments
	  (b) Tools
	  (c) Spares
	  (d) Vehicles
	  (e) Klystrons, tubes
	  (f) Fuel, lubrication
	3 Amortization

A complete analysis of the absolute and comparative costs of troposcatter and alternative facilities should be performed during the planning of a system. However, other parameters should be taken into account as well as economic considerations. Apart from cases in which a troposcatter system appears to be the only feasible solution in the given circumstances, there are considerations of privacy (a satellite could be not sufficiently reliable), security (repeater stations, satellites and cables may be vulnerable), greater independence etc. These factors may become more important than the economic considerations. Therefore generally the major users of troposcatter systems are governments (e.g. digital systems for military communications) and some private organizations (e.g. oil companies for communication with ocean drilling platforms).

More economical systems could be obtained by using larger antennas (for which only the initial outlay is larger than for small antennas) and solid state power amplifiers with relatively low power (say 100 W), thus drastically reducing operation and maintenance costs.

Equipment costs vary widely depending on the manufacturer, the specific system etc., but the ratios of the costs of the various components are perhaps a little more significant. To obtain at least a general idea of the apportioning of costs, let us consider a station equipped with a 1 kW quadruple diversity analog radio, a 60 channel FDM multiplex and two antennas of diameter 12 m with 10 m self-supporting towers. The breakdown of costs is roughly as follows:

radio equipment alone	
low power radio	40%
HPA 1	30%
HPA 2	30%
	100%
terminal	
complete radio with HPAs	55%
multiplex	15%
antenna plus tower 1	15%
antenna plus tower 2	15%
	100%

Note that 25% of the cost of the HPA is due to the cost of the klystron.

Costs and economic evaluation require detailed analysis, but this is inappropriate for the system designer who does not need an exact analysis but only a means for making the correct decision when comparing technical alternatives.

The *GAS 3 Manual* [11.5], which was written for the training of engineers worldwide, contains a number of chapters dealing with the economic problems of communication systems, and much of this discussion can also be applied to troposcatter systems. We advise the interested reader to refer to this manual.

11.14 Comments and suggestions for further reading

Some of the major manufacturers produce their own internal manuals or handbooks. References 11.6 and 11.7 are old but useful publications. Little other information is available in the literature. A good source, but accessible only to engineers working for manufacturing or consultant companies in this field, are the technical specifications annexed to requests for bids.

References

11.1 Ruze, J. Antenna tolerance theory—a review, *Proc. IEEE*, April 1966.
11.2 *International Civil Aviation Organization (ICAO) Convention on International Civil Aviation, Annex 10, Aeronautical Telecommunications*, ICAO, Montreal, July 1972.

11.3 Larsen, R. Measurements and predictions of multipath dispersion for troposcatter links, *AGARD Conf. Proc.*, **363**, 20–21, June 1984.
11.4 Shinn, D. H. Avoidance of radiation hazards from microwave antennas, *Marconi Rev.*, **39** (201), 61–80, 1976.
11.5 Transmission systems—economic and technical aspects of the choice of transmission systems, *GAS 3 Manual*, Vols 1 and 2, ITU, Geneva, 1976.
11.6 *USAF Handbook on Planning and Siting—Forward Propagation Tropospheric Scatter Communications Systems (RCA), TO 31R5-1-11*, Secretary of the Air Force, U.S. Department of Defense, August 1958.
11.7 *USAF Manual on Telecommunications Performance Standards, TO 31Z-10-1*, Vol. 5, Secretary of the Air Force, U.S. Department of Defense, 1960.

Appendix 1

CCIR Recommendations and Reports Concerning Transhorizon Systems

The CCIR books, which are updated and re-issued every 4 years after a Plenary Assembly, now (after the 16th Plenary Assembly, 1986) number 14 volumes, some of which split into two or three books. Only two of these books, Volumes 5 and 9, deal with radio links, and with troposcatter radio links in particular.

The documents of interest are listed below. We refer in particular to the 1986 edition. However, in new editions the identifying number of any Recommendation or Report is generally unchanged, the only variation being the final number which identifies the revision.

Volume 5: Propagation in non-ionized media

Recommendations

Recommendation 310-6: Definitions of terms relating to propagation in nonionized media. This contains various definitions.

Recommendation 617 (including *Report 238*): Propagation data required for the design of transhorizon radio relay systems. This provisionally approves the methods given in *Report 238*.

Recommendation 311-4: Presentation of data in studies of tropospheric wave propagation. This recommends that the results of transmission loss measurements are displayed on probability paper, and gives other advice.

Recommendation 526-1: Propagation by diffraction. This approves the methods given in *Report 715-2*.

Recommendation 452-4: Propagation data required for the evaluation of interference between stations on the surface of the Earth. This approves the methods given in *Report 569*.

Reports

Report 238-5 (included in *Recommendation 617*): Propagation data and prediction methods required for transhorizon radio relay systems. This includes methods and diagrams for calculating path loss. It deals with diversity, path antenna gain, fading, bandwidth and station siting, and defines the CCIR climates.

Report 715-2: Propagation by diffraction. This includes methods and diagrams for path loss calculations.

Report 1007: Statistical distributions in radio wave propagation. This gives formulae and diagrams for various distributions.

Report 563-3: Radiometeorological data. This gives definitions of refractivity, worldwide diagrams of refractivity and the refractivity gradient, and data on ducts and precipitation.

Report 718-2: Effects of tropospheric refraction on radiowave propagation. This deals, among other things, with the effective radius of the Earth, ducting, multipath, troposcatter propagation and distortion.

Report 723-2: Worst month statistics. This defines the worst month and applies the concept to practical problems, including transhorizon propagation.

Report 569-3: The evaluation of propagation factors in interference problems between stations on the surface of the Earth at frequencies above about 0.5 GHz. This also deals with interference due to troposcatter propagation.

Volume 9: Fixed service using radio relay systems

Recommendations

Recommendation 396-1: Hypothetical reference circuit for transhorizon radio relay systems for telephony using frequency division multiplex.

Recommendation 397-3: Allowable noise power in the hypothetical reference circuit of transhorizon radio relay systems for telephony using frequency division multiplex. This gives the noise performance objectives for the design of the links.

Recommendation 593: Noise in real circuits of multichannel transhorizon FM radio relay systems of less than 2500 km. This gives the noise performance for real circuits.

Recommendation 388: Radio frequency channel arrangements for transhorizon radio relay systems. This states that the frequency arrangement shall be agreed between administrations for individual cases.

Recommendation 302-1: Limitation of interference from transhorizon radio relay systems. This gives general recommendations.

Reports

Report 285-6: Propagation effects on the design and operation of transhorizon radio relay systems. This includes such topics as path antenna gain, noise, choice of some system parameters, optimum frequency deviation and the effects of multipath dispersion on digital transmission.

Report 376-5: Diversity techniques for radio relay systems. This deals with diversity, combiners, bandwidth, performance and digital systems.

Report 939: Interconnection of transhorizon radio links. This reviews the related problems.

Report 615-1: Transportable fixed radiocommunications equipment for relief operation. This includes the use of transhorizon equipment.

Report 932: Protection of radio relay stations against lightning discharges. This is of general importance.

Study programs

The CCIR has also produced the following study programs.

S.P. 7E/9: Preferred characteristics, permissible noise and signal distortion for the transmission of television signals over transhorizon radio relay systems.

S.P. 7F/9: Preferred frequency bands for tropospheric scatter radio relay systems.

S.P. 7G/9: Characteristics of digital transhorizon systems.

Appendix 2

Computer Calculations

It has been indicated in the text that some formulae can be calculated using a programmable pocket calculator or a desktop computer, and many designers have already computerized their calculations. Most of the formulae given in this book can be computerized without difficulty. The only problems likely to arise are the calculations of integrals and the calculation of implicit functions by successive approximation, and we discuss these further below.

Calculation of integrals

The calculation of integrals arises in connection with eqn (6.19) for the Rayleigh–Gaussian distribution with diversity and eqn (6.28) for the delay power spectrum. Suitable methods for calculating the integral are given in books on numerical calculations [A2.1, A2.2]. We used the Romberg method [A2.1] for which a calculator program was available in the Library of Programs, Hewlett-Packard, Geneva.

The Romberg method is based on the trapezoidal rule [A2.2]. In principle it approximates the exact value of the integral iteratively by dividing the integration interval a, b successively into $1, 2, 4, 8, 16, \ldots, 2^n$ equal intervals, evaluating the function at the extremes of each interval and calculating the area of the set of trapezoids approximating the given function. An algorithm for the rapid calculation and convergence of the results is applied. When the results of two successive iterations differ by less than the required accuracy the calculation is terminated.

Calculation of implicit functions by successive approximation

If in the function $F(x, y) = 0$ the dependent variable y is an implicit function of x, it is impossible to write the equation in the form $y = f(x)$ and to calculate y directly when x is given. If, in this case, we can put the equation in the form $x = f(y)$, we can calculate directly the x corresponding to any y and construct the curve of the function. We can therefore find graphically the value y_0 corresponding to a given x_0.

If we want to calculate y_0 using a computer we proceed as follows. We start with an approximate value of y_0 and calculate the corresponding x, which will be different from x_0. We then calculate at that point the increment Δx corresponding to a small increment Δy and, assuming a linear law, calculate by proportion the increment to be added to y to obtain x_0. Using the new incremented value of y we calculate a value of x nearer to x_0. We repeat the above steps iteratively until the difference $x - x_0$ is less than the required accuracy. This is Newton's method and corresponds to finding one of the zeros of the function $x - f(y)$.

If we have a second similar implicit function we can construct the second curve graphically. The solution of the system of the two equations is represented by the crossing point of the two curves. It can be calculated numerically only by successive approximations which approach the exact values of x and y indefinitely. Given an approximate value of y we find two values of x, one from each equation. Their difference should be brought to zero by appropriate variations of y. The problem is thus reduced to that of finding the zero of a function in the zone of interest and ignoring any other zeros.

An example of this problem arises when we want to compute two unknowns (latitude and longitude) of two equations (eqns (10.16)) for two stars. Given the longitude we cannot find the latitude directly. However, if we fix a value for the latitude, we can find two different values for the longitude directly from the two equations. We should therefore look for that value of latitude that makes the two longitudes equal (or makes their difference zero). The functions represent two circles which cross at two points, but the second point can easily be ignored as it is far from the zone of interest. Newton's method can also be used for this calculation.

References

A2.1 Davis, P. J. and Rabinowitz, P., *Numerical Integration*, Blaisdell, Waltham, MA, 1967.

A2.2 Smith, J. M., *Scientific Analysis on the Pocket Calculator*, Wiley, New York, 1975.

Bibliography

Books and monographs

Boithias, L. *Propagation des Ondes Radioélectriques dans l'Environnement Terrestre*, Dunod, Paris, 1984 (English translation (revised and updated): *Radiowave Propagation*, North Oxford Academic, London, 1987).

Bolt, F. D. Towers and masts for VHF and UHF transmitting aerials, *Tech. Monogr. 3103*, European Broadcasting Union Technical Center, Brussels, 1965.

du Castel, F. *Propagation Troposphérique et Faisceaux Hertziens Transhorizon*, Chiron, Paris, 1961.

Jacobsen, B. B. Thermal noise in multi-section radio links, *IEE Monogr. 262R*, Institution of Electrical Engineers, London, 1957.

Mackie, J. B. *The Elements of Astronomy for Surveyors*, Griffin, London, 1978.

Medhurst, R. G. and Hodgkinson, M. Intermodulation distortion due to fading in frequency modulation frequency division multiplex trunk radio systems, *Monogr. 240R*, Institution of Electrical Engineers, London, May 1957.

Panter, P. F. *Communication Systems Design—Line-of-sight and Troposcatter Systems*, McGraw-Hill, New York, 1972.

Schwartz, M., Bennett, W. R. and Stein, S. *Communication Systems and Techniques*, McGraw-Hill, New York, 1966.

Siegle, A. *Basic Plane Surveying*, Van Nostrand, Princeton, NJ, 1979.

Performance predictions for single tropospheric communication links and for several links in tandem, *NBS Tech. Note 102*, National Bureau of Standards, US Department of Commerce, August 1961.

Equipment characteristics and their relation to system performance for tropospheric communications circuits, *NBS Tech. Note 103*, National Bureau of Standards, US Department of Commerce, January 1963.

Transmission loss predictions for tropospheric communication circuits, *NBS Tech. Note 101* (two volumes), National Bureau of Standards, US Department of Commerce, January 1967.

International recommendations, conferences, symposia etc.

Advisory Group for Aerospace Research and Development Conf. Proc., **37**, August 1968.

Tropospheric radio wave propagation, *Advisory Group for Aerospace Research and Development Conf. Proc.*, **70**, September 1970.

Advisory Group for Aerospace Research and Development Conf. Proc., **244**, October 1977.

Propagation influences on digital transmission systems: problems and solutions, *Advisory Group for Aerospace Research and Development Conf. Proc.*, **363**, June 1984.

The American Ephemeris and Nautical Almanac, Naval Observatory, US Government Printing Office, Washington, DC; published annually.
The Astronomical Ephemeris, HMSO, London; published annually.
CCIR 16th Plenary Assembly, Dubrovnik, Vol. 5, Propagation in Nonionized Media; Vol. 9, Fixed Service Using Radio Relay Systems. ITU, Geneva, 1986.
CCITT 8th Plenary Assembly, ITU, Geneva, 1984.
Extended Range VHF Symposium, International Aeradio Limited, May 1963.
ICAO Bulletin, August 1973.
International Civil Aviation Organization (ICAO) Convention on International Civil Aviation, Annex 10, Aeronautical Telecommunications, ICAO, Montreal, July 1972.
Radio Regulations, Vols 1 and 2, ITU, Geneva, 1982.
Radio Transmission by Ionospheric and Tropospheric Scatter, Report of the Joint Technical Advisory Committee (JTAC); Preprint, Proc. IRE, 1960.
Transmission systems—economic and technical aspects of the choice of transmission systems, GAS 3 Manual, Vols 1 and 2, ITU, Geneva, 1976.
USAF Handbook on Planning and Siting—Foreward Propagation Tropospheric Scatter Communications Systems (RCA), TO 31R5-1-11, Secretary of the Air Force, US Department of Defense, August 1958.
USAF Manual on Telecommunications Performance Standards, TO 31Z-10-1, Vol. 5, Secretary of the Air Force, US Department of Defense, 1960.

Papers

Altman, F. Configurations for beyond-the-horizon diversity systems, Electr. Commun., 161–164, June 1956.
Altman, J. and Sichak, W. A simplified diversity communications system for beyond-the-horizon links, IRE Trans. Commun. Syst., 50, March 1956.
Arnstein, D. Correlated error statistics on troposcatter channels, IEEE Trans. Commun. Technol., 225–228, April 1971.
Baker, D. W. Modulation and signal-processing equipment for a new digital troposcatter link with military and general application, Telecommunication Transmission into the Digital Era, IEE Conf. Publ., **193**, 94–97.
Barrow, B. Error probabilities for telegraph signals transmitted on a fading FM carrier, Proc. IRE, 1613–1629, 1960.
Barrow, B. B. Diversity combination of fading signals with unequal mean strengths, IEEE Trans. Commun. Syst., 73–78, 1963.
Battesti, J. and Boithias, L. Propagation par les hétérogénéités de l'atmosphère et prévision des affaiblissements, AGARD Conf. Proc., **70**, 43-1–43-8, 1970.
Battesti, J., Boithias, L. and Misme, P. Calcul des affaiblissements en propagation transhorizon à partir des paramètres radiométéorologiques, Ann. Telecommun., **23** (5–6), 129–140, 1968.
Beach, C. D. and Trecker, J. M. A method for predicting interchannel modulation due to multipath propagation in FM and PM tropospheric radio systems, Bell Syst. Tech. J., **42** (1), 1–36, 1963.
Bello, P. A. A troposcatter channel model, IEEE Trans. Commun. Technol., **17** (2), 130–137, April 1969.
Bello, P. A. Selection of multichannel digital data systems for troposcatter channels, IEEE Trans. Commun. Technol., **17** (2), 138–161, April 1969.
Bello, P. A. A review of signal processing for scatter communications, AGARD Conf. Proc., **244**, 27-1–27-23, October 1977.

Bello, P. A. and Chase, D. A combined coding and modulation approach for high-speed data transmission over troposcatter channels, *Natl Telecommunications Conf., New Orleans, LA, 1–3 December 1975, Conf. Rec.*, Vol. 2, pp. 28-20–28-24, IEEE, New York, 1975.

Bello, P. A. and Crystal, T. H. A class of efficient high-speed digital modems for troposcatter links, *IEEE Trans. Commun. Technol.*, **17** (2), 162–183, April 1969.

Bello, P. A. and Ehrman, L. Error rates in diversity FDM–FM digital troposcatter transmission, *IEEE Trans. Commun. Technol.*, **17** (2), 183–191, April 1969.

Bello, P. A. and Ehrman, L. Performance of an energy detection FSK digital modem for troposcatter links, *IEEE Trans. Commun. Technol.*, **17** (2), 192–200, April 1969.

Bello, P. A., Ehrmann, L. and Alexander, P. Signal distortion and intermodulation with tropospheric scatter, *AGARD Conf. Proc.*, **70**, Part II, 36-1–36-17, 1970.

Bennet, W. R. Distribution of the sum of randomly phased components, *Q. Appl. Math.*, **5**, 385, January 1948.

Beverage, H., Laport, E. and Simpson, L. System parameters using tropospheric scatter propagation, *RCA Rev.*, 432–459, September 1955.

Birkemeier, W. P. and Sill, A. E. Precision alignment procedure for troposcatter systems, *Microwave J.*, 33–36, March 1977.

Boithias, L. and Battesti, J. Puissance moyenne de bruit dans les faisceaux hertziens transhorizon à modulation de fréquence, *Ann. Telecommun.*, **18** (5–6), 88–93, 1963.

Boithias, L. and Battesti, J. Etude expérimentale de la baisse du gain d'antenne dans les liaisons transhorizon, *Ann. Telecommun.*, **19** (9–10), 221–229, 1964.

Boithias, L. and Battesti, J. Les faisceaux hertziens transhorizon de haute qualité, *Ann. Telecommun.*, **20** (7–8), 11–12, 138–150, 237–254, 1965.

Boithias, L. and Battesti, J. Propagation due to tropospheric inhomogeneities, *Proc. IEEE, Part F*, **130** (7), 657–664, December 1983.

Booker, H. G. and Debettencourt, J. T. Theory of radio transmission by tropospheric scatter using very narrow beams, *Proc. IRE*, 281, March 1955.

Booker, H. G. and Sadon, W. E. A theory of radio scattering in the troposphere, *Proc. IRE*, April 1950.

Boyhan, J. W. A new forward acting predetection combiner, *IEEE Trans. Commun. Technol.*, **15** (5), 689–694, October 1967.

Boyle, A. W. A trans-horizon link in Arabia, an interim report, *Telecommun. J.*, **43**, 489–495, 1976.

Braine, M. R. Transmitting colour TV by troposcatter link, *Electron. Eng.*, 79–82, March 1978.

Brand, T. E., Connor, W. J. and Sherwood, A. R. AN/TRC 170 troposcatter communication system, *NATO Conf. on Digital Troposcatter, Brussels, March 1980*.

Brand, T. E., Connor, W. J., Sherwood, A. J., Unkauf, M. G., Tagliaferri, O. A., Liskov, N., Curtis, R., Boak, S., Bagnell, R., Abele, R. J., Smith, G. E. and Zawislan, F. *Technical Publications on Digital Troposcatter*, Raytheon Company, Sudburg, MA, June 1980.

Brayer, K. Error correction code performance on HF, troposcatter and satellite channels, *IEEE Trans. Commun. Technol.*, 781–789, October 1971.

Brennan, D. G. Linear diversity combining techniques, *Proc. IRE*, **47**, 1075–1102, June 1959.

Brose, J. 1 kW PA stage for wideband beyond-the-horizon radio link system, *AEG-Telefunken Prog.*, 100–101, 1970.

Cairns, J. B. S. A review of digital troposcatter links, *Conf. Publ. Telecom 79, 3rd World Telecommunications Forum, 24–26 September 1979*, Part 2, pp. 23.10.1–23.10.10, ITU, Geneva, 1979.

Carlton, B. F. Digital transmission over troposcatter links using independent sideband diversity, *Int. Conf. on Communications (ICC 75), San Francisco, CA, 16–18 June 1975*, Vol. 1, pp. 5-20–5-23, IEEE, New York, 1975.

du Castel, F. and Magnen, J. P. Etude de la qualité télégraphique dans les faisceaux hertziens transhorizon, *Ann. Telecommun.*, **14** (3–4), 93–103, 1959.

Chase, D. The application of error correction coding for troposcatter links, *Int. Conf. on Communications (ICC 75), San Francisco, CA, 16–18 June 1975*, pp. 5-11–5-14, IEEE, New York, 1975.

Chipp, R. D. and Cosgrove, T. Economic analysis of communication systems, *IRE Trans. Commun. Syst.*, 416–421, December 1962.

Colavito, C. Su un collegamento oltre orizzonte nel mare Mediterraneo centrale, *Rend. 19th Riunione Annuale AEI*, 1962.

Colavito, C. Progetto del ponte radio a diffusione troposferica Lampedusa—M. Cammarata, *Alta Freq.*, **38** (10), 808–815, 1969.

Collin, C. Evaluation empirique de la bande de cohérence en diffusion troposphérique, *Rev. Tech. Thomson-CSF*, **2** (3), 549–575, September 1979.

Connor, W. J. AN/TRC 170: A new digital troposcatter communication system, *Int. Conf. on Communications (ICC 78), Toronto, June 1978*, Vol. 3, IEEE, New York, 1978.

Crawford, A. B., Hogg, D. C. and Kummer, W. H. Studies in tropospheric propagation beyond-the-horizon, *Bell Syst. Tech. J.*, 1067, September 1959.

Crisholm, J. H., Rainville, L. P., Roche, J. F. and Root, H. G. Angular diversity reception at 2290 MHz over a 188 mile path, *IRE Trans. Commun. Syst.*, 195–201, September 1959.

Daniel, L. D. and Reinman, R. A. Modification of the Bello model for performance prediction of short range troposcatter links, *Int. Conf. on Communications (ICC 76), Philadelphia, PA, 14–16 June 1976*, Vol. 3, pp. 46-24–46-26, IEEE, New York, 1976.

Develet, J. A. An analytic approximation of phase-look receiver threshold, *IEEE Trans. Space Electron. Telem.*, 9–12, 1963.

Dougherty, H. T. A nomograph for predicting the performance of tropospheric scatter communication circuits, *IEEE Trans. Commun. Syst.*, 138–142, March 1963.

Elliott, J. C. Measurements of path intermodulation distortion over five tropospheric scatter paths, *IEEE Trans. Commun. Technol.*, 537–543, October 1970.

Field, C. The design and evaluation of a mobile digital troposcatter system, *GEC J. Res.*, **2** (3), 178–185, 1984.

Galpin, R. K. P. *Proposed Definition of a Spectral Emission Mask for Digital Tropospheric Scatter Transmission in Ace High (and Other Systems)*, SHAPE Technical Center, The Hague, January 1982.

Giordano, A. A., Lindholm, J. H. and Schonhoff, T. A. Error rate performance comparison of MLSE and decision feedback equalizer on Rayleigh fading multipath channels, *Int. Conf. on Communications (ICC 75), San Francisco, CA, 16–18 June 1975*, Vol. 1, pp. 5-1–5-5, IEEE, New York, 1975.

Gjessing, D. T. Scattering mechanisms and channel characterization in relation to broad band radio communications systems, *AGARD Conf. Proc.*, **244**, 1-1–1-15, October 1977.

Goeldner, J. and Schneider, W. IF protection switching system for wideband beyond-the-horizon radio links, *AEG-Telefunken Prog.*, 101–104, 1970.

Gough, M. W. The assessment of troposcatter radio relay systems in terms of CCIR Recommendations, *Marconi Rev.*, **29** (160), 1966.

Gough, M. W. The implications of certain CCIR objectives to troposcatter system design, *Marconi Rev.*, **31** (171), 240–253, 1968.

Gough, M. W. Angle diversity applied to tropospheric scatter systems, *AGARD Conf. Proc.*, **70**, 32-1–32-15, September 1970.

Gough, M. W. and Rider, G. C. Angle diversity in troposcatter communications, *Proc. IRE*, **122** (7), 713–719, July 1975.
Gough, M. W., Rider, G. C. and Larsen, R. Troposcatter angle diversity in practice, *Marconi Rev.*, 199–217, 1978.
Grisdale, G. L. The long-range aircraft communication problem, *World Aerosp. Syst.*, 74–78, February 1965.
Grzenda, C. J., Kern, D. R. and Monsen, P. Megabit digital troposcatter subsystem, *Natl Telecommunications Conf., New Orleans, LA, 1–3 December 1975, Conf. Rec.*, Vol. 2, pp. 28-15–28-19, IEEE, New York, 1975.
Gunther, F. A. Tropospheric scatter communications—past, present and future, *IEEE Spectrum*, **3** (9), 79–100, 1966.
Harris, D. P. An expanded theory for signal to noise performance of FM systems carrying frequency division multiplex, *IRE Natl Conv. Rec.*, Part 8, pp. 298–304, 1958.
Hill, S. J. British Post Office transhorizon radio links serving off-shore oil/gas production platforms, *Radio Electron. Eng.*, **50**, 397, August 1980.
Hirai, M., Nishikari, K., Fukushima, M., Kurihara, Y., Inone, R., Ikida, M., Neiva, S. and Kido, Y. Studies in UHF overland propagation beyond-the-horizon, *J. Radio Res. Lab., Tokyo*, 137, 1960.
Hoch, P. and Smith, N. F. Automated performance evaluation of troposcatter links, *10th Natl Communications Symp., Utica, NY, October 1964*.
Ince, A. N., Vogt, I. M. and Williams, H. P. A review of scatter communications, *AGARD Conf. Proc.*, **244**, 21-1–21-31, October 1977.
Johnson, J. K. Automatic power control of transmitters, *Electron. Eng.*, 69–77, December 1978.
Jost, R. and Gohlke, K. Gated subcarrier sideband diversity modem for digital transmission, *Natl Telecommunications Conf., New Orleans, LA, 1–3 December 1975, Conf. Rec.*, Vol. 2, pp. 28-6–28-9, IEEE, New York, 1975.
Junghans, H. and Weber, H. IF diversity combiner for a wideband beyond-the-horizon radio link system, *AEG-Telefunken Prog.*, 96–100, 1970.
Kennedy, D. J. A comparison of measured and calculated frequency correlation functions over 4.6 and 7.6 GHz troposcatter path, *IEEE Trans. Commun. Technol.*, 173–178, April 1972.
Kirk, F. W. and Osterholz, J. L. DCS digital transmission system performance, *Tech. Rep. 12-76*, Defense Communications Engineering Center, Reston, VA, November 1976.
Koono, T., Hirai, M., Inoue, R. and Ishizawa, Y. Antenna beam deflection loss and signal amplitude correlation in angle diversity reception in UHF beyond the horizon communications, *J. Radio Res. Lab. Tokyo*, **9** (41), 21–49, January 1962.
Krause, G. and Monsen, P. Results of an angle diversity field test experiment, *Natl Telecommunications Conf. (NTC 78), Birmingham, AL, 3–6 December 1978*, Vol. 2, pp. 17.2.1–17.2.6, IEEE, New York, 1978.
Kühn, U. and Dérer, I. Some experiments of tropospheric propagation beyond-the-horizon, *Telecommun. J.*, 149–155, March 1975.
Kühne, H. D. and Ramonat, R. D. Antenna and branching network technology for wideband beyond-the-horizon radio links, *AEG-Telefunken Prog.*, 104–107, 1970.
Lang, R. and Weber, H. System design of a 2 GHz wideband beyond-the-horizon radio link, *AEG-Telefunken Prog.*, 92–95, 1970.
Larsen, R. Quadruple space diversity in troposcatter systems, *Marconi Rev.*, 28–55, 1980.
Larsen, R. Measurements and predictions of multipath dispersion for troposcatter links, *AGARD Conf. Proc.*, **363**, 20–21, June 1984.

Larsen, R. and Cooke, T. R. F. Some measurements of dispersion and system performance on a digital troposcatter link, *IEE Colloq. on Digital Communications, May 1983*, IEE, London, 1983.

Lee, W. C. Y. Level crossing rates of an equal-gain predetection diversity combiner, *IEEE Trans. Commun. Technol.*, 417–426, August 1970.

Mack, A. Systems engineering of tactical multihop tropospheric scatter circuits, *Tech. Memo. M1872*, Signal Corps Engineering Laboratories, Redbank, NJ, May 1950.

Magnuski, H. A novel solution to troposcatter communications problems, *Telecommunications*, 7–13, January 1968.

Massaro, M. J. The distribution of error probability for Rayleigh fading and Gaussian noise, *IEEE Trans. Commum. Syst.*, 1856–1858, November 1974.

Medhurst, R. G. RF bandwidth of frequency division multiplex systems using frequency modulation, *Proc. IRE*, 189, February 1956.

Medhurst, R. G. Echo distortion in frequency modulation, *Electron. Radio Eng.*, **36** (7), 253, July 1959.

Medhurst, R. G. and Small, G. F. An extended analysis of echo distortion in the FM transmission of frequency division multiplex, *Proc. IRE, Part B*, 190, March 1956.

Meek, T. J. and Roda, G. Tropospheric scatter radio systems and their integration with public communication networks, *14th Int. Conv. on Communications, Genoa, 12–15 October 1966*, Istituto Internazionale delle Comunicazioni, Genoa, 1966.

Millen, G. L., Morrow, W. E., Pote, A. J., Radford, W. H. and Wiesmer, J. B. UHF long range communications, *Proc. IRE*, 1269, October 1955.

Monsen, P. Performance of an experimental angle-diversity troposcatter system, *IEEE Trans. Commun. Technol.*, 242–247, April 1972.

Monsen, P. Fading channel communications—adaptive processing can reduce the effect of fading on beyond-the-horizon digital radio links, *IEEE Commun. Mag.*, 16–25, January 1980.

Morita, S., Tachibana, H., Hoshino, T. and Kawasaki, H. Effect of angle diversity in troposcatter communication systems, *NEC (Nippon Electr. Co.) Res. Dev.*, **45**, 83–93, April 1977.

Morrow, W. E., Mack, C. L., Nichols, B. E. and Leonhard, J. Single sideband techniques in UHF long range communications, *Proc. IRE*, 1854, December 1956.

Nemirovskiy, A. S. Experimental study of the correlation of the slow variations of thermal and crosstalk noise in tropospheric links, *Telecommun. Radio Eng. (USSR)*, 61–62, May 1964.

Nicotra, G. Torri radio per telecomunicazioni, *Telecomunicazioni*, **51**, 3–12, 1974.

Norton, K. A., Vogler, L. E., Mansfield, W. V. and Short, P. S. The probability distribution of the amplitude of a constant vector plus a Rayleigh distributed vector, *Proc. IRE*, 1354, October 1955.

Osterholz, J. L. Design considerations for digital troposcatter communications systems, *AGARD Conf. Proc.*, **244**, 22-1–22-15, October 1977.

Osterholz, J. L. Megabit digital communications over a dispersive channel, *Natl Telecommunications Conf. (NTC 78), Birmingham, AL, 3–6 December 1978*, Vol. 2, IEEE, New York, 1978.

Parry, C. A. The ultimate long-haul capability of the tropospheric scatter mechanism, *Proc. 3rd Natl Conv. on Military Electronics*, pp. 121–127, IRE, New York, 1959.

Parry, C. A. A formalized procedure for the prediction and analysis of multichannel tropospheric scatter circuits, *Natl Conv. Record, September 1959*, Part 8, pp. 1–18, IRE, New York, 1959.

Parry, C. A. Optimum design considerations for radio relays utilizing the tropospheric scatter mode of propagation, *Commun. Electron.*, 71, March 1960.

Parry, C. A. Criteria for the ultimate capability of the optimized tropospheric scatter system, *IRE Trans. Commun. Syst.*, 187, September 1960.

Parry, C. A. On the prediction of the inherent bandwidth capability of the tropospheric scatter link, *IEEE Int. Conv. Rec., March 1963*, Part 8, pp. 215–232, IEEE, New York, 1963.

Parry, C. A. Predictions of intermodulation noise in an FM troposcatter system due to multipath, *IEEE Trans. Commun. Syst.*, 251–252, June 1964.

Perry, M. A. Digital data transmission via troposcatter channels, *IEEE Trans. Commun. Technol.*, 129, April 1969.

Pierce, J. N. Diversity improvement in frequency shift keying for Rayleigh fading conditions, *ASTIA Doc. N. AD-110105*, Air Force Cambridge Research Center, Bedford, MA, September 1956.

Pierce, J. N. Multiple diversity with non-independent fading, *Proc. IRE*, 427, February 1961.

Pusone, E. A troposcatter prediction model of long-term statistics of multipath dispersion and Doppler spread based on atmospheric parameters, *IEE Colloq. on Troposcatter Communications—Digital Development, London, October 1981, Dig. 1981/66*, p. 2/1.

Rev. Tech. Thomson Houston, Electron., **34**, June 1961 (special issue on tropospheric scatter).

Rice, S. O. Statistical fluctuations of radio field strength for beyond-the-horizon, *Proc. IRE*, 274, February 1953.

Rider, G. C., Gough, M. W., Larsen, R. and Clough, I. D. Troposcatter developments in the North Sea, *3rd World Telecommunications Forum, Geneva, September 1979*, Part 2, ITU, Geneva, 1979.

Rogers, J. D. Introduction to digital tropo for military tactical communication, *Commun. Broadcast.*, **6** (3), 3–9, 1981.

Rogers, J. D., Stears, M. H. and Baker, D. W. Radio equipment for digital tropo military tactical communication, *Commun. Broadcast.*, **7** (1), 61–71, 1981.

Ruze, J. Antenna tolerance theory—a review, *Proc. IEEE*, April 1966.

Schmitt, F. Statistics of troposcatter channels with respect to the applications of adaptive equalizing techniques, *AGARD Conf. Proc.*, **244**, 5-1–5-15, October 1977.

Sheffield, B. Nomograms for the statistical summation of noise in multihop communications systems, *IEEE Trans. Commun. Syst.*, 285–288, September 1963.

Sherwood, A. R. and Fantera, I. A. Multipath measurements over troposcatter paths with application to digital transmission, *Natl Telecommunications Conf., New Orleans, LA, 1–3 December 1975, Conf. Rec.*, Vol. 2, pp. 28-1–28-5, IEEE, New York, 1975.

Sherwood, A, Wenglin, G. and Wick, J. An experimental program leading to development of a tactical digital troposcatter system, *AGARD Conf. Proc.*, **244**, 30-1–30-16, October 1977.

Shinn, D. H. Avoidance of radiation hazards from microwave antennas, *Marconi Rev.*, **39** (201), 61–80, 1976.

Sing-Hsiung Lin. Statistical behavior of deep fades of diversity signals, *IRE Trans. Commun. Syst.*, 1100–1107, December 1972.

Skingley, B. S. Advances in tropospheric scatter techniques, *Communications 1974, Brighton.*

Skingley, B. S. Level control in tropospheric scatter systems, *AGARD Conf. Proc.*, **244**, 23-1–23-8, October 1977.

Skingley, B. S. Modern tropospheric scatter systems, *Electron. Eng.*, 37–40, May 1978.

Smith, D. A. and Garbutt, I. R. Percentage of time evaluation of tropospheric scatter links, *Marconi Instrum.*, **12** (1), 2–7.

Smith, G. E., Zawislan, F. and Abele, R. J. 2 GHz digital troposcatter terminal, *NATO*

Conf. on Digital Troposcatter, Brussels, March 1980.
Staras, H. Forward scattering of radio waves by anisotropic disturbance, *Proc. IRE*, 1374, October 1955.
Staras, H. Antenna to medium coupling loss, *IRE Trans. Antennas Propag.*, 228, April 1957.
Stewart, J. L. The power spectrum of a carrier frequency modulated by Gaussian noise, *Proc. IRE*, 1539, October 1954.
Sunde, E. D. Digital troposcatter transmission and modulation theory, *Bell Syst. Tech. J.*, 143–214, January 1964.
Sunde, E. D. Intermodulation distortion in analog FM troposcatter systems, *Bell Syst. Tech. J.*, 399–435, January 1964.
Tagliaferri, O. A. and Unkauf, M. G. Simplified performance estimate for an adaptive matched filter digital troposcatter demodulator, *Natl Telecommunications Conf., Los Angeles, CA, 1977, Conf. Rec.*, IEEE, New York, 1977.
Tremellen, K. W. and Cox, J. W. The influence of wave propagation on the planning of short wave communication, *J. Inst. Electr. Eng., Part 3A*, 212, 1947.
Unkauf, M. G. and Tagliaferri, O. A. Tactical digital troposcatter systems, *Natl Telecommunications Conf., Birmingham, AL, 3–6 December 1978*, Vol. 2, pp. 17.4.1–17.4.5, IEEE, New York.
Unkauf, M. G., Bagnell, R. and Tagliaferri, O. A. Tactical tropo performance of a distortion adaptive receiver, *Natl Telecommunications Conf., New Orleans, LA, 1–3 December 1975, Conf. Rec.*, Vol. 2, IEEE, New York, 1975.
Unkauf, M., Liskov, N., Curtis, R. and Boak, S. Advanced digital troposcatter modem technology, *Eastcon Conf. Rec., September 1977.*
Ural, A. T. *Multipath Measurements over Two Ace High Troposcatter Links*, SHAPE Technical Center, The Hague, January 1982.
Watt, A. D., Florman, E. F. and Plush, R. W. A note regarding the mechanism of UHF propagation beyond-the-horizon, *Proc. IRE*, 252, February 1960.
Willson, F. E. and Runge, W. A. Data transmission tests on tropospheric beyond-the-horizon radio systems, *IRE Trans. Commun. Syst.*, 40–43, March 1960.
Wood, H. B. Performance of VF/FM teleprinter circuits operating over a tropospheric scatter link, *Proc. IRE*, **110** (11), 1933–1939, November 1963.
Yeh, L. P. Basic analysis of controlled carrier operation of tropospheric scatter communication systems, *IRE Natl Conv. Rec.*, Part 8, pp. 261–283, 1958.
Yeh, L. P. New concepts in the statistical study of tropospheric scatter propagation data, *Wescon, August 1958.*
Yeh, L. P. Simple methods for designing troposcatter circuits, *IRE Trans. Commun. Syst.*, **8** (3), September 1960.
Yeh, L. P. Data transmission over FDM/FM troposcatter systems, *IEEE Trans. Commun. Technol.*, **18** (5), 490–501, October 1970.

Further references can be found in the literature listed above as well as in CCIR Reports.

Index

acceptance tests, 316–319
adaptive capability, systems with, 197–200
adaptive equalization, 101–102
adaptive modems, *see* Modems
advantages of troposcatter systems, 17
AGC, *see* Automatic gain control
air conditioning, 301
altazimuthal system, 269
altitude, 268
amplifiers, power, 6, 8, 160–163, 303
analog multiplex, 193–195
analog radio equipment, 167–175, 183–184, 185
analog troposcatter links, 13, 203–232
  performance objectives for, 203–205
angle diversity, 88–89
angular distance, 47
angular length, 46, 50
anomalous propagation, 78–79
antenna azimuth, 256–259, 271
antenna beamwidth, 43–44, 48, 70, 194, 236
antenna diameter, 8, 298
antenna gain, *see* Gain degradation
antennas, 188–193
  choice of, 298–300
  fixed, 188–191
  geometrical parameters of, 48
  mobile, 191, 192
  mounting of, 299, 305–311
  siting of, 274–280, 285–291, 330
application of troposcatter systems, 14
astronomical azimuth, 268
atmosphere, 2–3, 55
autocorrelation, 37–38
autocorrelation function, 38
automatic gain control (AGC), 91, 96, 97
azimuth, antenna, 256–259

background noise, 92
backward scatter, 64
bandwidth, 73–75
  CCIR Reports concerning, 330, 331
  coherence, 148
  correlation, 100–101, 150
  transmissible, 68, 72, 75
bandwidth compression, 8, 164
baseband combiner, 103–106, 157
BER, *see* Bit error rate
beyond-the-horizon radio paths, 1, 2–5, 55
billboard antennas, 188–190, 277–280
bit count integrity, 245
bit error rate (BER), 30, 148, 184, 231, 236
  curves of, 176, 177, 178, 180

carrier-to-noise ratio, 30–31, 178
carrier-to-threshold ratio, 246, 247, 248
CCIR, *see* International Radio Consultative Committee
CCITT, *see* International Telegraph and Telephone Consultative Committee
Chinese method for path loss, 132–133, 136
climate types, 64, 79–81, 285–288
  CCIR Reports concerning, 330
  function $Y(q\%)$ for, 126–127
  and path loss, 65, 128, 130–131, 134–135, 140–141
climatic regions, 11, 12–13, 79–81
coaxial feeders, 187–188
coherence bandwidth, 148
combination, *see* Linear combination of diversity branches
combiner, *see* Diversity combiner
common volume, 43, 47, 64, 68–69, 125, 132
  and antenna gain, 70–71
composed distributions, 35–37
computer calculations, 333–334

343

continental subtropical climate, 79
continental temperate climate, 80
continuous variables, 22
conventional radio equipment, 184–185
conventional troposcatter, 222–225, 250–252
convolutions, 35–37
correlation, 37–38
correlation bandwidth, 100–101, 150
correlation coefficient, 37
costs, 17, 18, 285–288, 324–326
crosspoint, 47, 50–51, 118, 119, 120
cumulative distributions, 22, 24

DAR, *see* Distortion adaptive receiver
DCS, *see* Defence Communications System
declination, 266, 268
deep fades, 243
Defence Communications System (DCS), 19, 243
degradation
 in transmission, 234–235
 see also Gain degradation
delay power spectrum, 74, 150, 151, 152
desert climate, 80
design, 285–288
 and feasibility, 11–13
 and reliability, 10–11
design calculations, 295–296
diffraction, 7, 55, 78
 CCIR Recommendations concerning, 329
 path loss due to, 142–146
diffraction path, 142–143
 knife-edge, 5, 50, 53
 smooth-earth, 5
diffraction zone, 2–3
digital multiplex, 195
digital radio equipment, 176–182, 184–186
digital transmission, modems for, 107–108, 185
digital troposcatter links, 13, 233–253
 design of, 248–252
 performance objectives for, 245–248
dicrete variables, 22
distortion adaptive receiver (DAR), 108–111
distributive curves, 23
diversity, 13, 155, 237, 238–242
 CCIR Reports concerning, 331
 order of, 9, 84, 101
 path loss with, 133–140
diversity branches (channels), 84

diversity combiner, 9, 84, 93, 103–111, 163
 baseband, 103–106, 157
 CCIR Reports concerning, 331
 equivalent circuit of, 91, 98
 i.f., 105, 106–107
diversity gain, 98–100
diversity systems, 84, 85–90, 155–157, 189
 explicit, 85, 87–100, 250
 implicit, 85, 100–111, 238
diversity techniques, 9, 83–112
Doppler spread, 73–74
double intersection from three stations, 277
dual diversity systems, 86–88
 distribution curves for, 94
ducts, 6, 57, 78–79, 147
 CCIR Reports concerning, 330
duplexers, 186–187
duplicated system, 87

Earth, effective radius of, *see* Effective radius
echoes, 218, 219–221
effective radius, 6, 42, 57–62, 117
 calculation of, 48–49
 CCIR Reports concerning, 330
elevation, horizon, 268
elevation angle, 11, 42–43, 46, 72, 140, 288
 calculation of, 49, 50
emphasis, 175
equal-gain combination, 93, 96
equatorial climate, 79
equipment, 155–201, 283, 285–288
 analog, 167–175, 183–184
 capability of, 196–200, 221–222
 choice of, 296–301
 digital, 176–182, 184
 failure of, 248
 testing of, 314–321
equivalent circuit
 of a diversity combiner, 91, 98
 for multipath transmission, 66–67
equivalent input r.f. level, 98
error function, 26
exciters, 158–160
exponential atmosphere, average, 56, 141
exponential distribution, 23, 28, 36

fading, 75, 133–140, 200, 238, 243
 CCIR Reports concerning, 330
 rate of, 84
fast fading, 75, 133–140, 200, 238, 243
fast–slow fading, 138–139
feasibility, 11–13, 282

feeders, 187–188, 303–305
feedhorn, 188
fixed systems, 14–15, 188–191, 290–291, 302–311
forward error-correction coding, 102–103, 185–186
forward scatter, 63–64
frequency bands, 8–9
　CCIR Reports concerning, 331
　choice of, 237, 291–293
　optimum, 9
frequency correlation function, 152
frequency diversity, 87–88
frequency hopping, 101
frequency modulation, 176–177
frequency planning, 293–295
frequency saving, 17
frequency-shift keying, 176

gain degradation, 8, 69–72, 193, 293
　CCIR Reports concerning, 330, 331
　measurements of, 71
gamma function, 31
Gaussian distribution, *see* Normal distribution
Gaussian wide sense stationary uncorrelated scattering, 76
geocentric latitude, 258, 259
geographic azimuth, 268, 269
geographic latitude, 258, 259, 268, 272–274
geographic longitude, 268, 272–274
geometrical data, definitions, 44–48
global reference circuit, 245, 246
grounding the plant, 311, 314

h.f. communications, 7
high quality troposcatter links, 225–228, 250, 251
horizon distance, 45, 47
　calculation of, 49
horizon skyline, determination of, 118, 119, 120, 260–263, 271–272
horizontal system, 269
hour angle, 266, 267, 268, 273
hydrometeors, 79, 147
hypothetical reference circuit, 204, 212, 330

i.f. combiner, 105, 106–107
illuminator, 188–190
implementation margin, 250
independent sideband diversity modem, 107–108

inhomogeneities in the troposphere, 3–4, 10–11, 55, 63
input thermal noise, 165–167
installation, 301–314
　of fixed systems, 302–311
　of mobile systems, 311–314
　tests of, 316–319
　of transportable systems, 311
instantaneous path loss, 133, 137
instantaneous signal, 78, 147
interference, 18, 147, 294, 295
　CCIR Recommendations concerning, 329, 330
　intersymbol, 148, 235, 249
intermodulation noise, 92, 169, 174, 205, 212–221, 293
International Radio Consultative Committee, 18–19
　climate types defined by, 79–81
　methods given by for calculating path loss, 117–133
　recommendations and reports of, 329–331
international recommendations, 18–19
International Telecommunications Union, 18–19
　Radio Regulations, 9, 19
International Telegraph and Telephone Consultative Committee, 18
ionoscatter, 3
ionosphere, 3
ionospheric reflections, 5, 7
ionospheric scatter, 2, 7
isolators, 186–187
ITU, *see* International Telecommunications Union

klystrons, 6, 161, 291, 322
knife-edge diffraction, 5
　calculations for, 50, 53, 143–144

latitude, determination of, 272–274
linear combination of diversity branches, 85, 90–91
　methods of, 93–98
　predetection or postdetection, 91–93
line-of-sight, *see* entries under LOS
local fixed system, 264
local sidereal time, 268
local zone time, 264, 268
log-normal distributions, 23, 26–27, 77–78
　convolution of, 36, 37, 78
longitude, determination of, 272–274

LOS radio links, 1, 2, 141
LOS repeaters, 4, 14
LOS zone, 2–3

maintenance, 296, 297, 323–324
maps, use of, 256, 259–260
maritime subtropical climate, 79–80
maritime temperate climate, 80–81
maximal-ratio combination, 93, 96–98
maximum likelihood detection, 102
mean value, 22
measurements, routine, 316–319
median value, 22
Mediterranean climate, 80
megabit transmission, 108–111
meridian, 267
meteor bursts, 7
military networks, 14, 17, 157
military standards, 19
mobile systems, 14, 16, 17, 191, 192, 284, 311–314
modems, 103, 185, 238, 249
   distortion adaptive, 108–111
   independent sideband diversity, 107–108
   multiple-subband, 107
modulation, 8, 176–177, 237
multipath analyzer, 315
multipath delay spread, 73–75, 148–152, 235–238, 315–316
multipath distortion, 100–101, 235–236, 248–249
   CCIR Reports concerning, 331
multipath propagation, 33–34, 66–68
   CCIR Reports concerning, 330
multipath signal, 72–73
multipath spectrum, 74, 148–152, 315–316
multiple-subband modem for digital transmission, 107
multiplex equipment, 193–195, 298

Nakagami distributions, 31–35
   m, 31–32, 33
   n, 32–35, 147
Nakagami–Rice distribution, 32–35, 147
National Bureau of Standards, 19
NBS, *see* National Bureau of Standards
noise, 173, 205–221
   intermodulation, 92, 169, 174, 205, 212–221, 293
   nonlinear, 10, 148, 175, 205, 217, 219, 226
   thermal, 92, 165–167, 169–171, 205, 206–212, 226
noise power ratio, 215

normal distributions, 23, 24–26
   convolution of, 36, 78

order of diversity, 9, 84
   total, 101
outage duration, 239, 242, 243
outage probability, 30, 77, 95–96, 97, 238, 240, 247
outage rate, 239, 241
outages, 244–245

parallel processing, 101
path, troposcatter, *see* Troposcatter path
path antenna gain, 193
path attenuation, *see* Path loss
path beamwidth, 43–44
path calculation, 221–222, 248–252
path differences, calculation of, 52–53
path dispersion parameter, 184, 235–237
path distortion, 148–153, 184
   calculation of, 116–117
path gain, 191, 193, 194
path geometry, 44–46
   calculation of, 117–120
path loss, 8, 113–153, 196–197, 221–222, 235, 293, 315
   calculation of, 116–133
   CCIR Reports concerning, 330, 331
   dependence of on main parameters, 140–141
   diffraction, 142–148
   instantaneous, 133, 137
   between isotropic antennas, 114–116
   for LOS links, 141
path tests, 314–316
performance, evaluation of, 319–321
phase modulation, 176, 177
phase-shift keying, 176
pilot tone, 156–157, 160
point-to-point links, 2
polar climate, 81
polarization discrimination, 86, 88
polarization diversity, 88
postdetection combination, 91–93, 163
power amplifiers, 160–163
power spectrum, 150
power supply, 162, 285–286, 300–301
predetection combination, 91–93, 163
profile, troposcatter path, 42–44, 66
propagation, 68–69
   anomalous, 78–79
   CCIR Recommendations concerning, 329

by diffraction, 2, 33–35, 55
  tests of, 314–316
  by troposcatter, 2, 10–11

quadruple diversity systems, 86–87, 88, 156, 250
  distribution curves for, 95

radar, 6
radio equipment, 296–298
radio horizon, 42
radiometeorological method for path loss, 132
radiometeorological parameters, 65
  CCIR Reports concerning, 330
Radio Regulations, 9
radius of the Earth, effective, *see* Effective radius
rake technique, 315
random variables, 21–23
Rayleigh distributions, 23, 27–31, 75–78
  convolution of, 36, 37, 78
Rayleigh fading, 137
receivers, 163–165
receiver threshold, 8, 168–169, 179–182
receiving system, 163–167
reciprocity theorem, 70
reflections
  from aircraft, 78
  ionospheric, 5, 7
  specular, 78
refraction, 6
  CCIR Reports concerning, 330
refractive index, 6, 56, 68
refractivity, 11, 56–65
  CCIR Reports concerning, 330
  see also Surface refractivity
refractivity gradient, 11, 57, 65, 132
  CCIR Reports concerning, 330
reliability, and design, 10–11
repeaters, 4, 14, 17
r.f. filters, 186
right ascension, 266, 268
routine measurements, 316–319

satellites, 2, 7, 324
scatter angle, 47, 50, 118, 140, 236
scatterers, 63, 70, 72
scattering theory, 64
scatter points, 63
scintillation, 68
selector, 93, 94–96
service channels, 157, 158, 160, 183–184

service probability, 125
shadow zone, 44
sidelobes, 79, 147, 190
sidereal time, 267–268
signal recirculation with time gating, 102
signal selection, 94–96
signal-to-noise ratio, 169–171, 177, 179, 222, 231
  in telephone channels, 234
site surveys, 263–264, 282–283, 284–288
siting of stations, 288–291
slot ratio, 215, 221
slot fading, 75
smooth-earth diffraction, 145–146
space diversity, 87–88, 189
spherical excess, 257, 259
spurious frequencies, 293, 295
squint angle, 89
standard deviation, 23
station layout, 302–303, 304
statistical distributions, 21–39
  CCIR Reports concerning, 330
stratosphere, 2–3
subrefraction, 79
super-refraction, 79, 141
supervisory channels, 157, 158, 160
surface refractivity, 56, 62, 65
survey problems, 255–281
symmetry factor, 47, 118, 119, 120, 121
systems, troposcatter, *see* Troposcatter systems

takeoff angle, *see* Elevation angle
telegraphic threshold, 169
telegraphy, 9, 228–231
telephone channels, 8, 9, 10, 169–171
telephone threshold, 168–169
terminal, composition of, 155–158, 159
testing, 314–321
test instruments, 316, 317
tetrodes, 161, 291
thermal noise, 92, 165–167, 169–171, 205, 206–212, 226
  CCIR conditions for, 228
threshold, 30–31, 173–174
  receiver, 8, 168–169, 179–182
threshold extension, 92, 164, 168
time delay spread, 235
time diversity, 90
time gating, signal recirculation with, 102
topographic problems, 255–281
traffic capacity, 8, 238
traffic parameters, 182

transhorizon radio links, 1
transmission quality, 244
transmitters, 158–160
transmitting system, 158–163
transmultiplexers, 186
transportable systems, 14–17, 284, 311
   CCIR Reports concerning, 331
traveling-wave tubes, 161, 291
triodes, 161, 291
troposcatter, 1, 7
   CCIR Reports concerning, 330
   propagation by, 2, 55–81
   see also Troposcatter systems
troposcatter path, 2–5, 17
   feasibility of, 11–13
   and gain loss, 71, 72
   geometry of, 41–53
   profile of, 42–44, 118, 123
   tests of, 314–316
troposcatter signal, 75–78
troposcatter systems
   conventional, 222–225, 250–252
   design of, 10–11, 281–301
   equipment for, 155–201, 296–301
   fixed, 14–15, 188–191, 290–291, 302–311
   high quality, 225–228, 250, 251
   implementation of, 281–327
   installation of, 301–314
   mobile systems, 14, 16, 17, 284, 311–314
   performance of, 227, 229, 319–321
   testing of, 316–321
   transportable, 14–17, 284, 311
troposcatter terminal, 155–158, 159
troposcatter, 2–3
troposphere, 2, 55
   inhomogeneities in, 3–4, 10–11, 55, 63
tropospheric scatter, *see* Troposcatter

unavailability, 246
uncertainty zone, 124
universal time, 267
uranographic equatorial system, 264–265

variance, 22
vernal point, 266

water vapor, 63
waveguide feeders, 187–188
worst month, 115

Yeh's method, 137

zone time correction, 267

The Artech House Radar Library

Adamy, David L., **Preparing and Delivering Effective Technical Presentations**

Alison, W.B.W., **A Handbook for the Mechanical Tolerancing of Waveguide Components**

Arams, Frank R., ed., **Infrared-to-Millimeter Wavelength Detectors**

Arbenz, Kurt, and Alfred Wohlhauser, **Advanced Mathematics for Practicing Engineers**

Banakh, V.A., and V.L. Mironov, **Lidar in a Turbulent Atmosphere**

Barton, David K., **Modern Radar System Analysis**

Barton, David K. and Harold R. Ward, **Handbook of Radar Measurement**

Beckmann, Peter, and Andre Spizzichino, **The Scattering of Electromagnetic Waves from Rough Surfaces**

Blackman, Samuel S., **Multiple-Target Tracking with Radar Applications**

Blake, Lamont V., **Radar Range-Performance Analysis**

Brookner, Eli, ed., **Radar Technology**

Currie, Nicholas C., and Charles E. Brown, eds., **Principles and Applications of Millimeter-Wave Radar**

DiFranco, J.V., and W.L. Rubin, **Radar Detection**

Erst, Stephen J., **Receiving Systems Design**

Fielding, John E., and Gary D. Reynolds, **RGCALC: Radar Range Detection Software and User's Manual**

Hansen, R.C., ed., **Significant Phased Array Papers**

Hovanessian, S.A., **Radar System Design and Analysis**

Hughes, Richard Smith, **Logarithmic Amplification**

Knott, Eugene F., John F. Shaeffer, and Michael T. Tuley, **Radar Cross Section**

Kolosov, A.A., et al., **Over-The-Horizon Radar**

Leonov, A.I., and K.I. Fomichev, **Monopulse Radar**

Lewis, Bernard, Frank Kretschmer, and Wesley Shelton, **Aspects of Radar Signal Processing**

Maksimov, M.V., et al., **Radar Anti-Jamming Techniques**

Meeks, M.L., **Radar Propagation at Low Altitudes**

Mensa, Dean L., **High Resolution Radar Imaging**

Ostroff, Edward D., et al., **Solid-State Radar Transmitters**

Ostrovityanov, R.V., and F.A. Basalov, **Statistical Theory of Extended Radar Targets**

Schleher, D. Curtis, **Introduction to Electronic Warfare**

Sherman, Samuel M., **Monopulse Principles and Techniques**

Skillman, William A., **SIGCLUT: Surface and Volumetric Clutter-to-Noise, Jammer, and Target Signal-to-Noise Radar Calculation Software and User's Manual**

Stevens, Michael C., **Secondary Surveillance Radar**

Torrieri, Don J., **Principles of Secure Communication Systems**

Wehner, Donald R., **High Resolution Radar**

Whalen, Timothy, **Writing and Managing Winning Technical Proposals**

Wiley, Richard G., **Electronic Intelligence: The Analysis of Radar Signals**

Wiley, Richard G., **Electronic Intelligence: The Interception of Radar Signals**

Wiley, Richard G., and Michael B. Szymanski, **Pulse Train Analysis Using Personal Computers**